Leckie
the education publisher
for Scotland

Primary Maths for Scotland

2nd Level Maths
Teacher Guide

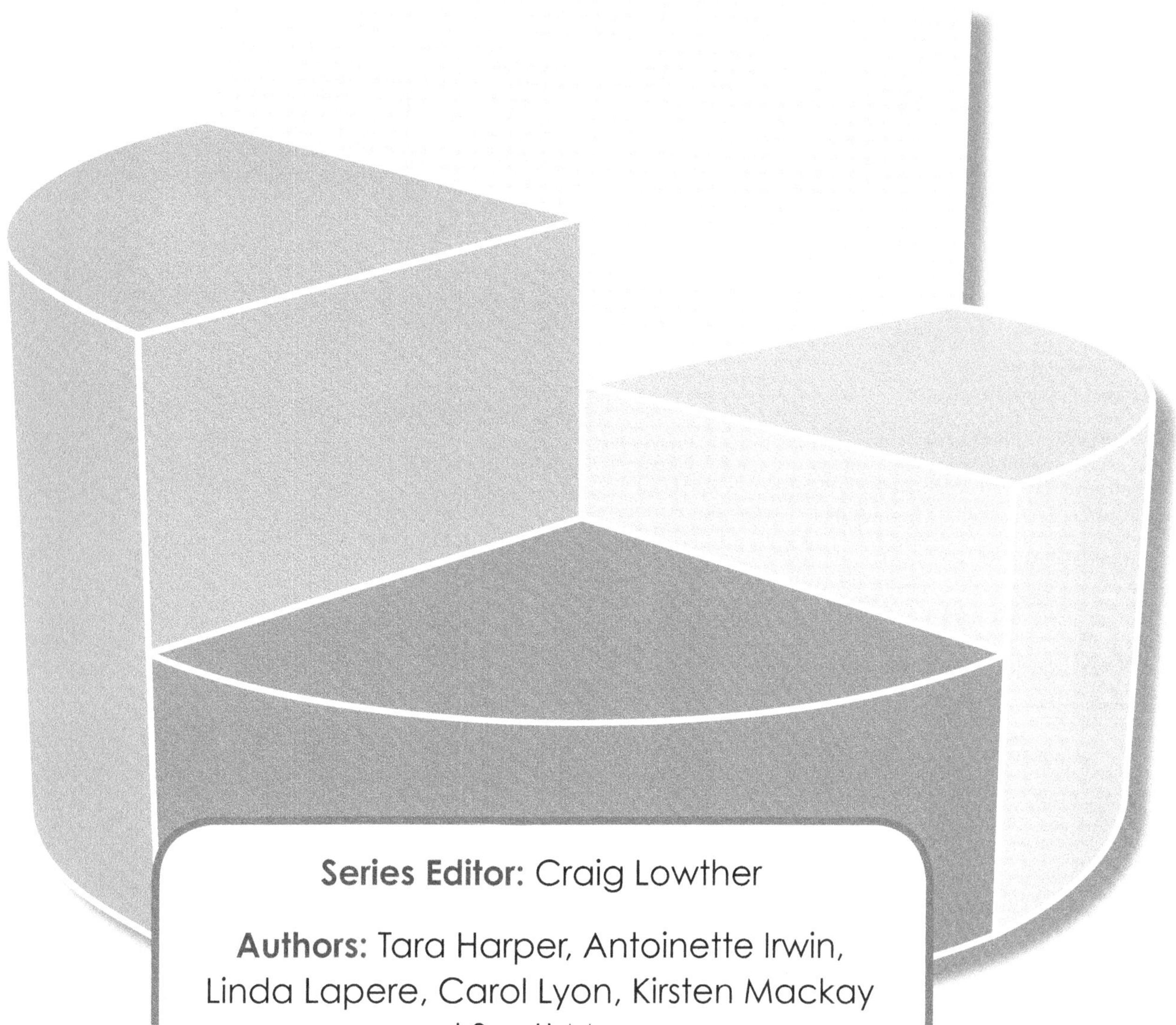

Series Editor: Craig Lowther

Authors: Tara Harper, Antoinette Irwin, Linda Lapere, Carol Lyon, Kirsten Mackay and Scott Morrow

001/04122019

11

ISBN 9780008348953

Published by
Leckie
An imprint of HarperCollins *Publishers*
1 Robroyston Gate, Glasgow G33 1JN
T: 0844 576 8126 F: 0844 576 8131
leckiescotland@harpercollins.co.uk www.leckiescotland.co.uk

HarperCollins *Publishers*
Macken House, 39/40 Mayor Street Upper, Dublin 1, D01 C9W8, Ireland

Publisher: Fiona McGlade
Managing editor: Craig Balfour
Project editor: Peter Dennis

Special thanks
Copy editor: Peter Dennis
Layout: Jouve
Proofreader: Louise Robb

A CIP Catalogue record for this book is available from the British Library.

Acknowledgements
Images © Shutterstock.com

Whilst every effort has been made to trace the copyright holders, in cases where this has been unsuccessful, or if any have inadvertently been overlooked, the Publishers would gladly receive any information enabling them to rectify any error or

Printed by Ashford Colour Ltd.

FSC www.fsc.org
MIX
Paper | Supporting responsible forestry
FSC® C011748

Contents

Contents

Introduction

Primary Maths for Scotland programme overview

This Primary Maths for Scotland Second Level Teacher Guide has been published in conjunction with the Primary Maths for Scotland Second Level Maths Textbooks 2A, 2B and 2C, and the Second Level Maths Assessment Guide and Problem Solving Pack. These titles complete a whole-school mathematics programme which also includes Early Level and First Level.

This series has been written by a team of Scottish numeracy experts specially for Scottish schools and has been written to the Curriculum for Excellence Benchmarks to plan routes through the Experiences and Outcomes for Numeracy and Mathematics. They have been designed to develop conceptual understanding and procedural fluency whilst being easy to implement alongside existing plans and resources.

The team used research-informed pedagogy to guide the creation of these resources. The principle of social constructivism was used to inform the teaching, and how resources and teaching ideas have been phrased. Pupils should be active participants in the learning process and be challenged and encouraged to talk about their ideas and share their strategies for solving mathematical problems. We want all pupils to enjoy learning mathematics. This mind-set and motivation makes learning and teaching a truly positive experience for everyone.

Primary Maths for Scotland programme components

Primary Maths for Scotland Second Level Teacher Guide

At the beginning of the Teacher Guide is a suggested teaching plan for second level. The curriculum has been organised into three parts: 2A, 2B and 2C, each of which correspond to a year's worth of work for children learning at the expected rate. The teaching plan can be used by teachers as their planning document or can be ignored in favour of an existing or different plan. The Teacher Guide and the Textbooks have been organised by Experiences and Outcomes, and not by suggested teaching plan, so that each topic is easily found, and teachers can use the resources as they see fit.

Suggested Teaching Plan

2A Block	Textbook 2A Chapter	Textbook 2A Units	Benchmark/E&O
1	Number – order and place value	2.1 Reading and writing whole numbers 2.2 Representing and describing whole numbers 2.3 Place value partitioning of numbers with up to five digits 2.4 Comparing and ordering numbers in the range 0–10 000	MNU 2-02a • Reads, writes and orders whole numbers to 1 000 000, starting from any number in the sequence. • Explains the link between a digit, its place and its value for whole numbers to 1 000 000. • Partitions a wide range of whole numbers and decimal fractions to three decimal places, for example, 3·6 = 3 ones and 6 tenths = 36 tenths.

This provides a suggested order of topics to follow. The resources are designed to be flexible and will fit with an established curriculum design, but if you are looking for a revised order then this is your starting point. The order is designed to support the pupils to progressively build knowledge and skills, while being able to return to previously learned topics as part of a spiral curriculum of deliberate spaced practice. This helps pupils to remember and practice old skills that they need to continue to learn. If you are working with a multi-stage class it would be prudent to choose one sequence of topics for the class and then to differentiate to meet the needs of groups and individuals.

Chapters

Each chapter in the Teacher Guide is divided into three sections that correspond to the contents of Textbooks 2A, 2B and 2C. For each of the three sections within a topic, the following guidance is provided:

Section overview

Textbook 2A units	Review units	Prior learning	Benchmarks/ Es&Os
Unit 1.1 Rounding whole numbers to the nearest 10	**1C** Units 1.1 and 1.2	• I can read and write numbers with up to 4 digits. • I understand that the place of a digit within a number shows its value. • I know the meaning of each digit in the ones, tens, hundreds and thousands columns.	**MNU 2-01a** • Rounds whole numbers to the nearest 1000, 10 000 and 100 000.

This table shows the Curriculum for Excellence Experiences and Outcomes and Benchmarks covered by each unit in this section. It is not our expectation that a child who successfully completes a single unit has demonstrated completion of a benchmark, rather it is through regularly returning to ideas and building skills and knowledge that ultimately a child will be able to demonstrate their understanding of each benchmark in new and unfamiliar contexts. This is when they, and you, know they are secure.

- Column 1 lists the Teacher Guide units in this section, which correspond to the Textbook units
- Column 2 directs teachers to units that children can review if they are not yet ready to complete the work in the unit and do not possess the required pre-requisite knowledge
- Column 3 describes the prior learning that children are expected to have mastered before embarking on the unit
- Column 4 shows the coverage of the CfE and Benchmarks within this section.

Conceptual context

Conceptual context

In daily life, we often round numbers to allow us to easily work with them and to check answers. It is important that pupils have a clear understanding of the numerals they are working with, and their value, to be able to estimate and round effectively. Pupils require a wide range of opportunities in meaningful contexts to evaluate the accuracy of calculations by estimating, and to explore different levels of accuracy when estimating. In this way, they will develop the ability to habitually estimate if answers are reasonable, and be able to use rounding in ways that are relevant to the context. For instance, if working out a problem such as:

This section provides an overview of the topic. It will include a brief explanation of key aspects of learning covered, including prior knowledge, calculation methods and next steps in learning.

Resources

Resources

- Number lines 0-100, 0-1000 in a range of increments – ones, tens, hundreds, tenths
- Numeral cards 0-100
- Decimats
- Arrow cards
- Physical materials to represent numbers such as cubes or place value sticks
- Money – mock £10, £100 notes and £1 coins

This section lists the physical resources which can be used to support the teaching and learning of the units in this topic.

We have listed resources that we use regularly with our own classes. We do not expect you to buy specific items of equipment. Many objects, such as counters or cubes, will be available in a number of different forms. We do believe strongly that when pupils first encounter new ideas that they should have access to and be encouraged to use concrete materials. As pupils become more confident, they can move onto imaging the materials in their minds and then move onto using the number facts and skills without the help of manipulatives.

Key Vocabulary

The main mathematical terms relating to the units in this section are listed.

Key vocabulary

- **Approximately**
- **Estimate**
- **Guess**

A **full glossary** of all the key mathematical terms and definitions can be found at collins.co.uk/PrimaryMathsforScotland/freeresources

The key vocabulary is listed and defined to promote the use of correct mathematical words and consistency across classes.

Misconceptions

Potential misconception	Addressing the misconception
Units 1.1, 1.2 and 1.3 The pupil might not understand which digit in the number to look at to determine whether to round up or down. For example, 3821 rounded to the nearest 1000 is 3000 (pupil is using last digit to round incorrectly).	• Write number on an empty number line and ask them which multiple of 1000 the number is between and how they decide which multiple is nearer to the number. Ask them which digit helps them decide.

We have listed many of the common misconceptions that pupils will reveal to you when they have an opportunity to talk about their ideas. The left-hand column indicates the units where the potential misconceptions are most likely to occur.

Where errors or misconceptions are highlighted it is important to address them immediately to support the pupil to change their thinking.

Teaching ideas

Unit 1.1 Rounding whole numbers to the nearest 10, 100 or 1000

- Use numeral cards to make a 5-digit number, e.g. 23 709. Ask questions such as: **What is this number? What does the 3 represent? How many hundreds in the whole number?** to check understanding of place value.

These suggested activities add breadth, depth and challenge to the textbook resource. They can be used to introduce a topic before pupils practise new skills or as part of the reinforcement of learning. The team have highlighted key questions in bold that you can use with individuals or groups of pupils. These questions are carefully constructed to provoke thought and to help reveal the pupils' current understanding and any misconceptions they may have. The answers and explanations that the pupils offer will provide feedback that will help you and them to decide on the next steps in their learning.

Resource sheets

Where Resource sheets are listed in the text, you can download them from collins.co.uk/PrimaryMathsforScotland/freeresources to use with your pupils.

Glossary of Key Vocabulary

A full glossary of all the key mathematical terms and definitions from each topic can be found at collins.co.uk/PrimaryMathsforScotland/freeresources

Primary Maths for Scotland Second Level Textbooks

We have organised the curriculum into three parts: 2A, 2B and 2C, each of which correspond to a year's worth of work for children learning at the expected rate. There are three textbooks covering Second Level: 2A, 2B and 2C. Each textbook is organised by Experiences and Outcomes, and not by suggested teaching plan. The textbooks are progressive and should be used in sequence until the child has mastered the learning from each book.

The features of the textbooks include:

What we are learning

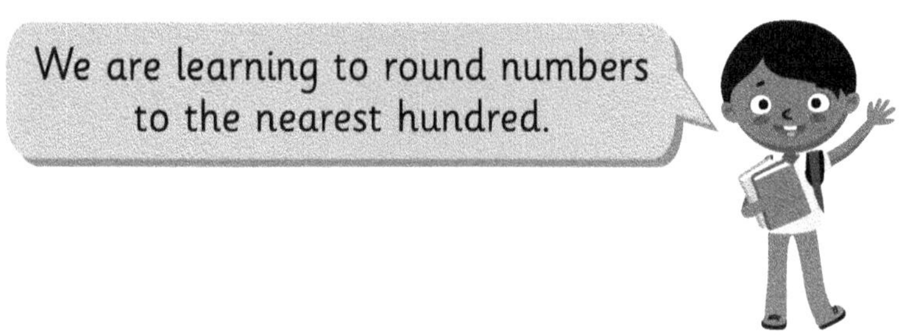

Familiar characters introduce the learning intention so that the pupil knows what they are learning and can use this to reflect on their progress.

Before we start

Before we start

a) What is this number rounded to the nearest hundred?

b) What number is 100 more than this number?

824

This contains a diagnostic question that will help you find out what knowledge and strategy the pupil knows and can remember. This will help you to pitch the start-point of your teaching.

Let's learn

Let's learn

We can use a number line to help us round numbers to the nearest hundred.

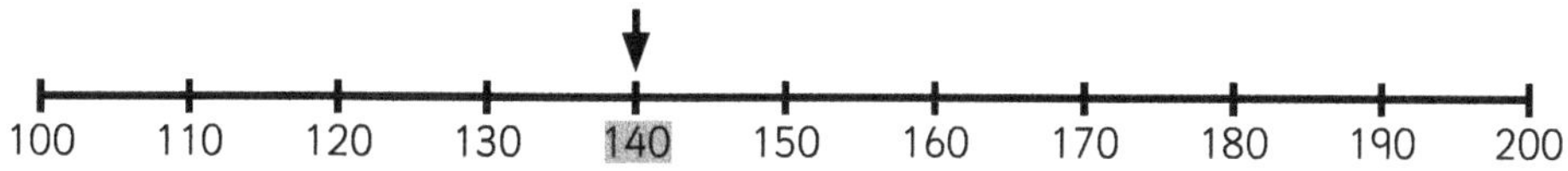

The number 140 is between 100 and 200, but it is closer to the number 100, so we round **down**.

These worked examples can be used by you and/or the pupils to help understand the intended learning and the skills and knowledge required for the unit. The teaching ideas section in the Teacher Guide provides more ideas for introducing and extending this initial learning phase.

Let's practise

Let's practise

1) Which is the closest hundred to these numbers?

Use the number lines to help you.

a) 337

b) 458

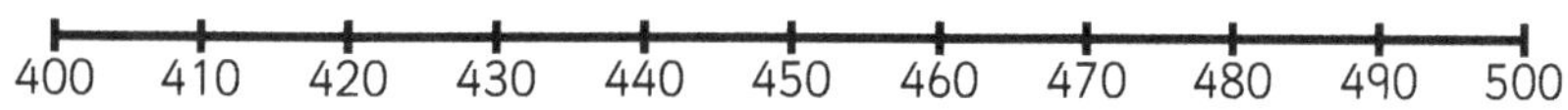

These exercises can be used for group, pair or individual practice. The exercises are not long and repetitive; rather, they are designed to help the pupil to embed the learning in the unit in the initial phase of learning. Deep memory comes from being motivated to learn and through regular, deliberate spaced practice over days, weeks and months. We advocate that pupils regularly return to, and build on, previously learned skills and knowledge as part of a spiral curriculum. We are also happy for our pupils to choose the questions they want to start at, if they feel confident with the topic. There are answers at the back of the Textbooks so that pupils can get immediate feedback on their solutions and find out what they need to do next to improve their learning.

Challenge

CHALLENGE!

How many different three-digit numbers can you make with these digit cards? Write each number you make in numerals and in words.

3 0 5

This final section can be used as an extension to the Let's practise section or as the main part of the learning. These questions are usually written for pairs of pupils to work on together so that they are supported to talk and share their ideas.

Further resources and ideas are available from collins.co.uk/PrimaryMathsforScotland/freeresources

Primary Maths for Scotland Second Level Problem Solving Pack

This pack is available to support your pupils to develop the numeracy and mathematics skills from the Education Scotland Benchmarks. We advise that after a topic has been covered with a class or group that you then uses the relevant problem with that class/group a week or so after the topic has ended. This is ideal for promoting memory and metacognition. Completing the mathematics topic prior to the problem-solving activity should mean that the pupils can focus on the problem-solving skills and they are not trying to learn a new maths skill or process at the same time. The Problem Solving Pack is designed to be used flexibly to fit with the curriculum order the pupils are following.

Primary Maths for Second First Level Assessment Pack

Assessment Packs are available to support you to make assessment decisions on each pupil's progress through the numeracy and mathematics curriculum. The Second Level Assessment pack includes Yearly Progress checks that can be used throughout the year to formatively assess ongoing progress and End of Level assessments that cover all the Experiences and Outcomes of the Broad General Education Numeracy and Mathematics curriculum.

We welcome your thoughts and feedback, please contact the team here: leckiescotland@harpercollins.co.uk

Suggested Teaching Plans

Primary Maths for Scotland 2A scope and sequence

Chapter	Block	Units
Number – order and place value	1	2.1, 2.2, 2.3, 2.4
Number – addition and subtraction	2	3.1, 3.2, 3.3, 3.4, 3.5, 3.6
Time	3	8.1, 8.2
Mathematics, its impact on the world, past, present and future	4	10.1
Estimation and rounding	5	1.1, 1.2
Number – multiplication and division	6	4.1, 4.2, 4.3, 4.4, 4.5
Measurement	7	9.1, 9.2, 9.3, 9.4
Number – addition and subtraction	8	3.7, 3.8, 3.9, 3.10, 3.11, 3.12
Angles, symmetry and transformation	9	14.1, 14.2, 14.3
Number – multiplication and division	10	4.6, 4.7, 4.8, 4.9, 4.10, 4.11, 4.12
Measurement	11	9.5, 9.6, 9.7, 9.8
Number – order and place value	12	2.5, 2.6, 2.7, 2.8
Estimation and rounding	13	1.3, 1.4
Money	14	7.1, 7.2, 7.3, 7.4
Fractions, decimal fractions and percentages	15	6.1, 6.2, 6.3, 6.4
Time	16	8.3, 8.4, 8.5
Ideas of chance and uncertainty	17	16.1
Data handling and analysis	18	15.1, 15.2, 15.3, 15.4
Fractions, decimal fractions and percentages	19	6.5, 6.6, 6.7
Multiplies, factors and primes	20	5.1
Patterns and relationships	21	11.1
Expressions and equations	22	12.1
Angles, symmetry and transformation	23	14.4, 14.5, 14.6, 14.7
2D shapes and 3D objects	24	13.1. 13.2. 13.3. 13.4

Note: sections of the benchmarks/Es&Os not covered in a 2A block have been highlighted grey in the table. These will be covered in 2B and 2C.

2A Block	Textbook 2A Chapter	Textbook 2A Units	Benchmark/E&O
1	Number – order and place value	2.1 Reading and writing whole numbers 2.2 Representing and describing whole numbers 2.3 Place value partitioning of numbers with up to five digits 2.4 Comparing and ordering numbers in the range 0–10 000	MNU 2-02a • Reads, writes and orders whole numbers to 1 000 000, starting from any number in the sequence. • Explains the link between a digit, its place and its value for whole numbers to 1 000 000. • Partitions a wide range of whole numbers and decimal fractions to three decimal places, for example, 3·6 = 3 ones and 6 tenths = 36 tenths.
2	Number – addition and subtraction	3.1 Mental addition and subtraction of two- and three-digit numbers 3.2 Adding and subtracting 1, 10, 100 and 1000 3.3 Adding and subtracting multiples of 100 3.4 Adding and subtracting by making 10s or 100s 3.5 Adding and subtracting multiples of 1000 3.6 Adding and subtracting multiples of 10 and 100	MNU 2-02a MNU 2-03a • Adds and subtracts multiples of 10, 100 and 1000 to and from whole numbers and decimal fractions to two decimal places. • Adds and subtracts whole numbers and decimal fractions to two decimal places, within the number range 0 to 1 000 000.
3	Time	8.1 Telling the time to the minute – 12-hour clock 8.2 Converting between 12-h and 24-h time	MNU 2-10a MNU 2-10b MNU 2-10c • Reads and records time in both 12-hour and 24-hour notation and converts between the two.
4	Mathematics, its impact on the world, past, present and future	10.1 Mathematical inventions and different number systems	MTH 2-12a • Researches and presents examples of the impact mathematics has in the world of life and work. • Contributes to discussions and activities on the role of mathematics in the creation of important inventions, now and in the past.
5	Estimation and rounding	1.1 Rounding whole numbers to the nearest 10 1.2 Rounding whole numbers to the nearest 100	MNU 2-01a • Rounds whole numbers to the nearest 1000, 10 000 and 100 000.

2A Block	Textbook 2A Chapter	Textbook 2A Units	Benchmark/E&O
6	Number – multiplication and division	4.1 Recalling multiplication and division facts for 2, 5 and 10 4.2 Recalling multiplication and division facts for 3 4.3 Solving multiplication problems 4.4 Multiplying whole numbers by 10 and 100 4.5 Dividing whole numbers by 10 and 100	MTH 2-03c • Uses multiplication and division facts to the 10th multiplication table. • Multiplies and divides whole numbers by multiples of 10, 100 and 1000. • Multiplies whole numbers by two-digit numbers.
7	Measurement	9.1 Using familiar objects to estimate length, mass, area and capacity 9.2 Estimating and measuring length 9.3 Estimating and measuring mass 9.4 Converting units of length	MNU 2-11a MNU 2-11b MNU 2-11c • Uses the comparative size of familiar objects to make reasonable estimations of length, mass, area and capacity. • Estimates to the nearest appropriate unit, then measures accurately: length, height and distance in millimetres (mm), centimetres (cm), metres (m) and kilometres (km); mass in grams (g) and kilograms (kg); and capacity in millilitres (ml) and litres (l). • Converts between common units of measurement using decimal notation, for example, 550 cm = 5·5 m; 3·009 kg = 3009 g. • Chooses the most appropriate measuring device for a given task and carries out the required calculation, recording results in the correct unit. • Reads a variety of scales accurately.
8	Number – addition and subtraction	3.7 Solving word problems 3.8 Representing word problems in different ways 3.9 Using non-standard place value partitioning 3.10 Adding using a semi-formal written method 3.11 Adding three-digit numbers using standard algorithms 3.12 Representing and solving word problems	MNU 2-03a MTH 2-03c • Adds and subtracts multiples of 10, 100 and 1000 to and from whole numbers and decimal fractions to two decimal places. • Adds and subtracts whole numbers and decimal fractions to two decimal places, within the number range 0 to 1 000 000.

2A Block	Textbook 2A Chapter	Textbook 2A Units	Benchmark/E&O
9	Angles, symmetry and transformation	14.1 Identifying angles 14.2 Using an eight-point compass 14.3 Plotting points using coordinates	MTH 2-17a MTH 2-17b MTH 2-17c MTH 2-17d MTH 2-18a MTH 2-19a • Uses mathematical language including acute, obtuse, straight and reflex to describe and classify a range of angles identified within shapes in the environment. • Uses knowledge of the link between the eight compass points and angles to describe, follow and record directions. • Describes, plots and records the location of a point, in the first quadrant, using coordinate notation.
10	Number – multiplication and division	4.6 Solving multiplication problems by partitioning 4.7 Solving multiplication and division problems 4.8 Recalling multiplication and division facts for 4 4.9 Solving division problems 4.10 Multiplying by changing the order of factors 4.11 Solving multiplication problems 4.12 Using known facts and halving to solve division problems	MTH 2-03c • Uses multiplication and division facts to the 10th multiplication table. • Multiplies and divides whole numbers by multiples of 10, 100 and 1000. • Multiplies whole numbers by two-digit numbers.

2A Block	Textbook 2A Chapter	Textbook 2A Units	Benchmark/E&O
11	Measurement	9.5 Calculating the perimeter of simple shapes 9.6 Finding the area of regular shapes in square cm/m 9.7 Finding the volume of cubes and cuboids by counting cubes 9.8 Estimating and measuring capacity	MNU 2-11a MNU 2-11b MNU 2-11c • Uses the comparative size of familiar objects to make reasonable estimations of length, mass, area and capacity. • Estimates to the nearest appropriate unit, then measures accurately: length, height and distance in millimetres (mm), centimetres (cm), metres (m) and kilometres (km); mass in grams (g) and kilograms (kg); and capacity in millilitres (ml) and litres (l). • Calculates the perimeter of simple straight sided 2D shapes in millimetres (mm), centimetres (cm) and metres (m). • Calculates the area of squares, rectangles and right-angled triangles in square millimetres (mm^2), square centimetres (cm^2) and square metres (m^2). • Calculates the volume of cubes and cuboids in cubic centimetres (cm^3) and cubic metres (m^3). • Chooses the most appropriate measuring device for a given task and carries out the required calculation, recording results in the correct unit. • Reads a variety of scales accurately.
12	Number – order and place value	2.5 Reading and writing decimal fractions 2.6 Representing and describing decimal fractions 2.7 Comparing and ordering decimal fractions 2.8 Recognising the context for negative numbers	MNU 2-02a MNU 2-03b MNU 2-04a • Reads, writes and orders sets of decimal fractions to three decimal places. • Explains the link between a digit, its place and its value for numbers to three decimal places. • Partitions a wide range of whole numbers and decimal fractions to three decimal places, for example, 3·6 = 3 ones and 6 tenths = 36 tenths. • Identifies familiar contexts in which negative numbers are used

2A Block	Textbook 2A Chapter	Textbook 2A Units	Benchmark/E&O
13	Estimation and rounding	1.3 Rounding decimal fractions to the nearest whole number 1.4 Estimating the answer using rounding	MNU 2-01a • Rounds decimal fractions to the nearest whole number, to one decimal place and two decimal places. • Applies knowledge of rounding to give an estimate to a calculation appropriate to the context
14	Money	7.1 Writing amounts using digital notation 7.2 Budgeting 7.3 Saving Money 7.4 Profit and loss	MNU 2-09a MNU 2-09b MNU 2-09c • Carries out money calculations involving the four operations. • Compares costs and determines affordability within a given budget. • Calculates profit and loss accurately, for example, when working with a budget for an enterprise activity.
15	Fractions, decimal fractions and percentages	6.1 Comparing fractions 6.2 Identifying equivalent fractions pictorially 6.3 Identifying and creating equivalent fractions 6.4 Simplifying fractions	MNU 2-07a MNU 2-07b MTH 2-07c • Uses knowledge of equivalent forms of common fractions, decimal fractions and percentages (for example, $\frac{3}{4} = 0{\cdot}75 = 75\%$) to solve problems. • Creates equivalent fractions and uses this knowledge to put a set of the most commonly used fractions in order. • Expresses fractions in their simplest form.

2A Block	Textbook 2A Chapter	Textbook 2A Units	Benchmark/E&O
16	Time	8.3 Converting minute intervals to fractions of an hour 8.4 Calculating time intervals or durations 8.5 Speed, time and distance calculations	MNU 2-10a MNU 2-10b MNU 2-10c • Reads and records time in both 12-hour and 24-hour notation and converts between the two. • Knows the relationships between commonly used units of time and carries out simple conversion calculations, for example, changes 1 ¾ hours into minutes. • Uses and interprets a range of electronic and paper-based timetables and calendars to plan events or activities and solve real life problems. • Calculates durations of activities and events including situations bridging across several hours and parts of hours using both 12-hour clock and 24-hour notation. • Estimates the duration of a journey based on knowledge of the link between speed, distance and time. • Partitions a wide range of whole numbers and decimal fractions to three decimal places, for example, 3·6 = 3 ones and 6 tenths = 36 tenths.
17	Ideas of chance and uncertainty	16.1 Predicting and explaining simple chance situations	MNU 2-22a • Uses the language of probability accurately to describe the likelihood of simple events occurring, for example equal chance; fifty-fifty; one in two, two in three; percentage chance; and 1/6. • Plans and carries out simple experiments involving chance with repeated trials, for example, 'what is the probability of throwing a six if you throw a die fifty times?'. • Uses data to predict the outcome of a simple experiment.

2A Block	Textbook 2A Chapter	Textbook 2A Units	Benchmark/E&O
18	Data handling and analysis	15.1 Reading and interpreting information 15.2 Organising and displaying data 15.3 Reading and interpreting pie charts 15.4 Collecting data	MNU 2-20a MNU 2-20b MTH 2-21a • Devises ways of collecting data in the most suitable way for the given task. • Collects, organises and displays data accurately in a variety of ways including through the use of digital technologies, for example, creating surveys, tables, bar graphs, line graphs, frequency tables, simple pie charts and spreadsheets. • Analyses, interprets and draws conclusions from a variety of data. • Displays data appropriately making effective use of technology and chooses a suitable scale when creating graphs.
19	Fractions, decimal fractions and percentages	6.5 Writing decimal equivalents to tenths 6.6 Comparing numbers with one decimal place 6.7 Calculating a simple fraction of a value	• Uses knowledge of equivalent forms of common fractions, decimal fractions and percentages (for example, ¾ = 0·75 = 75%) to solve problems. • Calculates simple fractions of quantity and uses this knowledge to solve problems, for example, find 3/5 of 60.
20	Multiples, factors and primes	5.1 Recognising multiples and factors	MTH 2-05a • Identifies multiples and factors of whole numbers and applies knowledge and understanding of these when solving relevant problems in number, money and measurement.
21	Patterns and relationships	11.1 Exploring and extending number sequences	MTH 2-13a • Explains and uses a rule to extend well known number sequences including square numbers, triangular numbers and Fibonacci sequence.
22	Expressions and equations	12.1 Solving simple equations using known number facts	MTH 2-15a • Solves simple algebraic equations with one variable, for example, a - 30 = 40 and 4b = 20.

2A Block	Textbook 2A Chapter	Textbook 2A Units	Benchmark/E&O
23	Angles, symmetry and transformation	14.4 Lines of symmetry 14.5 Creating designs with lines of symmetry 14.6 Measuring angles up to 180° 14.7 Understanding scale	MTH 2-17a MTH 2-17b MTH 2-17c MTH 2-17d MTH 2-18a MTH 2-19a • Uses mathematical language including acute, obtuse, straight and reflex to describe and classify a range of angles identified within shapes in the environment. • Measures and draws a range of angles to within ± 2°. • Interprets maps, models or plans with simple scales, for example, 1 cm:2 km. • Identifies and illustrates line symmetry on a wide range of 2D shapes and applies this understanding to complete a range of symmetrical patterns, with and without the use of digital technologies.
24	2D shapes and 3D Objects	13.1 Drawing 2D shapes 13.2 Naming and sorting 2D shapes 13.3 Drawing 2D shapes and 3D objects 13.4 Describing and sorting prisms and pyramids	MTH 2-16a MTH 2-16b MTH 2-16c • Describes 3D objects and 2D shapes using specific vocabulary including regular, irregular, diagonal, radius, diameter and circumference. Applies this knowledge to demonstrate understanding of the relationship. between 3D objects and their nets. • Identifies and describes 3D objects and 2D shapes within the environment and explains why their properties match their function. Uses digital technologies and mathematical instruments to draw 2D shapes and make representations of 3D objects, understanding that not all parts of the 3D object can be seen.

Primary Maths for Scotland 2B scope and sequence

Chapter	Block	Units
Number – order and place value	1	2.1, 2.2, 2.3, 2.4, 2.5, 2.6
Number – addition and subtraction	2	3.1, 3.2, 3.3, 3.4
Time	3	8.1, 8.2, 8.3
Mathematics, its impact on the world, past, present and future	4	10.1
Estimation and rounding	5	1.1
Number – multiplication and division	6	4.1, 4.2, 4.3, 4.4
Measurement	7	9.1, 9.2, 9.3
Number – addition and subtraction	8	3.5, 3.6, 3.7, 3.8, 3.9
Angles, symmetry and transformation	9	14.1, 14.2, 14.3, 14.4
Number – multiplication and division	10	4.5, 4.6, 4.7, 4.8, 4.9
Fractions, decimal fractions and percentages	11	6.1, 6.2, 6.3
Patterns and relationships	12	11.1
Multiplies, factors and primes	13	5.1, 5.2
Number – order and place value	14	2.7, 2.8, 2.9
Estimation and rounding	15	1.2
2D shapes and 3D objects	16	13.1, 13.2, 13.3, 13.4
Estimation and rounding	17	1.3
Number – addition and subtraction	18	3.10, 3.11, 3.12
Fractions, decimal fractions and percentages	19	6.4, 6.5
Number – addition and subtraction	20	3.13, 3.14
Number – multiplication and division	21	4.10, 4.11
Money	22	7.1, 7.2, 7.3, 7.4, 7.5
Measurement	23	9.4, 9.5, 9.6, 9.7, 9.8
Angles, symmetry and transformation	24	14.5, 14.6, 14.7, 14.8
Number – multiplication and division	25	4.12, 4.13
Fractions, decimal fractions and percentages	26	6.6, 6.7
Time	27	8.4, 8.5, 8.6
Expressions and equations	28	12.1
Fractions, decimal fractions and percentages	29	6.8, 6.9, 6.10, 6.11
Data handling and analysis	30	15.1, 15.2, 15.3, 15.4
Ideas of chance and uncertainty	31	16.1

Note: sections of the benchmarks/Es&Os not covered in a 2B block have been highlighted grey in the table. These will be covered in 2A and 2C.

2B Block	Textbook 2B Chapter	Textbook 2B Units	Benchmark/E&O
1	Number – order and place value	2.1 Reading and writing whole numbers 2.2 Representing and describing whole numbers 2.3 Place value partitioning of whole numbers 2.4 Number sequences 2.5 Comparing and ordering whole numbers 2.6 Identifying and placing positive and negative numbers	MNU 2-02a MNU 2-03b MNU 2-04a • Reads, writes and orders whole numbers to 1 000 000, starting from any number in the sequence. • Explains the link between a digit, its place and its value for whole numbers to 1 000 000. • Explains the link between a digit, its place and its value for numbers to three decimal places. • Partitions a wide range of whole numbers and decimal fractions to three decimal places, for example, 3·6 = 3 ones and 6 tenths = 36 tenths.
2	Number – addition and subtraction	3.1 Mental addition and subtraction 3.2 Adding and subtracting a string of numbers 3.3 Adding and subtracting multiples of 10, 100 and 1000 3.4 Using place value partitioning	MNU 2-03a MTH 2-03c • Adds and subtracts multiples of 10, 100 and 1000 to and from whole numbers and decimal fractions to two decimal places. • Adds and subtracts whole numbers and decimal fractions to two decimal places, within the number range 0 to 1 000 000.
3	Time	8.1 Reading and writing 12-h and 24-h time 8.2 Converting units of time using fractions 8.3 Calculating time intervals using timetables	MNU 2-10a MNU 2-10b MNU 2-10c • Reads and records time in both 12-hour and 24-hour notation and converts between the two. • Knows the relationships between commonly used units of time and carries out simple conversion calculations, for example, changes 1 hours into minutes. • Uses and interprets a range of electronic and paper-based timetables and calendars to plan events or activities and solve real life problems. • Calculates durations of activities and events including situations bridging across several hours and parts of hours using both 12-hour clock and 24-hour notation.

2B Block	Textbook 2B Chapter	Textbook 2B Units	Benchmark/E&O
4	Mathematics, its impact on the world, past, present and future	10.1 Mathematical inventions and different number systems	MTH 2-12a • Researches and presents examples of the impact mathematics has in the world of life and work. • Contributes to discussions and activities on the role of mathematics in the creation of important inventions, now and in the past.
5	Estimation and rounding	1.1 Rounding whole numbers to the nearest 10, 100 or 1000	MNU 2-01a • Rounds whole numbers to the nearest 1000, 10 000 and 100 000.
6	Number – multiplication and division	4.1 Recalling multiplication and division facts for 6 4.2 Recalling multiplication and division facts for 9 4.3 Multiplying multiples of 10, 100 or 1000 4.4 Dividing by multiples of 10, 100 or 1000	MTH 2-03c • Uses multiplication and division facts to the 10th multiplication table. • Multiplies and divides whole numbers by multiples of 10, 100 and 1000.
7	Measurement	9.1 Estimating and measuring length 9.2 Estimating and measuring mass 9.3 Estimating and measuring capacity	MNU 2-11a MNU 2-11b MNU 2-11c • Uses the comparative size of familiar objects to make reasonable estimations of length, mass, area and capacity. • Estimates to the nearest appropriate unit, then measures accurately: length, height and distance in millimetres (mm), centimetres (cm), metres (m) and kilometres (km); mass in grams (g) and kilograms (kg); and capacity in millilitres (ml) and litres (l). • Converts between common units of measurement using decimal notation, for example, 550 cm = 5·5 m; 3·009 kg = 3009 g. • Reads a variety of scales accurately.

2B Block	Textbook 2B Chapter	Textbook 2B Units	Benchmark/E&O
8	Number – addition and subtraction	3.5 Adding four-digit numbers using standard algorithms 3.6 Column subtraction: three-digit numbers 3.7 Column subtraction: four-digit numbers 3.8 Subtracting three-digit numbers using standard algorithms 3.9 Subtracting four-digit numbers using standard algorithms	MNU 2-03a MTH 2-03c • Adds and subtracts multiples of 10, 100 and 1000 to and from whole numbers and decimal fractions to two decimal places. • Adds and subtracts whole numbers and decimal fractions to two decimal places, within the number range 0 to 1 000 000.
9	Angles, symmetry and transformation	14.1 Identifying and sorting angles 14.2 Measuring and drawing angles 14.3 Finding missing angles 14.4 Locating objects using bearings	MTH 2-17a MTH 2-17b MTH 2-17c MTH 2-17d MTH 2-18a MTH 2-19a • Uses mathematical language including acute, obtuse, straight and reflex to describe and classify a range of angles identified within shapes in the environment. • Measures and draws a range of angles to within $\pm 2°$ • Uses knowledge of the link between the eight compass points and angles to describe, follow and record directions.
10	Number – multiplication and division	4.5 Solving division problems 4.6 Solving multiplication problems 4.7 Using partitioning to solve division problems 4.8 Using rounding and compensating to solve division problems 4.9 Solving problems involving addition, subtraction, multiplication and division	MTH 2-03c • Uses multiplication and division facts to the 10th multiplication table. • Multiplies and divides whole numbers by multiples of 10, 100 and 1000. MNU 2-03a MTH 2-03c • Adds and subtracts multiples of 10, 100 and 1000 to and from whole numbers and decimal fractions to two decimal places. • Adds and subtracts whole numbers and decimal fractions to two decimal places, within the number range 0 to 1 000 000.

2B Block	Textbook 2B Chapter	Textbook 2B Units	Benchmark/E&O
11	Fractions, decimal fractions and percentages	6.1 Identifying equivalent fractions 6.2 Calculating equivalent fractions 6.3 Comparing and ordering fractions	MNU 2-07a MNU 2-07b MTH 2-07c • Uses knowledge of equivalent forms of common fractions, decimal fractions and percentages (for example, ¾ = 0·75 = 75%) to solve problems. • Creates equivalent fractions and uses this knowledge to put a set of the most commonly used fractions in order.
12	Patterns and relationships	11.1 Exploring and extending number sequences	MTH 2-13a • Explains and uses a rule to extend well known number sequences including square numbers, triangular numbers and Fibonacci sequence. • Applies knowledge of multiples, square numbers and triangular numbers to generate number patterns.
13	Multiples, factors and primes	5.1 Recognising all factors of a number 5.2 Identifying multiples of numbers	MTH 2-05a • Identifies multiples and factors of whole numbers and applies knowledge and understanding of these when solving relevant problems in number, money and measurement.
14	Number – order and place value	2.7 Reading and writing decimal fractions 2.8 Representing and describing decimal fractions 2.9 Comparing and ordering decimal fractions	MNU 2-02a MNU 2-03b MNU 2-04a • Reads, writes and orders sets of decimal fractions to three decimal places. • Explains the link between a digit, its place and its value for numbers to three decimal places. • Partitions a wide range of whole numbers and decimal fractions to three decimal places, for example, 3·6 = 3 ones and 6 tenths = 36 tenths.
15	Estimation and rounding	1.2 Rounding decimal fractions to two places	MNU 2-01a • Rounds decimal fractions to the nearest whole number, to one decimal place and two decimal places. • Applies knowledge of rounding to give an estimate to a calculation appropriate to the context.

2B Block	Textbook 2B Chapter	Textbook 2B Units	Benchmark/E&O
16	2D shapes and 3D Objects	13.1 Naming and sorting polygons 13.2 Describing and drawing circles 13.3 Constructing 3D objects 13.4 Constructing nets	MTH 2-16a MTH 2-16b MTH 2-16c • Describes 3D objects and 2D shapes using specific vocabulary including regular, irregular, diagonal, radius, diameter and circumference. Applies this knowledge to demonstrate understanding of the relationship between 3D objects and their nets. • Identifies and describes 3D objects and 2D shapes within the environment and explains why their properties match their function. • Knows that the radius is half of the diameter. • Uses digital technologies and mathematical instruments to draw 2D shapes and make representations of 3D objects, understanding that not all parts of the 3D object can be seen.
17	Estimation and rounding	1.3 Using rounding to estimate the accuracy of a calculation	MNU 2-01a • Rounds whole numbers to the nearest 1000, 10 000 and 100 000. • Rounds decimal fractions to the nearest whole number, to one decimal place and two decimal places. • Applies knowledge of rounding to give an estimate to a calculation appropriate to the context.
18	Number – addition and subtraction	3.10 Mental and written strategies 3.11 Representing and solving word problems 3.12 Solving multi-step word problems	MNU 2-03a MTH 2-03c • Adds and subtracts multiples of 10, 100 and 1000 to and from whole numbers and decimal fractions to two decimal places. • Adds and subtracts whole numbers and decimal fractions to two decimal places, within the number range 0 to 1 000 000.
19	Fractions, decimal fractions and percentages	6.4 Decimal equivalents to tenths and hundredths 6.5 Decimal equivalents to simple fractions	MNU 2-07a MNU 2-07b MTH 2-07c • Uses knowledge of equivalent forms of common fractions, decimal fractions and percentages (for example, ¾ = 0·75 = 75%) to solve problems.

2B Block	Textbook 2B Chapter	Textbook 2B Units	Benchmark/E&O
20	Number – addition and subtraction	3.13 Adding decimal fractions and whole numbers 3.14 Adding and subtracting decimal fractions	MNU 2-03a MTH 2-03c • Adds and subtracts multiples of 10, 100 and 1000 to and from whole numbers and decimal fractions to two decimal places. • Adds and subtracts whole numbers and decimal fractions to two decimal places, within the number range 0 to 1 000 000.
21	Number – multiplication and division	4.10 Multiplying decimal fractions 4.11 Dividing whole numbers by 10, 100 and 1000	MTH 2-03c • Uses multiplication and division facts to the 10th multiplication table. • Multiplies and divides whole numbers by multiples of 10, 100 and 1000. • Multiplies and divides decimal fractions to two decimal places by 10, 100 and 1000.
22	Money	7.1 Money problems using the four operations 7.2 Budgeting 7.3 Profit and loss 7.4 Discounts 7.5 Credit, debit and debt	MNU 2-09a MNU 2-09b MNU 2-09c • Carries out money calculations involving the four operations. • Compares costs and determines affordability within a given budget. • Demonstrates understanding of the benefits and risks of using bank cards and digital technologies. • Calculates profit and loss accurately, for example, when working with a budget for an enterprise activity. MTH 2-05a • Identifies multiples and factors of whole numbers and applies knowledge and understanding of these when solving relevant problems in number, money and measurement.

2B Block	Textbook 2B Chapter	Textbook 2B Units	Benchmark/E&O
23	Measurement	9.4 Converting metric units 9.5 Imperial measurement 9.6 Calculating perimeter 9.7 Calculating the area of regular shapes 9.8 Calculating Volume	MNU 2-11a MNU 2-11b MNU 2-11c • Uses the comparative size of familiar objects to make reasonable estimations of length, mass, area and capacity. • Estimates to the nearest appropriate unit, then measures accurately: length, height and distance in millimetres (mm), centimetres (cm), metres (m) and kilometres (km); mass in grams (g) and kilograms (kg); and capacity in millilitres (ml) and litres (l). • Calculates the perimeter of simple straight sided 2D shapes in millimetres (mm), centimetres (cm) and metres (m). • Calculates the area of squares, rectangles and right-angled triangles in square millimetres (mm^2), square centimetres (cm^2) and square metres (m^2). • Calculates the volume of cubes and cuboids in cubic centimetres (cm^3) and cubic metres (m^3). • Converts between common units of measurement using decimal notation, for example, 550 cm = 5·5 m; 3·009 kg = 3009 g. • Chooses the most appropriate measuring device for a given task and carries out the required calculation, recording results in the correct unit. • Reads a variety of scales accurately. • Shows awareness of imperial units used in everyday life, for example, miles or stones.

2B Block	Textbook 2B Chapter	Textbook 2B Units	Benchmark/E&O
24	Angles, symmetry and transformation	14.5 Reading coordinates 14.6 Line symmetry 14.7 Symmetrical pictures and diagrams 14.8 Reading scale maps	MTH 2-17a MTH 2-17b MTH 2-17c MTH 2-17d MTH 2-18a MTH 2-19a • Interprets maps, models or plans with simple scales, for example, 1 cm:2 km. • Describes, plots and records the location of a point, in the first quadrant, using coordinate notation. • Identifies and illustrates line symmetry on a wide range of 2D shapes and applies this understanding to complete a range of symmetrical patterns, with and without the use of digital technologies.
25	Number – multiplication and division	4.12 Solving division problems with remainders 4.13 Solving multiplication and division problems	MTH 2-03c • Uses multiplication and division facts to the 10th multiplication table. • Multiplies and divides whole numbers by multiples of 10, 100 and 1000.
26	Fractions, decimal fractions and percentages	6.6 Adding and subtracting fractions 6.7 Calculating the fraction of a value	MNU 2-07a MNU 2-07b MTH 2-07c • Uses knowledge of equivalent forms of common fractions, decimal fractions and percentages (for example, ¾ = 0·75 = 75%) to solve problems. • Calculates simple fractions of quantity and uses this knowledge to solve problems, for example, find 3/5 of 60.

2B Block	Textbook 2B Chapter	Textbook 2B Units	Benchmark/E&O
27	Time	8.4 Measuring time 8.5 Speed, distance and time calculations 8.6 Time problems	MNU 2-10a MNU 2-10b MNU 2-10c • Reads and records time in both 12-hour and 24-hour notation and converts between the two. • Knows the relationships between commonly used units of time and carries out simple conversion calculations, for example, changes 1 hours into minutes. • Calculates durations of activities and events including situations bridging across several hours and parts of hours using both 12-hour clock and 24-hour notation. • Estimates the duration of a journey based on knowledge of the link between speed, distance and time. • Chooses the most appropriate timing device in practical situations and records using relevant units, including hundredths of a second.
28	Expressions and equations	12.1 Solving equations using mathematical rules	MTH 2-15a • Solves simple algebraic equations with one variable, for example, $a - 30 = 40$ and $4b = 20$.
29	Fractions, decimal fractions and percentages	6.8 Comparing numbers with two decimal places 6.9 Percentage 6.10 Converting fractions to percentages 6.11 Percentage calculation	MNU 2-07a MNU 2-07b MTH 2-07c • Uses knowledge of equivalent forms of common fractions, decimal fractions and percentages (for example, $\frac{3}{4} = 0{\cdot}75 = 75\%$) to solve problems. • Calculates simple percentages of a quantity and uses this knowledge to solve problems in everyday contexts, for example, calculates the sale price of an item with a discount of 15%.

2B Block	Textbook 2B Chapter	Textbook 2B Units	Benchmark/E&O
30	Data handling and analysis	15.1 Working with a range of graphs 15.2 Using pie charts 15.3 Creating and interpreting graphs 15.4 Drawing conclusions from graphs	MNU 2-20a MNU 2-20b MTH 2-21a • Devises ways of collecting data in the most suitable way for the given task. • Collects, organises and displays data accurately in a variety of ways including through the use of digital technologies, for example, creating surveys, tables, bar graphs, line graphs, frequency tables, simple pie charts and spreadsheets. • Analyses, interprets and draws conclusions from a variety of data. • Displays data appropriately making effective use of technology and chooses a suitable scale when creating graphs.
31	Ideas of chance and uncertainty	16.1 Investigating the possible outcomes of random events	MNU 2-22a • Uses the language of probability accurately to describe the likelihood of simple events occurring, for example equal chance; fifty-fifty; one in two, two in three; percentage chance; and 1/6. • Plans and carries out simple experiments involving chance with repeated trials, for example, 'what is the probability of throwing a six if you throw a die fifty times?'.

Primary Maths for Scotland 2C scope and sequence

Chapter	Block	Units
Number – order and place value	1	2.1, 2.2, 2.3, 2.4, 2.5, 2.6
Number – addition and subtraction	2	3.1, 3.2, 3.3
Mathematics, its impact on the world, past, present and future	3	10.1
Estimation and rounding	4	1.1
Number – multiplication and division	5	4.1, 4.2
Time	6	8.1, 8.2
Measurement	7	9.1, 9.2
Estimation and rounding	8	1.3
Number – addition and subtraction	9	3.4, 3.5, 3.6, 3.7, 3.8
Angles, symmetry and transformation	10	14.1, 14.2, 14.3, 14.4, 14.5
Multiplies, factors and primes	11	5.1, 5.2
Patterns and relationships	12	11.1
Fractions, decimal fractions and percentages	13	6.1, 6.2, 6.3, 6.4
Number – order and place value	14	2.7, 2.8, 2.9, 2.10, 2.11
Estimation and rounding	15	1.2
Number – addition and subtraction	16	3.9, 3.10, 3.11, 3.12, 3.13, 3.14, 3.15
2D shapes and 3D objects	17	13.1, 13.2, 13.3, 13.4
Number – multiplication and division	18	4.3, 4.4, 4.5, 4.6
Fractions, decimal fractions and percentages	19	6.5, 6.6, 6.7, 6.8, 6.9, 6.10
Measurement	20	9.3, 9.4, 9.5
Data handling and analysis	21	15.1, 15.2, 15.3, 15.4
Number – multiplication and division	22	4.7, 4.8, 4.9, 4.10
Expressions and equations	23	12.1
Measurement	24	9.6, 9.7, 9.8, 9.9, 9.10, 9.11, 9.12
Money	25	7.1, 7.2, 7.3, 7.4, 7.5
Time	26	8.3, 8.4, 8.5
Ideas of chance and uncertainty	27	16.1
Number – multiplication and division	28	4.11, 4.12, 4.13
Angles	29	14.6, 14.7, 14.8, 14.9

Note: sections of the benchmarks/Es&Os not covered in a 2C block have been highlighted grey in the table. These will be covered in 2A and 2B.

2C Block	Textbook 2C Chapter	Textbook 2C Units	Benchmark/E&O
1	Number – order and place value	2.1 Reading and writing whole numbers 2.2 Representing and describing whole numbers 2.3 Place value partitioning of whole numbers 2.4 Number sequences 2.5 Comparing and ordering whole numbers 2.6 Negative numbers	MNU 2-02a MNU 2-03b MNU 2-04a • Reads, writes and orders whole numbers to 1 000 000, starting from any number in the sequence. • Explains the link between a digit, its place and its value for whole numbers to 1 000 000. • Explains the link between a digit, its place and its value for numbers to three decimal places. • Partitions a wide range of whole numbers and decimal fractions to three decimal places, for example, 3·6 = 3 ones and 6 tenths = 36 tenths. • Identifies familiar contexts in which negative numbers are used. • Orders numbers less than zero and locates them on a number line.
2	Number – addition and subtraction	3.1 Mental addition and subtraction 3.2 Adding and subtracting a string of numbers 3.3 Using place value partitioning to add and subtract	MNU 2-03a MTH 2-03c • Adds and subtracts multiples of 10, 100 and 1000 to and from whole numbers and decimal fractions to two decimal places. • Adds and subtracts whole numbers and decimal fractions to two decimal places, within the number range 0 to 1 000 000.
3	Mathematics, its impact on the world, past, present and future	10.1 Mathematical inventions and different number systems	MTH 2-12a • Researches and presents examples of the impact mathematics has in the world of life and work. • Contributes to discussions and activities on the role of mathematics in the creation of important inventions, now and in the past.
4	Estimation and rounding	1.1 Rounding whole numbers	MNU 2-01a • Rounds whole numbers to the nearest 1000, 10 000 and 100 000.

2C Block	Textbook 2C Chapter	Textbook 2C Units	Benchmark/E&O
5	Number – multiplication and division	4.1 Recalling multiplication and division facts for 7 4.2 Recalling multiplication and division facts for 8	MTH 2-03c • Uses multiplication and division facts to the 10th multiplication table
6	Time	8.1 Investigating how long a journey will take 8.2 Calculating the duration of activities and events	MNU 2-10a MNU 2-10b MNU 2-10c • Reads and records time in both 12-hour and 24-hour notation and converts between the two. • Knows the relationships between commonly used units of time and carries out simple conversion calculations, for example, changes 1 ¾ hours into minutes. • Uses and interprets a range of electronic and paper-based timetables and calendars to plan events or activities and solve real life problems. • Calculates durations of activities and events including situations bridging across several hours and parts of hours using both 12-hour clock and 24-hour notation. • Chooses the most appropriate timing device in practical situations and records using relevant units, including hundredths of a second. • Selects the most appropriate unit of time for a given task and justifies choice.

2C Block	Textbook 2C Chapter	Textbook 2C Units	Benchmark/E&O
7	Measurement	9.1 Estimating and measuring length 9.2 Estimating and measuring mass	MNU 2-11a MNU 2-11b MNU 2-11c • Uses the comparative size of familiar objects to make reasonable estimations of length, mass, area and capacity. • Estimates to the nearest appropriate unit, then measures accurately: length, height and distance in millimetres (mm), centimetres (cm), metres (m) and kilometres (km); mass in grams (g) and kilograms (kg); and capacity in millilitres (ml) and litres (l). • Converts between common units of measurement using decimal notation, for example, 550 cm = 5·5 m; 3·009 kg = 3009 g. • Chooses the most appropriate measuring device for a given task and carries out the required calculation, recording results in the correct unit. • Reads a variety of scales accurately.
8	Estimation and rounding	1.3 Using rounding to estimate the answer	MNU 2-01a • Rounds whole numbers to the nearest 1000, 10 000 and 100 000. • Applies knowledge of rounding to give an estimate to a calculation appropriate to the context.
9	Number – addition and subtraction	3.4 Adding whole numbers using standard algorithms 3.5 Subtracting whole numbers using standard algorithms 3.6 Mental and written calculation strategies 3.7 Representing word problems 3.8 Solving multi-step word problems	MNU 2-03a MTH 2-03c • Adds and subtracts multiples of 10, 100 and 1000 to and from whole numbers and decimal fractions to two decimal places. • Adds and subtracts whole numbers and decimal fractions to two decimal places, within the number range 0 to 1 000 000.

2C Block	Textbook 2C Chapter	Textbook 2C Units	Benchmark/E&O
10	Angles, symmetry and transformation	14.1 Drawing triangles 14.2 Reflex angles 14.3 Finding missing angles 14.4 Using bearings 1 14.5 Using bearings 2	MTH 2-17a MTH 2-17b MTH 2-17c MTH 2-17d MTH 2-18a MTH 2-19a • Uses mathematical language including acute, obtuse, straight and reflex to describe and classify a range of angles identified within shapes in the environment. • Measures and draws a range of angles to within ± 2°. • Knows that complementary angles add up to 90° and supplementary angles add up to 180° and uses this knowledge to calculate missing angles. • Uses knowledge of the link between the eight compass points and angles to describe, follow and record directions. • Interprets maps, models or plans with simple scales, for example, 1 cm:2 km.
11	Multiples, factors and primes	5.1 Using knowledge of multiples and factors to work out divisibility rules 5.2 Using knowledge of multiples and factors to solve problems	MTH 2-05a • Identifies multiples and factors of whole numbers and applies knowledge and understanding of these when solving relevant problems in number, money and measurement.
12	Patterns and relationships	11.1 Applying knowledge of multiples, square numbers and triangular numbers to generate number patterns	MTH 2-13a • Explains and uses a rule to extend well known number sequences including square numbers, triangular numbers and Fibonacci sequence. • Applies knowledge of multiples, square numbers and triangular numbers to generate number patterns.

2C Block	Textbook 2C Chapter	Textbook 2C Units	Benchmark/E&O
13	Fractions, decimal fractions and percentages	6.1 Converting fractions 6.2 Comparing and ordering fractions 6.3 Simplifying fractions 6.4 Adding and subtracting fractions	MNU 2-07a MNU 2-07b MTH 2-07c • Uses knowledge of equivalent forms of common fractions, decimal fractions and percentages (for example, ¾ = 0·75 = 75%) to solve problems. • Creates equivalent fractions and uses this knowledge to put a set of the most commonly used fractions in order. • Expresses fractions in their simplest form.
14	Number – order and place value	2.7 Reading and writing decimal fractions 2.8 Representing and describing decimal fractions 2.9 Zero as a placeholder in decimal fractions 2.10 Partitioning decimal fractions 2.11 Comparing and ordering decimal fractions	MNU 2-02a MNU 2-03b MNU 2-04a • Reads, writes and orders whole numbers to 1 000 000, starting from any number in the sequence. • Explains the link between a digit, its place and its value for whole numbers to 1 000 000. • Reads, writes and orders sets of decimal fractions to three decimal places. • Explains the link between a digit, its place and its value for numbers to three decimal places. • Partitions a wide range of whole numbers and decimal fractions to three decimal places, for example, 3·6 = 3 ones and 6 tenths = 36 tenths.
15	Estimation and rounding	1.2 Rounding decimal fractions	MNU 2-01a • Rounds whole numbers to the nearest 1000, 10 000 and 100 000. • Rounds decimal fractions to the nearest whole number, to one decimal place and two decimal places. • Applies knowledge of rounding to give an estimate to a calculation appropriate to the context.

2C Block	Textbook 2C Chapter	Textbook 2C Units	Benchmark/E&O
16	Number – addition and subtraction	3.9 Adding whole numbers and decimal fractions 3.10 Adding decimal fractions using standard algorithms 3.11 Subtracting decimal fractions 3.12 Subtracting decimal fractions using standard algorithms 3.13 Adding and subtracting decimal fractions 3.14 Representing word problems involving decimal fractions 3.15 Multi-step word problems	MNU 2-03a MTH 2-03c • Adds and subtracts multiples of 10, 100 and 1000 to and from whole numbers and decimal fractions to two decimal places. • Adds and subtracts whole numbers and decimal fractions to two decimal places, within the number range 0 to 1 000 000.
17	2D shapes and 3D Objects	13.1 Describing and sorting triangles 13.2 Drawing 2D shapes 13.3 Making representations of 3D objects 13.4 Nets of prisms	MTH 2-16a MTH 2-16b MTH 2-16c • Describes 3D objects and 2D shapes using specific vocabulary including regular, irregular, diagonal, radius, diameter and circumference. Applies this knowledge to demonstrate understanding of the relationship between 3D objects and their nets. • Identifies and describes 3D objects and 2D shapes within the environment and explains why their properties match their function. • Uses digital technologies and mathematical instruments to draw 2D shapes and make representations of 3D objects, understanding that not all parts of the 3D object can be seen.

2C Block	Textbook 2C Chapter	Textbook 2C Units	Benchmark/E&O
18	Number – multiplication and division	4.3 Multiplication with decimal fractions 4.4 Division with decimal fractions 4.5 Solving multiplication problems 4.6 Solving multiplication problems involving decimal fractions	MTH 2-03c • Uses multiplication and division facts to the 10th multiplication table • Multiplies and divides whole numbers by multiples of 10, 100 and 1000. • Multiplies and divides decimal fractions to two decimal places by 10, 100 and 1000. • Multiplies decimal fractions to two decimal places by a single digit. • Divides whole numbers and decimal fractions to two decimal places, by a single digit, including answers expressed as decimal fractions, for example, $43 \div 5 = 8{\cdot}6$.
19	Fractions, decimal fractions and percentages	6.5 Converting decimal fractions to fractions 6.6 Calculating a fraction of a fraction 6.7 Dividing a fraction by a whole number 6.8 Dividing a whole number by a fraction 6.9 Equivalents: fractions, decimals and percentages 6.10 Solving fraction, decimal and percentage problems	MNU 2-07a MNU 2-07b MTH 2-07c • Uses knowledge of equivalent forms of common fractions, decimal fractions and percentages (for example, $\frac{3}{4} = 0{\cdot}75 = 75\%$) to solve problems. • Calculates simple fractions of quantity and uses this knowledge to solve problems, for example, find 3/5 of 60. • Calculates simple percentages of a quantity and uses this knowledge to solve problems in everyday contexts, for example, calculates the sale price of an item with a discount of 15%. • Creates equivalent fractions and uses this knowledge to put a set of the most commonly used fractions in order. • Expresses fractions in their simplest form.

2C Block	Textbook 2C Chapter	Textbook 2C Units	Benchmark/E&O
20	Measurement	9.3 Estimating and measuring area 9.4 Estimating and measuring capacity 9.5 Estimating imperial measurements	MNU 2-11a MNU 2-11b MNU 2-11c • Uses the comparative size of familiar objects to make reasonable estimations of length, mass, area and capacity. • Estimates to the nearest appropriate unit, then measures accurately: length, height and distance in millimetres (mm), centimetres (cm), metres (m) and kilometres (km); mass in grams (g) and kilograms (kg); and capacity in millilitres (ml) and litres (l). • Calculates the area of squares, rectangles and right-angled triangles in square millimetres (mm^2), square centimetres (cm^2) and square metres (m^2). • Converts between common units of measurement using decimal notation, for example, 550 cm = 5·5 m; 3·009 kg = 3009 g. • Chooses the most appropriate measuring device for a given task and carries out the required calculation, recording results in the correct unit. • Reads a variety of scales accurately.

2C Block	Textbook 2C Chapter	Textbook 2C Units	Benchmark/E&O
21	Data handling and analysis	15.1 Sampling 15.2 Interpreting graphs 15.3 Understand sampling bias 15.4 Pie charts	MNU 2-20a MNU 2-20b MTH 2-21a • Devises ways of collecting data in the most suitable way for the given task. • Collects, organises and displays data accurately in a variety of ways including through the use of digital technologies, for example, creating surveys, tables, bar graphs, line graphs, frequency tables, simple pie charts and spreadsheets. • Analyses, interprets and draws conclusions from a variety of data. • Draws conclusions about the reliability of data, taking into account, for example, the author, the audience, the scale and sample size used. • Displays data appropriately making effective use of technology and chooses a suitable scale when creating graphs.
22	Number – multiplication and division	4.7 Solving division problems using place value 4.8 Solving multiplication problems using proportional adjustment 4.9 Solving division problems 4.6 Solving multiplication problems	MTH 2-03c • Uses multiplication and division facts to the 10th multiplication table • Multiplies and divides whole numbers by multiples of 10, 100 and 1000. • Multiplies whole numbers by two-digit numbers. • Multiplies decimal fractions to two decimal places by a single digit. • Divides whole numbers and decimal fractions to two decimal places, by a single digit, including answers expressed as decimal fractions, for example, $43 \div 5 = 8{\cdot}6$.
23	Expressions and equations	12.1 Solving equations with inequalities	MTH 2-15a • Solves simple algebraic equations with one variable, for example, $a - 30 = 40$ and $4b = 20$.

2C Block	Textbook 2C Chapter	Textbook 2C Units	Benchmark/E&O
24	Measurement	9.6 Converting imperial measurements 9.7 Calculating perimeter (rectilinear shapes) 9.8 Calculating area: triangles 9.9 Measuring area (composite shapes) 9.10 Calculating volume 9.11 Volume: composite shapes 9.12 Capacity problems	MNU 2-11a MNU 2-11b MNU 2-11c • Uses the comparative size of familiar objects to make reasonable estimations of length, mass, area and capacity. • Estimates to the nearest appropriate unit, then measures accurately: length, height and distance in millimetres (mm), centimetres (cm), metres (m) and kilometres (km); mass in grams (g) and kilograms (kg); and capacity in millilitres (ml) and litres (l). • Calculates the perimeter of simple straight sided 2D shapes in millimetres (mm), centimetres (cm) and metres (m). • Calculates the area of squares, rectangles and right-angled triangles in square millimetres (mm^2), square centimetres (cm^2) and square metres (m^2). • Calculates the volume of cubes and cuboids in cubic centimetres (cm^3) and cubic metres (m^3). • Converts between common units of measurement using decimal notation, for example, 550 cm = 5·5 m; 3·009 kg = 3009 g. • Chooses the most appropriate measuring device for a given task and carries out the required calculation, recording results in the correct unit. • Draws squares and rectangles accurately with a given perimeter or area.

2C Block	Textbook 2C Chapter	Textbook 2C Units	Benchmark/E&O
25	Money	7.1 Money problems using the four operations 7.2 Budgeting 7.3 Profit and loss 7.4 Discounts 7.5 Hire purchase	MNU 2-09a MNU 2-09b MNU 2-09c • Carries out money calculations involving the four operations. • Compares costs and determines affordability within a given budget. • Demonstrates understanding of the benefits and risks of using bank cards and digital technologies. • Calculates profit and loss accurately, for example, when working with a budget for an enterprise activity.
26	Time	8.3 Investigating ways speed, time and distance can be measured 8.4 Calculating time accurately using a stopwatch 8.5 Converting between units of time	MNU 2-10a MNU 2-10b MNU 2-10c • Reads and records time in both 12-hour and 24-hour notation and converts between the two. • Knows the relationships between commonly used units of time and carries out simple conversion calculations, for example, changes 1 ¾ hours into minutes. • Uses and interprets a range of electronic and paper-based timetables and calendars to plan events or activities and solve real life problems. • Calculates durations of activities and events including situations bridging across several hours and parts of hours using both 12-hour clock and 24-hour notation. • Estimates the duration of a journey based on knowledge of the link between speed, distance and time. • Chooses the most appropriate timing device in practical situations and records using relevant units, including hundredths of a second. • Selects the most appropriate unit of time for a given task and justifies choice.

2C Block	Textbook 2C Chapter	Textbook 2C Units	Benchmark/E&O
27	Ideas of chance and uncertainty	16.1 Predicting and explaining the outcomes of simple chance situations and experiments	MNU 2-22a • Uses the language of probability accurately to describe the likelihood of simple events occurring, for example equal chance; fifty-fifty; one in two, two in three; percentage chance; and 1/6. • Plans and carries out simple experiments involving chance with repeated trials, for example, 'what is the probability of throwing a six if you throw a die fifty times?'. • Uses data to predict the outcome of a simple experiment.
28	Number – multiplication and division	4.11 Solving division problems using an algorithm 4.12 Solving division problems involving decimal fractions 4.13 Solving multi-step problems using the order of operations	MTH 2-03c • Uses multiplication and division facts to the 10th multiplication table • Multiplies and divides whole numbers by multiples of 10, 100 and 1000. • Multiplies and divides decimal fractions to two decimal places by 10, 100 and 1000. • Multiplies whole numbers by two-digit numbers. • Divides whole numbers and decimal fractions to two decimal places, by a single digit, including answers expressed as decimal fractions, for example, $43 \div 5 = 8{\cdot}6$. • Applies the correct order of operations in number calculations when solving multi-step problems.

2C Block	Textbook 2C Chapter	Textbook 2C Units	Benchmark/E&O
29	Angles, symmetry and transformation	14.6 Using coordinates 14.7 Symmetry 1 14.8 Symmetry 2 14.9 Using scale	MTH 2-17a MTH 2-17b MTH 2-17c MTH 2-17d MTH 2-18a MTH 2-19a • Uses mathematical language including acute, obtuse, straight and reflex to describe and classify a range of angles identified within shapes in the environment. • Measures and draws a range of angles to within ± 2° • Uses knowledge of the link between the eight compass points and angles to describe, follow and record directions. • Interprets maps, models or plans with simple scales, for example, 1 cm:2 km. • Describes, plots and records the location of a point, in the first quadrant, using coordinate notation. • Identifies and illustrates line symmetry on a wide range of 2D shapes and applies this understanding to complete a range of symmetrical patterns, with and without the use of digital technologies.

1 Estimation and rounding

Estimation and rounding 2A

Section overview

Textbook 2A units	Review units	Prior learning	Benchmarks/ Es&Os
Unit 1.1 Rounding whole numbers to the nearest 10	**1C** Units 1.1 and 1.2	• I can read and write numbers with up to 4 digits. • I understand that the place of a digit within a number shows its value. • I know the meaning of each digit in the ones, tens, hundreds and thousands columns.	**MNU 2-01a** • Rounds whole numbers to the nearest 1000, 10 000 and 100 000. • Rounds decimal fractions to the nearest whole number, to one decimal place and two decimal places. • Applies knowledge of rounding to give an estimate to a calculation appropriate to the context.
Unit 1.2 Rounding whole numbers to the nearest 100	**1C** Units 1.1 and 1.2	• I can read and write numbers with up to 4 digits. • I understand that the place of a digit within a number shows its value. • I know the meaning of each digit in the ones, tens, hundreds and thousands columns.	
Unit 1.3 Rounding decimal fractions to the nearest whole number	**1C** Units 1.1 and 1.2 **2A** Unit 2.6	• I can read and write decimal fractions to one decimal place. • I can count in decimal amounts to one decimal place forwards and backwards. • I understand the value of each digit in decimal fractions up to one decimal place. • I can round whole numbers to the nearest ten and the nearest hundred.	
Unit 1.4 Estimating the answer using rounding	**1C** Units 1.1 and 1.2	• I can read and write numbers with up to 4 digits. • I understand that the place of a digit within a number shows its value. • I know the meaning of each digit in the ones, tens, hundreds and thousands columns. • I can round numbers to the nearest ten or hundred.	

Conceptual context

In daily life, we often round numbers to allow us to easily work with them and to check answers. It is important that pupils have a clear understanding of the numerals they are working with, and their value, to be able to estimate and round effectively. Pupils require a wide range of opportunities in meaningful contexts to evaluate the accuracy of calculations by estimating, and to explore different levels of accuracy when estimating. In this way, they will develop the ability to habitually estimate if answers are reasonable, and be able to use rounding in ways that are relevant to the context. For instance, if working out a problem such as:

You need 1·8 metres each to make a shelf. How many shelves can you make with 50 metres of wood?

pupils need to understand that rounding to the nearest whole number (2m) then dividing 50 by 2 will give a reasonable estimate of 25, with some left over.

It is important that pupils use number lines (as well as a range of other materials to represent numbers visually if needed) to allow them to explore how and why we round up or down. In this way, they can also learn to visualise these materials to help them make an active decision on rounding, rather than simply following a rule.

For example, if we want to round 382 to the nearest hundred:

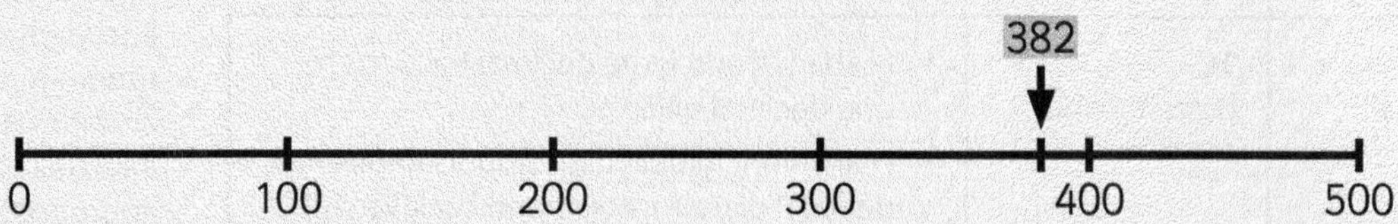

Pupils need to learn to identify multiples of hundred on either side, find the halfway point between them, and use this understanding to decide whether to round up or down. Wide experience of using number lines will allow them to develop strong understanding of the relative size of numbers, and the ability to make reasonable estimates.

They also need to understand that the context of the problem will affect the way we round numbers. For example, if we are working out how much string we would need to get nine 27·5cm lengths of string, then it would be sensible to round up to 30cm and estimate 9×30 to make sure that we have enough string with some left over. However if we were estimating lengths of wood that needed to fit between two posts 27·5cm apart then it would be more sensible to round down to 27cm to make sure they are not too long to fit! Pupils need a range of experiences, estimating problems relevant to daily life, so they understand that the context of a problem affects how we estimate, how accurate we need to be, and what form rounding should take.

Resources

- Number lines 0-100, 0-1000 in a range of increments – ones, tens, hundreds, tenths
- Numeral cards 0-100
- Decimats
- Arrow cards
- Physical materials to represent numbers such as cubes or place value sticks
- Money – mock £10, £100 notes and £1 coins

Key vocabulary

- **Approximately**
- **Estimate**
- **Guess**
- **Roughly**
- **Round**

Misconceptions

Potential misconception	Addressing the misconception
Units 1.1, 1.2, 1.3 and 1.4 The pupil might not understand which digit in the number to look at to determine whether to round up or down. For example, 829 rounded to the nearest 100 is 900 (pupil is using last digit to round incorrectly).	• Show number on a number line and ask them which multiples of 10 the number is between and how they decide which multiple is nearer to the number. Ask them which digit helps them decide.
Units 1.1, 1.2, 1.3 and 1.4 The pupil might not understand the convention for rounding up if a digit is 5 or more.	• Show number on a number line and allow them to explore which multiple is nearer on either side. Explain that we usually round up when the digit is five or more.
Units 1.1, 1.2, 1.3 and 1.4 The pupil always rounds up, e.g. 62 rounded to the nearest ten is 70.	• Show number on a number line and ask them which multiples the number is between and how they decide which multiple is nearer to the number. Ask them which digit helps them decide.
Units 1.1, 1.2, 1.3 and 1.4 The pupil thinks rounding is just changing the numbers to zero, e.g. 587 rounded to the nearest 100 is 500.	• Show number on a number line and ask them which multiples the number is between and how they decide which multiple is nearer to the number. Ask them which digit helps them decide.
Unit 1.4 The pupil doesn't understand that the context matters when using rounding.	• Give a problem based on a real-life context, such as 'Amman has £56 and is estimating how many friends he can take to a concert if tickets cost £10 each.' In this problem he can't use the extra £6 as it's not enough for a ticket, so rounds down to £50 rather than the usual convention of rounding up for numbers ending in 5 or over.

Teaching ideas

Unit 1.1 Rounding whole numbers to the nearest 10

- Take a pack of numeral cards to 100 and shuffle them. Ask a pupil to pick one and tell you the number. Find the numeral on a number line to 100. Ask: **What multiples are on either side of the number – can you find them on the number line?** For example, if you picked 32, the multiples of 30 and 40 are the multiples of ten either side. **Which is nearer to the number?** In this case 32 is nearer to 30, so we round down.
- Repeat with lots of different examples, rounding up and down.
- Discuss what happens if we have a number ending with 5. Explain that we usually round up in this case, although it depends on the particular context. For example, if we were measuring materials to build a fence, there may be instances where we would always round down to make sure the length would not be too big to fit between posts. Or if we wanted to buy tickets that cost £10 each and a group had £65, we would need to round down to the nearest ten because we can't use the £5 to buy another ticket. So there may be some particular contexts where we would always round down or round up, but otherwise the rule is that for numbers 5 and over, we round up.
- Extend to 3-digit and 4-digit numbers once confident.
- Ask for examples why we might round to the nearest ten in daily life, e.g. using money, directions.

Unit 1.2 Rounding whole numbers to the nearest 100

- Check understanding of rounding to the nearest 10 for a range of numbers. Ask questions such as: **What is 88 rounded to the nearest ten? How about 739? What number do I look at to decide? If I think of a number and round it up to 50, what could all the possible numbers be?**

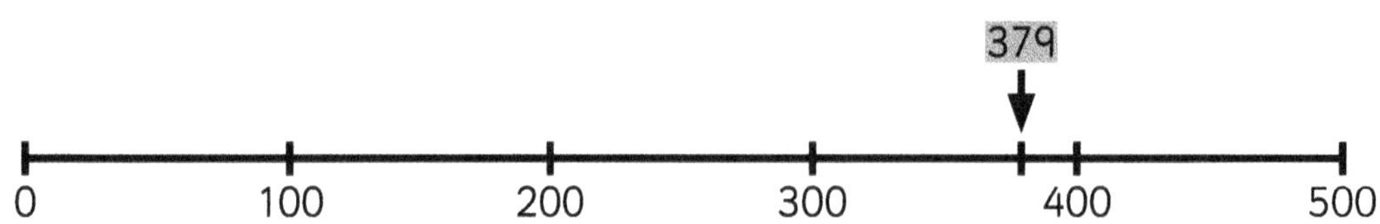

- Ask pupils what the multiples of 100 on either side would be. Ask them to write them on the number line. **Which multiple of 100 is nearer to 379?** Discuss how they know. Explain that we round to the nearest 100 by finding the closest multiple of 100. Repeat with different examples.
- Discuss what happens when a number ends in 50; explain that we usually round up in this case, although it depends on the particular context.
- Extend to 4-digit numbers once confident.

Unit 1.3 Rounding decimal fractions to the nearest whole number

- Check understanding of decimal fractions by showing decimats for tenths and asking what number they represent.
- Draw an empty number line on the whiteboard from 0 to 1. Ask pupils in turns to estimate and write where various tenths would be on the number line, e.g. 0·6, 0·2, 0·9. Now ask: **Which are closer to 1 and which are closer to zero? What about 0·5?** Explain that we usually round up for numbers that end in 0·5 or higher, although it depends on the particular context.
- Now give examples such as 21·3 – ask pupils to draw empty number lines on whiteboards that show which whole numbers are on either side. **Which whole number is closer? How do they know?** Repeat with a wide range of examples for numbers up to 4 digits.
- Discuss examples of when we might round decimal fractions to the nearest whole in daily life, e.g. rounding centimetres to the nearest millimetre or metre when estimating lengths, rounding prices to the nearest pound to estimate costs.

Unit 1.4 Estimating the answer using rounding

- Check understanding of rounding to the nearest 10, 100 and whole number.
- Show a number problem in context, e.g. Isla and Nuria are working out the cost of new camping gear. The tent costs £179 and the sleeping bags cost £52. Isla works out they need £699. Ask the pupils to roughly estimate if that seems likely to be correct. Ask the pupils to round each number to the nearest ten and write their answers on the board. **What is £180 and £50? Was Isla's answer reasonable?**
- Isla's answer of £699 is not reasonable; using rounding we can see the total should be roughly £230.
- Explain that we often use rounding to quickly estimate if an answer is reasonable.
- Give a wide range of examples for pupils to round and estimate, allowing them to explore rounding to the nearest ten, hundred or whole number for different problems.

Estimation and rounding 2B

Section overview

Textbook 2B units	Review units	Prior learning	Benchmarks/Es&Os
Unit 1.1 Rounding whole numbers to the nearest 10, 100 or 1000	**2A** Units 1.1 and 1.2	• I can read and write numbers with up to 5 digits. • I understand that the place of a digit within a number shows its value. • I know the meaning of each digit in the ones, tens, hundreds, thousands and ten thousands columns.	**MNU 2-01a** • Rounds whole numbers to the nearest 1000, 10 000 and 100 000. • Rounds decimal fractions to the nearest whole number, to one decimal place and two decimal places. • Applies knowledge of rounding to give an estimate to a calculation appropriate to the context.
Unit 1.2 Rounding decimal fractions to two places	**2A** Units 1.1, 1.2 and 2.6	• I can read and write decimal fractions to two decimal places. • I can count in decimal amounts to two decimal places forwards and backwards. • I understand the value of each digit in decimal fractions up to two decimal places • I can round whole numbers to the nearest 10, 100 or 1000.	
Unit 1.3 Using rounding to estimate the accuracy of a calculation	**2A** Units 1.1, 1.2 and 2.6	• I can read and write numbers with up to 5 digits. • I understand that the place of a digit within a number shows its value. • I know the meaning of each digit in the ones, tens, hundreds, thousands and ten thousands columns • I can round numbers to the nearest 10, 100 or 1000.	

Conceptual context

In Textbook 2B, knowledge of rounding larger numbers to the nearest 10, 100 or 1000 is extended to allow us to easily work with them and to develop skills in estimating the reasonableness of answers. We also extend knowledge of rounding decimal fractions to the nearest tenth. It is important that pupils have a clear understanding of the numerals they are working with and their value to be able to estimate and round effectively. Materials such as decimats are very useful to check pupils' understanding of reading, writing and representing decimal fractions.

Pupils should use empty number lines, (as well as a range of other materials to represent numbers visually if needed) to allow them to explore how and why we round up or down. In this way, they can also learn to visualise these materials to help them make an active decision on rounding for themselves, rather than simply following a rule.

At this stage, it is important that pupils explore a wide range of contexts involving estimating to develop the ability to assess what levels of accuracy are most useful when rounding. For example, if we were working out how many visitors have visited an event in a month in order to compare attendance over different years, and the numbers we are looking at are 38 498 and 42 940, then for these purposes rounding to the nearest thousand would be accurate enough to give us reasonable estimates to compare. However, if we were estimating measurements for curtains and wanted to know how much material to buy for two windows with widths of 187·5cm then rounding to the nearest ten would be a more useful estimate.

It is also important that pupils begin to understand how useful it is to estimate calculations as a matter of course using their understanding of rounding, operations and the relative size of the numbers they are working with, to evaluate if answers to a problem are reasonable.

Resources

- Number lines 0-100, 0-100, 0-10 000 in a range of increments – ones, tens, hundreds, thousands tenths, hundredths
- Numeral cards 0-9 that can be used to make a range of numbers up to 5 digits
- Decimats
- Arrow cards
- Place value materials – sticks, counters, base 10 materials
- Money – mock £10, £100 notes and a range of coins

Key vocabulary

- **Estimate**
- **Round**

Misconceptions

Potential misconception	Addressing the misconception
Units 1.1, 1.2 and 1.3 The pupil might not understand which digit in the number to look at to determine whether to round up or down. For example, 3821 rounded to the nearest 1000 is 3000 (pupil is using last digit to round incorrectly).	• Write number on an empty number line and ask them which multiple of 1000 the number is between and how they decide which multiple is nearer to the number. Ask them which digit helps them decide.
Units 1.1, 1.2 and 1.3 The pupil might not understand the convention for rounding up if a digit is 5 or more.	• Show number on a number line and allow them to explore which multiple is nearer on either side. Explain that we usually round up when the digit is five or more.
Units 1.1, 1.2 and 1.3 The pupil always rounds up, e.g. 6.32 rounded to the nearest ten is 6.4.	• Show number on a number line and ask them which multiples the number is between and how they decide which multiple is nearer to the number. Ask them which digit helps them decide.
Units 1.1, 1.2 and 1.3 The pupil thinks rounding is just changing the numbers to zero, e.g. 15 687 rounded to the nearest 1000 is 15 000.	• Show number on a number line and ask them which multiples the number is between and how they decide which multiple is nearer to the number. Ask them which digit helps them decide.
Unit 1.3 The pupil doesn't understand that the context matters when using rounding.	• Give a problem based on a real-life context, such as 'Amman has £56 and is estimating how many friends he can take to a concert if tickets cost £10 each.' In this problem he can't use the extra £6 as it's not enough for a ticket, so rounds down to £50 rather than the usual convention of rounding up for numbers ending in 5 or over.

Teaching ideas

Unit 1.1 Rounding whole numbers to the nearest 10, 100 or 1000

- Use numeral cards to make a 5-digit number, e.g. 23 709. Ask questions such as: **What is this number? What does the 3 represent? How many hundreds in the whole number?** to check understanding of place value.
- Draw an empty number line on the whiteboard and mark on 23 709 about half way. Ask the pupils: **What are the multiples of ten on either side of this number?** Ask someone to write on the number line the multiple of ten on either side. Revise learning from previous units where we learned that we usually round up if a number ends in 5 or more. **What is 23 709 rounded to the nearest 10?**

- Now ask the pupils if they can find the multiples of 100 on either side of 23 709. Ask someone to write the multiples of 100 on the number line. Ask: **What digit do we look at to decide whether to round up or down?** Revise learning from previous units where we learned that when a number ends in 50 or more we usually round up, and when a number is less than 50 we usually round down. Ask the pupils if we round up or down in this case.
- Now repeat for multiples of 1000 on either side.
- Repeat for a wide range of examples and discuss examples when we might use rounding in daily life, such as estimating amounts of money for budgets or comparing distances.

Unit 1.2 Rounding decimal fractions to two places

- Write a decimal fraction to two places on the board, for example 3·72, and ask the pupils to read the number. Ask the pupils to show this number using decimats. Ask questions such as: **What does the 3 represent? What about the 2? How many tenths in the whole number?** to check understanding of place value.
- Draw an empty number line on the board from 3 to 4 and ask: **Where would this number be on the number line?** Mark 3·72 on the number line and ask: **What would this number be rounded to the nearest whole number? How do you know?** to check understanding of learning in previous units.
- Now ask: **What would the tenths be on either side of this number?** Ask someone to mark them on the number line. Ask: **Which tenth is this number closer to, 3·7 or 3·8?** Explain that we usually round up for numbers that end in 0·05 or higher, although it depends on the particular context.
- Now give examples such as 0·39 – ask pupils to draw empty number lines on whiteboards that show which tenths are on either side. **Which tenth is closer? How do they know?**
- Repeat with a wide range of examples for decimal fractions to two places.
- Discuss examples of when we might round decimal fractions to the nearest tenth in daily life, e.g. rounding measurements such as lengths or weights or comparing temperatures.

Unit 1.3 Using rounding to estimate the accuracy of a calculation

- Check understanding of rounding to the nearest 10, 100 and 1000.
- Check understanding of rounding decimal fractions to the nearest tenth or whole number.
- Show a number problem in context, e.g. Finlay and his dad are comparing the costs of holidays. It would cost £2398 for a family of four to go to Cyprus. A holiday to Portugal costs £1847 for a family of four. Finlay works out the difference in cost is £1091.
- Explain that we often use rounding to quickly estimate if an answer is reasonable. Ask the pupils to round each number to the nearest hundred and write their answers on the board. **What is £2400 – £1800? Was Finlay's answer reasonable?**
- Rounding is a useful skill to check our answers. Give a wide range of examples for pupils to round and estimate, allowing them to explore rounding to the nearest ten, hundred, thousand, tenth or whole number for different problems.

Estimation and rounding 2C

Section overview

Textbook 2C units	Review units	Prior learning	Benchmarks/Es&Os
Unit 1.1 Rounding whole numbers to the nearest 1000, 10 000 or 100 000	**2B** Units 1.2 and 1.2	• I can read and write numbers with up to 6 digits. • I understand that the place of a digit within a number shows its value. • I know the meaning of each digit in each column of numbers up to 6 digits.	**MNU 2-01a** • Rounds whole numbers to the nearest 1000, 10 000 and 100 000. • Rounds decimal fractions to the nearest whole number, to one decimal place and two decimal places. • Applies knowledge of rounding to give an estimate to a calculation appropriate to the context.
Unit 1.2 Rounding decimal fractions to three places	**2B** Units 1.1, 1.2 and 2.7	• I can read and write decimal fractions to three decimal places. • I can count in decimal amounts to three decimal places forwards and backwards. • I understand the value of each digit in decimal fractions up to three decimal places. • I can round decimal fractions to the nearest whole number or 1/10.	
Unit 1.3 Using rounding to estimate the answer	**2B** Units 1.1 and 1.2	• I can read and write numbers with up to 6 digits. • I understand that the place of a digit within a number shows its value. • I know the meaning of each digit in numbers of up to 6 digits. • I can round numbers to the nearest 10, 100, 1000, 10 000 or 100 000.	

Conceptual context

In Textbook 2C we extend knowledge of rounding larger numbers to the nearest 1000, 10 000 or 100 000 to allow pupils to easily work with them and to evaluate the accuracy of answers. We also extend knowledge of rounding decimal fractions to the nearest hundredth. It is very important that pupils have a clear understanding of the numerals they are working with, and their value, to be able to estimate and round effectively.

At this stage, pupils need to be developing the ability to estimate calculations habitually, and make sensible decisions on how to estimate based on the context of a problem. It is therefore vital that pupils are given a wide range of opportunities to estimate calculations based on different contexts. They need to understand what levels of accuracy will be most useful based on the problem they are working out, when rounding will be the most useful way to estimate, and by what multiples, and also times when rounding may not be the most sensible way of estimating if an answer is reasonable. For example, if estimating the answer to $0{\cdot}58 \times 298$ it would be more useful to round to think of this problem as approximately half of 300 to reach a rough estimate of 150, rather than rounding to the nearest whole number, with a much less accurate estimate of 300.

Resources

- Empty number lines
- Numeral cards 0-9 that can be used to make a range of numbers up to 6 digits
- Decimats – tenths, hundredths and thousandths
- Arrow cards
- Place value materials
- Money – mock £10, £100 notes and a range of coins

Key vocabulary

- **Estimate**
- **Round**

Misconceptions

Potential misconception	Addressing the misconception
Units 1.1, 1.2 and 1.3 The pupil might not understand which digit in the number to look at to determine whether to round up or down. For example, 37 294 rounded to the nearest 10 000 is 30 000 (pupil is using last digit to round incorrectly).	• Write number on an empty number line and ask them which multiple of 10 000 the number is between and how they decide which multiple is nearer to the number. Ask them which digit helps them decide.
Units 1.1, 1.2 and 1.3 The pupil might not understand the convention for rounding up if a digit is 5 or more.	• Show number on a number line and allow them to explore which multiple is nearer on either side. Explain that we usually round up when the digit is five or more.
Units 1.1, 1.2 and 1.3 The pupil always rounds up, e.g. 0·231 rounded to the nearest hundredth is 0·24.	• Show number on a number line and ask them which multiples the number is between and how they decide which multiple is nearer to the number. Ask them which digit helps them decide.
The pupil thinks rounding is just changing the numbers to zero, e.g. 915 687 rounded to the nearest 100 000 is 900 000.	• Show number on a number line and ask them which multiples the number is between and how they decide which multiple is nearer to the number. Ask them which digit helps them decide.
Unit 1.3 The pupil doesn't understand that the context matters when using rounding.	• Give a problem based on a real-life context, such as 'Amman has £56 and is estimating how many friends he can take to a concert if tickets cost £10 each.' In this problem he can't use the extra £6 as it's not enough for a ticket, so rounds down to £50 rather than the usual convention of rounding up for numbers ending in 5 or over.

Teaching ideas

Unit 1.1 Rounding whole numbers to the nearest 1000, 10 000 or 100 000

- Use numeral cards to make a 6-digit number, e.g. 384 298. Ask questions such as: **What is this number? What does the 3 represent? How many thousands in the whole number?** to check understanding of place value.

- Draw an empty number line on the whiteboard and mark on 384 298 about half way. Ask the pupils: **What are the multiples of thousand on either side of this number?** Ask someone to write on the number line the multiple of thousand on either side. Revise learning from previous units where we learned that we usually round up if a number ends in 500 or more. Ask the pupils if we round up or down in this case.
- Now ask the pupils if they can find the multiples of 10 000 on either side of 384 298. Ask someone to write the multiples of 10 000 on the number line. Ask: **What number do we look at to decide whether to round up or down?** Ask if we round up or down in this case.
- Now repeat for multiples of 100 000 on either side.
- Repeat for a wide range of examples and discuss examples when we might use rounding in daily life, such as estimating numbers of votes or the cost of housing.
- Ask questions such as: **What number do I look at to decide? If I think of a number and round it up to 50 000 – what could all the possible numbers be?**

Unit 1.2 Rounding decimal fractions to two places

- Write a decimal fraction to three places on the board, for example 18·714, and ask the pupils to read the number. Ask questions such as: **What does the 1 represent? What about the 4? How many hundredths in the whole number?** to check understanding of place value.
- Draw an empty number line on the board from 18 to 19 and ask: **Where would this number be on the number line?**

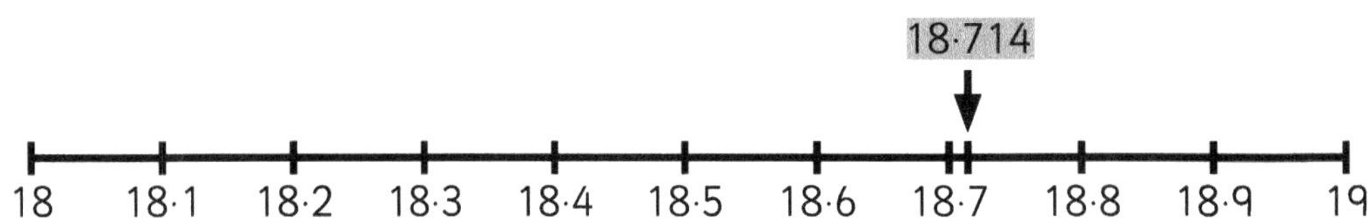

- Mark 18·714 on the number line and ask: **What would this number be rounded to the nearest whole number? What about the nearest tenth? How do you know? What digit do you look at to decide?** to check understanding of learning in previous units.
- Now ask: **What would the hundredths be on either side of this number?** Ask someone to mark them on the number line. Ask: **Which hundredth is this number closer to, 18·71 or 18·72?** Explain that we usually round up for numbers that end in 0·005 or higher, although it depends on the particular context.
- Now give a wide range of examples of decimal fractions to 3 places – ask pupils to draw empty number lines on whiteboards that show which hundredths are on either side. **Which hundredth is closer? How do you know?**
- Repeat with a wide range of examples for decimal fractions to three places.
- Discuss examples of when we might round decimal fractions to the nearest hundredth in daily life, e.g. rounding measurements such as distances or weights.

Unit 1.3 Using rounding to estimate the accuracy of a calculation

- Check understanding of rounding to the nearest 10, 100, 1000, 10 000 or 100 000.
- Check understanding of rounding decimal fractions to the nearest tenth, hundredth or whole number.
- Show a number problem in context, for example:
- Amman earns £29·75 each week during the summer holidays. He wants to work out how much money he will have after the 6 weeks of holidays. He works out this will be £139·75.
- Ask the pupils to quickly estimate if an answer is reasonable by rounding to the nearest whole number. **Was Amman's answer reasonable? How do you know?**
- Rounding is a useful skill to check our answers. Give a wide range of examples for pupils to round and estimate, allowing them to explore rounding to different multiples for different problems and using different processes.

2 Number - order and place value

Number - order and place value 2A

Section overview

Textbook 2A units	Review units	Prior learning	Benchmarks/ Es&Os
Unit 2.1 Reading and writing whole numbers	**1B** Unit 2.3 **1C** Units 2.1 and 2.2	• I can read 3-digit numbers written in both words and numerals. • I can write 3-digit numbers in both words and numerals. • I know that zero is a place holder and can use it correctly when writing 3-digit numbers.	**MNU 2-02a** Reads, writes and orders whole numbers to 1 000 000, starting from any number in the sequence.
Unit 2.2 Representing and describing whole numbers	**1B** Unit 2.10 **1C** Unit 2.6	• I can build 3-digit numbers with concrete materials. • I can describe the position of each digit in a 3-digit number using the words hundreds, tens and ones. • I can explain the value of each digit in a 3-digit number using 100s, 10s and 1s.	**MNU 2-02a** Explains the link between a digit, its place and its value for whole numbers to 1 000 000.
Unit 2.3 Place value partitioning of numbers with up to five digits	**1C** Units 2.7 and 2.8	• I can partition any 3-digit number into 100s, 10s and 1s. For example, 867 = 800 + 60 + 7. • I can partition 2-digit numbers in different ways. For example, I know that 53 = 50 + 3 but it can also be 40 + 13, 30 + 23 or 10 + 43. • I can count forwards and backwards in tens and hundreds.	**MNU 2-02a** Partitions a wide range of whole numbers and decimal fractions to three decimal places, for example, 3·6 = 3 ones and 6 tenths = 36 tenths.
Unit 2.4 Comparing and ordering numbers in the range 0-10 000	**1B** Units 2.12 and 2.13 **1C** Units 2.9 and 2.10	• I can compare two 3-digit numbers by looking at the position of the digits and thinking about their value. • I can use place value to order a set of 3-digit numbers from smallest to largest. • I can use place value to order a set of 3-digit numbers from largest to smallest.	**MNU 2-02a** Reads, writes and orders whole numbers to 1 000 000, starting from any number in the sequence.
Unit 2.5 Reading and writing decimal fractions	**1B** Units 5.1, 5.2, 5.3 and 5.5	• I know that a fraction is an equal part of one whole. • I know that the number under the line in a fraction tells us the number of equal parts that make up the whole. • I know that the number above the line in a fraction tells us the number of equal parts being used. • I know that ten tenths make one whole. • I can read and write common fractions in tenths, for example 1/10 (one tenth) and 4/10 (four tenths).	**MNU 2-02a** Reads, writes and orders sets of decimal fractions to three decimal places.

Unit 2.6 Representing and describing decimal fractions	**2A** Unit 2.5	• I know that ten tenths make one whole. • I know that a decimal point separates a whole number and a fraction part. • I can read decimal fractions with one decimal place. • I can write decimal fractions with one decimal place.	**MNU 2-02a** Explains the link between a digit, its place and its value for numbers to three decimal places.
Unit 2.7 Comparing and ordering decimal fractions	**1C** Units 2.9 and 2.10	• I can position whole numbers on a number line. • I can arrange whole numbers in order, from smallest to largest and from largest to smallest. • I can use the symbols =, ≠, < and > to compare whole numbers.	**MNU 2-02a** Reads, writes and orders sets of decimal fractions to three decimal places.
Unit 2.8 Recognising the context for negative numbers	**1C** Unit 2.3	• I understand the term 'whole number'. • I understand that whole numbers are greater than zero. • I know that temperature is recorded in degrees Celsius (°C). • I know that freezing point is *zero degrees Celsius* (0°C). • I know that height and depth can be measured in metres (m) and I can read simple scales graduated in metres.	**MNU 2-04a** Identifies familiar contexts in which negative numbers are used. Orders numbers less than zero and locates them on a number line.

Conceptual context

Place value *underpins* effective mental and written calculation. Deep *conceptual* place value understanding allows us to manipulate numbers flexibly and choose the most efficient method to solve a number problem. Central to this is the ability to *partition* numbers in different ways according to their place values. For example, 46 = 40 + 6 (standard place value partitioning, also referred to as canonical form) or 30 + 16, 20 + 26, 10 + 36 (non-standard place value partitioning or non-canonical form).

Prior to naming the places, pupils need lots of opportunities to represent whole numbers pictorially and with concrete materials. Through these experiences they will begin to develop a sense of the relative sizes of numbers and come to understand that the value of each digit is determined by its place.

Pupils encounter decimal fractions in their daily lives, most commonly within the context of money. However, it is important that experiences designed to develop pupils' understanding of the relationships between the places in whole numbers (i.e. 10 ones = 1 ten, 10 tens = 1 hundred, 10 hundreds = 1 thousand) is extended to include decimal fractions. Through such experiences, pupils will come to realise that decimal fractions are numbers that lie in between two whole numbers (e.g. the number 2·7 comes in-between the numbers 2 and 3) and that the purpose of the decimal point is to separate a number of whole units from their fractional parts.

Through experiences such as those described above, pupils will begin to appreciate the relative sizes of whole and decimal numbers, enabling them to compare and order the numbers and position them on a number line.

Resources

- Digit cards
- Place value houses
- Dot squares and strips
- Graduated number line
- Place value arrow cards
- Blank ten frames
- Base 10 blocks
- Place value counters
- Fraction tiles
- Mini whiteboards
- Calculators
- 0-9 spinners or dice
- Post-it notes
- Squared paper

Key vocabulary

- **Adjacent**
- **Canonical form**
- **Celsius**
- **Decimal fraction**
- **Decimal point**
- **Digit**
- **Negative number**
- **Non-canonical form**
- **Numeral**
- **Partition**
- **Place holder**
- **Positive number**
- **Symbol**
- **Value**

Misconceptions

Potential misconception	Addressing the misconception
Unit 2.1 The pupil may include extra zeros when writing numbers in numerals. For example, writes three thousand, four hundred and twenty-six as 3000426.	Represent 4-digit numbers with base 10 blocks and link representations to place value houses. Initially, representations should include *at least* 1 thousand, 1 hundred, 1 ten and 1 one before building up to numbers that include zeros as place holders.
Units 2.1, 2.2, 2.3 and 2.4 The pupil may be able to name the places as 'thousands, hundreds, tens and ones' without fully understanding the multiplicative relationship between them. For example, may read the digit in the thousands column as 'five thousands' but does not equate this with 'five lots of one thousand'.	It is important that pupils see, and are helped to visualise, the individual units of one that make up one unit of ten, one unit of one hundred and one unit of one thousand, respectively. They need lots of opportunities to build 4-digit numbers with base 10 blocks, where the individual ones can be seen or easily visualised, before progressing to work with more abstract units such as place value counters.

Potential misconception	Addressing the misconception
Unit 2.3 The pupil can work out a non-standard partition (e.g. 408 = 200 + 200 + 8) but has difficulty seeing this as 20 tens + 20 tens + 8 ones.	Provide concrete materials grouped in tens (for example, base 10 blocks) and challenge pupils to build as many different representations of the same number as they can. For example, 46 could be represented by 4 tens and 6 ones, 3 tens and 16 ones, 2 tens and 26 ones, 1 ten and 36 ones or 46 ones. Ask pupils to justify their answers by counting in tens and ones. Extend the activity to include 3-digit numbers.
Units 2.5 and 2.6 The pupil may have experienced decimals within the context of money or measure and so wrongly believe that the decimal point separates two whole units (e.g. pounds and pence) rather than whole units and fractional parts.	Provide pupils with lots of opportunities to cut objects into ten equal parts to create tenths, then reassemble some of the 10 tenths to create whole units. For example, three circles each cut into ten equal parts gives us 30 tenths. To represent the decimal fraction 2·4 we need two whole circles and 4 tenths of a circle.
Units 2.5 and 2.6 The pupil may understand that the decimal point separates whole units and fractional parts but wrongly believes that, for example, 0·5 = one fifth.	Repeat the above activity but focus on equivalence. For example, take two identical circles, cut one into tenths and halve the other. Lay the five tenths and one half on top of each other to 'prove' they are equivalent.
Unit 2.7 The pupil may believe that the number after 0·9 is 0·10.	Ask pupils to use the constant function on their calculator to count from zero by 0·1. Pause at 0·9 and ask pupils to predict what number will come next. Ask them to check their predictions on the calculator and discuss any conflicts. Invite those who made a correct prediction to use concrete materials to convince their peers.

Teaching ideas

Unit 2.1 Reading and writing whole numbers

- Ask pupils to enter 999 + 1 into a calculator. **Can anyone tell me what this number says?** We say 'one thousand' and we write 1000.
- Show the number 1000 on a place value house. One thousand has 1 thousand, 0 hundreds, 0 tens and 0 ones.
- Show 2000, 4000, 7000, etc. on place value houses. Discuss the purpose of the zeros, i.e. they keep the thousands digit in place. **How would we write the numbers represented by these place value houses in words and in numerals?**
- **What about these numbers?** Represent a series of 4-digit numbers with none, one or two zeros, on place value houses. Ask pupils to say the number represented. Where appropriate, discuss the position of the 'zero' place holder(s).
- Show pupils a range of 4-digit numbers written in words and ask them to 'show you' the number in numerals on a mini whiteboard or calculator display.
- Give individuals/pairs of pupils a numeral dice or spinner and ask them to generate a list of ten 4-digit numbers. They should write each number down in numerals and in words. Swap with another individual/pair. **Has *'and'* been used correctly? Has *'zero'* been used correctly?**

- Discuss numbers such as 504, 1034 and 3434. **Is '*and*' used the same way each time?**
- **Are answers such as 0679 acceptable? Justify your answer.**

Unit 2.2 Representing and describing whole numbers

- Model the relationship between the places with base 10 blocks. Show the smallest base 10 block (1). **What number is this?** Then show the next largest base 10 block (10). **What number is this? How much bigger is it than the first block?** 1 ten = 10 ones.
- Show the next largest base 10 block (100). **What number is this? How much bigger is it than the first/second block?** 1 hundred = 100 ones. 1 hundred = 10 tens.
- Show the next largest base 10 block (1000). **What number is this? How many hundreds are there? How do you know?** 10 hundreds = 1000.
- Show a range of 4-digit numbers on place value houses and invite pupils to represent them with base 10 blocks. **Can you 'see' the individual ones that make up each thousand/hundred/ten?**
- Represent the same set of numbers with place value counters. **Can you 'see' the individual ones that make up each thousand/hundred/ten this time? Each 1000 counter has a value of one thousand. It is worth one thousand ones.**
- Challenge pupils to work in small groups to represent a 4-digit number using everyday objects such as paper clips, pasta shapes, beads, straws, plastic cups, rubber bands, etc. Their model should clearly show the values of the digits. **Can you work out the number represented by other groups' models? Can they guess your number?**

Unit 2.3 Place value partitioning of numbers with up to five digits

- Display a 4-digit number written in numerals; for example, 9826. Discuss the value of each digit, i.e. 9826 is made up of 9 thousands, 8 hundreds, 2 tens and 6 ones. **How can we make the number 9826 using place value arrow cards? Which cards do we need? How do you know? How can we check?** Represent the same number with place value arrow cards. Draw out that 9826 = 9000 + 800 + 20 + 6.
- Present pupils with a range of 4-digit numbers written in numerals. Invite them to model each number using place value arrow cards and record the partitions on a mini whiteboard. For example, 1653 = 1000 + 600 + 50 + 3. **Is it possible to represent a 4-digit number using only one, two or three place value arrow cards? Why is that?**
- Using base 10 blocks, model and discuss different representations of the same 3-digit number. For example, 362 can be 3 hundreds, 6 tens and 2 ones (300 + 60 + 2). It can also be 2 hundreds, 16 tens and 2 ones (200 + 160 + 2) or 3 hundreds 5 tens and 12 ones (300 + 50 + 12). **How many other ways can we find to partition the number 362? How can we check our solutions?**

- Ask pupils to work in pairs to find as many different ways as they can to partition the same 3-digit number (for example 205) then check their solutions by counting in 100s, 10s and 1s. **Compare your solutions with another pair. Have you found them all? How can you be sure?**
- Provide each pair with a set of cards showing standard and non-standard place value partitions for two different 3-digit numbers, for example 139 and 391. Include a few red herrings. Challenge pairs to find and record the cards that can be used to represent each number. For example, 139 could match with these cards:

1 hundred, 3 tens and 9 ones	13 tens and 9 ones	12 tens and 19 ones

391 could be matched to these cards:

3 hundred and 91 ones	3 hundred, 9 tens and 1 one	391 ones

Unit 2.4 Comparing and ordering numbers in the range 0–10000

- Discuss the 'Let's learn' section on page 20. Display a different set of 4-digit numbers and work together to compare and order them. **Which number will come first/last? Why? How many numbers are there in-between ______ and ______? What effect will changing the first digit of the number ______ (to a ______) have on the order? For example, 9825 changed to 4825 in the following sequence: 4258, 4528, 4852, 9258, 9582, 9825, 9852.**
- Ask a group of pupils to each choose a 4-digit number within a given range and write it on a post-it note. Ask each pupil to estimate the position of their number on an empty number line and attach it. **Are the numbers in the correct order? If not, what do we need to change?** Challenge all pupils to write and show a number that would fit in-between a given pair.
- Ask all pupils to each write a 4-digit number within a given range on their mini whiteboard. They should then work together to compare numbers and arrange themselves in a given order around the room. **How did you work out the correct order? What strategy did you use?**
- Repeat the above activities for a mixed set of 2-, 3- and 4-digit numbers.

Unit 2.5 Reading and writing decimal fractions

- Discuss the 'Let's learn' section on page 22. Practise reading *different* decimal fractions written in numerals. **What does this decimal fraction say? What does it mean?**
- Order a set of common fraction cards from 0/10 to 10/10 on a number line. Discuss how to write each fraction as a decimal fraction. **What does zero mean when we write, for example, 0·7? What is the same and what is different about the numbers 1·0 and 1?**

- Present pupils with decimal fractions written in words, for example 'one point six'. Invite pupils to draw a diagram to represent each decimal fraction on squared paper. **What have you drawn? Why does your diagram show one point six? Is there another way to show this decimal fraction on squared paper? How many different ways can you find to represent the same decimal fraction on squared paper?** For example, one point six could be shown as:

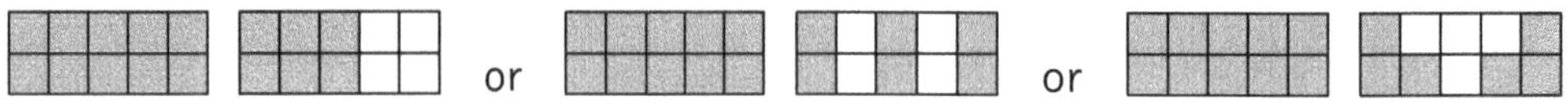

- Look for examples of first place decimal fractions at home and in the environment. For example, 2·5 litres, 1·2 kilograms, 0·6 metres. **What do they mean?**

Unit 2.6 Representing and describing decimal fractions

- Discuss the models in the 'Let's learn' section on page 24. **What is the same and what is different? In which models is it possible to see the ten tenths that make up the one whole? How many tenths altogether in 1·2? How many biscuits altogether in 6·5 packets?**
- Use base 10 blocks (rods and small cubes) to represent first place decimal fractions smaller than 10. **Which blocks represent the ones? Which blocks represent the tenths? How is this different to the way in which we have used base 10 blocks previously? Would a decimal point make the difference clearer? For example:**

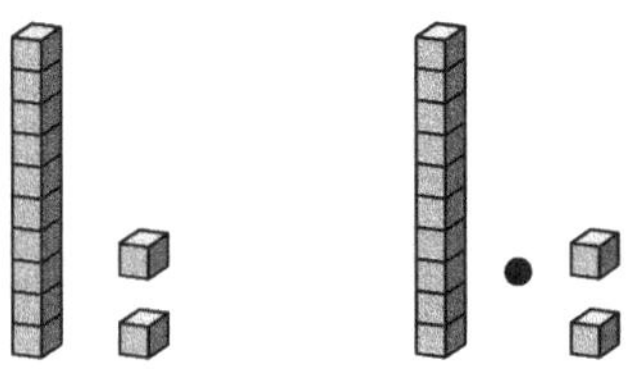

12 twelve   1·2 one point two

- Challenge pupils to work in small groups to represent a decimal fraction with one decimal place using materials of their choice (for example, base 10 materials, straws, cups and counters, paper strips, ten frames). Models should clearly show the whole units and the fractional parts (tenths). **Can you work out the decimal fraction represented by other groups' models? Can they guess your number? Are the tenths obvious or is a decimal point needed?**

Unit 2.7 Comparing and ordering decimal fractions

- Remind pupils of the symbols >, <, = and ≠. **What helps you to remember them?** Write pairs of numbers (whole numbers and decimal fractions) on the board. Invite pupils to insert the correct symbol to make each statement true. **Is there more than one correct answer? Is this always or sometimes true?**
- Provide the symbol and invite pupils to write two decimal fractions to make the number statement true. **Is there more than one correct answer? Is this always or sometimes true?**
- Display two cards showing two different whole numbers at either end of a washing line/empty number line. Show pupils some other cards, each with a different decimal fraction within this range written on it. **Where should this number go? Why?**
- Give each pupil a blank card and invite them to write a decimal fraction to fit in-between two numbers on the washing line/empty number line. **What does this number mean? Where should it go? What makes you say that?**
- Ask pupils to count by 0·1/0·2/0·5 on a calculator from various starting numbers. **Which keys will you need to use? Why? What happens if you miss out the zero? What happens if you miss out the decimal point?**
- Record each sequence generated on a number line. Ask pupils to say each sequence out loud, forwards and backwards. **Why can't the number after 0·9 be 0·10?**
- **Which calculator keys would we need to press to count in halves? Why is that? What about counting in fifths?**
- Practise counting in decimal fractions, forwards and backwards. Focus on bridging a whole number, for example, *zero point eight, zero point nine, one, one point one, one point two…*

Unit 2.8 Recognising the context for negative numbers

- Ask pupils to start at 5 and keep subtracting 1 on their calculator. **What happened when you subtracted 1 from zero? Have you seen numbers like this before? If so, where? How do we say this number? What happened when you subtracted 1 from −1?**
- Discuss the 'Let's' learn' section on page 29. Call out positive and negative numbers and ask the pupils to record them on their mini whiteboards.
- Discuss different types of thermometer. Record the daily temperature for one week/month. **Was the temperature above or below zero degrees Celsius each day? How do you know?**
- Practise counting forwards and backwards along a graduated number line showing both positive and negative numbers.
- Challenge pupils to picture the number line in their heads. Walk forwards to count forwards aloud. Walk backwards to count backwards aloud. Take one step for each number (a large open space such as a gym hall or playground is advisable!).

Number – order and place value 2B

Section overview

Textbook 2B units	Review units	Prior learning	Benchmarks/Es&Os
Unit 2.1 Reading and writing whole numbers	**2A** Unit 2.1	• I can read 4-digit numbers written in both words and numerals. • I can write 4-digit numbers in both words and numerals. • I know that zero is a place holder and can use it correctly when writing 4-digit numbers.	**MNU 2-02a** Reads, writes and orders whole numbers to 1 000 000, starting from any number in the sequence.
Unit 2.2 Representing and describing whole numbers	**2A** Unit 2.2	• I can represent 4-digit numbers in different ways. For example, by using base 10 blocks, place value counters or place value arrow cards. • I can describe the position of each digit in a 4-digit number using the words thousands, hundreds, tens and ones. • I can explain the value of each digit in a 4-digit number using 1000s, 100s, 10s and 1s.	**MNU 2-02a** Explains the link between a digit, its place and its value for whole numbers to 1 000 000.
Unit 2.3 Place value partitioning of whole numbers	**2A** Unit 2.3	• I know that our place value system is based on tens: 10 ones = 1 ten, 10 tens = 1 hundred, 10 hundreds = 1 thousand, 10 thousands = 10 thousand. • I can partition any 4-digit number into 1000s, 100s, 10s and 1s. For example, 3861 = 3000 + 800 + 60 + 1. • I can partition 3-digit numbers in different ways. For example, I know that 453 = 400 + 50 + 3 but it can also be 300 + 150 + 3.	**MNU 2-02a** Partitions a wide range of whole numbers and decimal fractions to three decimal places, for example, 3·6 = 3 ones and 6 tenths = 36 tenths.
Unit 2.4 Number sequences	**2A** Unit 3.2	• I can count forwards and backwards in ones from any 4-digit number, including those that bridge a ten, hundred or thousand. • I can say and write the number that comes before or after any 4-digit number. • I can count forwards and backwards in tens, hundreds and thousands. • I can say and write the number that is 10 more/less, 100 more/less or 1000 more/less than any 4-digit number. • I can add or subtract multiples of 10 and 100 to or from a 4-digit number.	**MNU 2-02a** Reads, writes and orders whole numbers to 1 000 000, starting from any number in the sequence.
Unit 2.5 Comparing and ordering whole numbers	**2A** Unit 2.4	• I can compare two 4-digit numbers by looking at the position of the digits and thinking about their value. • I can use place value to order a set of 4-digit numbers from smallest to largest. • I can use place value to order a set of 4-digit numbers from largest to smallest. • I can use the symbols =, ≠, < and > to compare numbers.	**MNU 2-02a** Reads, writes and orders whole numbers to 1 000 000, starting from any number in the sequence.

Unit 2.6 Identifying and placing positive and negative numbers	**2A** Unit 2.8	• I can identify and place whole numbers within the range 0-10 000 on a number line. • I know that negative numbers are whole numbers below zero and positive numbers are whole numbers above zero. • I can give examples of where negative numbers are used in everyday life. • I can use the words plus and minus, and the + and – symbols, to talk and write about positive and negative numbers.	**MNU 2-04a** Identifies familiar contexts in which negative numbers are used. Orders numbers less than zero and locates them on a number line.
Unit 2.7 Reading and writing decimal fractions	**2A** Units 2.5 and 6.5	• I know that a decimal point separates a whole number and a fraction part. • I can read decimal fractions with one decimal place. • I can write decimal fractions with one decimal place.	**MNU 2-02a** Reads, writes and orders sets of decimal fractions to three decimal places.
Unit 2.8 Representing and describing decimal fractions	**2A** Unit 2.6	• I can represent decimal fractions with one decimal place in different ways. • I can describe decimal fractions with one decimal place in words and numerals. • I can read and write decimal fractions with two decimal places. • I can write mixed numbers by combining whole numbers and hundredths.	**MNU 2-02a** Explains the link between a digit, its place and its value for numbers to three decimal places.
Unit 2.9 Comparing and ordering decimal fractions	**2A** Units 2.7 and 6.6	• I can position decimal fractions with one decimal place on a number line. • I can arrange decimal fractions written to one decimal place in order, from smallest to largest and from largest to smallest. • I can use the symbols =, ≠, < and > to compare decimal fractions with one decimal place. • I can read, write, build and describe decimal fractions with two decimal places.	**MNU 2-02a** Reads, writes and orders sets of decimal fractions to three decimal places.

Conceptual context

Place value *underpins* effective mental and written calculation. Deep *conceptual* place value understanding allows us to manipulate numbers flexibly and choose the most efficient method to solve a number problem. Previously, pupils learned that numbers can be partitioned in different ways according to their place values. For example, 382 = 300 + 80 + 2 but can also be 200 + 180 + 2, 100 + 280 + 2, 300 + 82, 300 + 70 + 12, etc. In Book 2B, this idea is extended to include 4-digit numbers.

Prior to naming the places, pupils need lots of opportunities to represent whole numbers with concrete materials and pictorially. Through these experiences they will develop their understanding of **abstract composite units**. For example, they come to realise that 8 tens = 80 ones, 3 hundreds = 30 tens, etc. This is important because it allows pupils to see, for example, that whilst the number 2146 is made up of 2 thousands, 1 hundred, 4 tens and 6 ones it can also be 1 thousand, 11 hundreds, 3 tens and 16 ones, a key factor in understanding how written algorithms work.

Previously, pupils learned that 10 ones = 1 ten, 10 tens = 1 hundred and 10 hundreds = 1 thousand. They may therefore naturally assume that 10 thousands = 1 million. Place value houses will help them see the cyclical pattern in the way we say, read and write whole numbers with the 'ones, tens and hundreds pattern' being repeated for the thousands. This will enable them to read, write, compare and order increasingly large numbers.

Previous work on decimal numbers expressed to the first decimal place is extended to include numbers to the second decimal place. The distinction between how we say the numbers on either side of the decimal point should be made clear, e.g. for 21·21 we say *twenty-one point two one*, **not** *twenty-one point twenty-one*. Pupils need ample opportunities to read decimal fractions correctly in this way because permitting 'bad habits' may exacerbate the common misconception that the decimal point separates two whole numbers.

As for whole numbers, pupils also need lots of opportunities to build, describe, compare and order decimal fractions, using a range of concrete and pictorial representations such as base 10 blocks and number lines, so that they can begin to develop an understanding of the relative sizes of the places.

Resources

- Digits cards 0-9
- Number line showing 100 graduations
- Place value arrow cards
- Place value houses
- Base 10 blocks
- Place value counters
- Mini whiteboard
- Calculators
- Squared paper
- Post-it notes
- Washing line and pegs
- One-millimetre squared paper

Key vocabulary

- **Celsius**
- **Decimal**
- **Decimal fraction**
- **Digit**
- **Mixed number**
- **Numeral**
- **Partition**
- **Positive number**
- **Sequence**
- **Symbol**
- **Value**

Misconceptions

Potential misconception	Addressing the misconception
Units 2.1, 2.2, 2.3, 2.4 and 2.5 The pupil may think that the number after 9999 is one million.	Practice in reading numbers beyond 9999, represented by place value houses, will help pupils see the repeating 'ones, tens and hundreds' pattern in the way we read and write whole numbers. Pupils also need lots of practice in saying and writing number word sequences that bridge the thousands or ten thousands. For example: 43 998, 43 999, 44 000, 44 001, 44 002…… 69 998, 69 999, 70 000, 70 001, 70 002……
Units 2.3, 2.4 and 2.5 The pupil may be able to partition numbers 'additively', e.g. 13 212 = 10 000 + 3000 + 200 + 10 + 2 but may not fully understand the multiplicative relationship between the places.	Demonstrate the relationship between the places using base 10 blocks where it is possible to 'see' the individual units of one that make up each unit of ten, one hundred and one thousand respectively. Use mathematical language such as: 10 ones = 1 ten. Ten lots of one. One ten times. $10 \times 1 = 10$ 10 tens = 1 hundred. Ten lots of ten. Ten ten times. $10 \times 10 = 100$ 10 hundreds = 1 thousand Ten lots of 100. 100 ten times. $10 \times 100 = 1000$ Make a tower with ten 1000 blocks. Ten lots of 1000. 1000 ten times. $10 \times 1000 = 10\,000$ Can the pupils use one-millimetre squared paper to show that 10 is ten times bigger than 1, 100 is ten times bigger than 10, 1000 is ten times bigger than 100 and 10 000 is ten times bigger than 1000?
Units 2.7, 2.8 and 2.9 The pupil may think that the decimal point separates two whole units, e.g. metres and centimetres.	Provide pupils with lots of opportunities to represent decimal fractions in ways that highlight the fractional parts that make up the whole. For example, by using concrete materials such as base 10 blocks, or representing the numbers pictorially on squared paper.
Unit 2.9 The pupil may think that the more digits a number has, the larger it is. For example, may believe that 7·24 is greater than 9.	Check that pupils understand the meaning of such numbers by asking them to represent each with concrete materials or pictorially. Follow this with lots of practice in identifying/positioning decimal fractions between whole numbers on a number line.

Teaching ideas

Unit 2.1 Reading and writing whole numbers

- Ask pupils to enter 9999 + 1 into a calculator. **Can anyone tell me what this number says?** We say 'ten thousand'. Show the number 10 000 written in numerals and ask pupils to compare the written version with the version on their calculator screen. **What is the same and what is different? Which is easier to read? Why? How might we write the number 10 000 in words?**
- Show the number 10 000 on place value houses. Ten thousand is a five-digit number. It is made up of tens of thousands, thousands, hundreds, tens and ones.
- We read a 5-digit number by saying the number of thousands first, followed by the rest of the number. For example, 45 607 says forty-five thousand, six hundred and seven.
- Represent a series of 5-digit numbers on place value houses. Ask the pupils to say the numbers represented. Ask: '**How many thousands?'** each time. For example, 45 607 has forty-five thousands.
- Call out a range of 5-digit numbers and ask pupils to 'show you' each number in numerals on a mini whiteboard. Compare answers with a partner. **Are they the same? If not, why not? Who is correct?** Discuss zero as a place holder where appropriate.
- Demonstrate how to write the same set of 5-digit numbers in words. Draw pupils' attention to the comma in each case; for example, twenty-six thousand, three hundred and eleven. **When reading, what do we do when we see a comma? What might the purpose of the comma be here?** Stress that the comma serves the same purpose as the space when the number is written in numerals.
- Ask pupils to work in pairs. One calls out a 5-digit number and writes it on their mini whiteboard in numerals. The other writes the same number in words. Pairs compare what they have written. **Do the words and numerals mean the same? If not, what do you think has happened? Is the small space/comma in the correct place?** Swap roles.

Unit 2.2 Representing and describing whole numbers

- Show a range of 5-digit numbers on place value houses and discuss the value of the digits. For example, **What is the value of the 2? What about the 4? Which digit has the greatest value? What is it worth? Which digit is worth least? Why? How can the same digit have different values?**

Thousands		
H	T	O
	1	4

Ones		
H	T	O
2	3	3

- If available, introduce pupils to 'tens of thousands' place value counters. Challenge them to work in pairs to create the 5-digit number shown by each place value house, then check their answers by counting. For example, 10000, 11000, 12000, 13000, 14000, 14100, 14200, 14210, 14220, 14230, 14231, 14232, 14233. **Can you write this number in words and in numerals?**
- Display a range of 5-digit numbers on a flipchart or whiteboard, some of which contain zeros. Ask pupils to represent each number with place value arrow cards. **Is it possible to represent a 5-digit number using only one, two, three or four place value arrow cards? Why is that?**

Unit 2.3 Place value partitioning of whole numbers

- Present pupils with a range of 5-digit numbers written in numerals. Invite them to partition each number into tens of thousands, thousands, hundreds, tens and ones and record the partitions on a mini whiteboard.
- Use base 10 blocks to demonstrate the multiplicative relationship between the digits in 4-digit numbers, i.e. 10 ones = 1 ten, 10 tens = 1 hundred, 10 hundreds = 1 thousand. Model and discuss different representations of the same 4-digit number. For example, 7841 can be 7 thousands, 8 hundreds, 4 tens and 1 one. It can also be 6 thousands, 18 hundreds, 4 tens and 1 one, or 7 thousands, 7 hundreds, 14 tens and 1 one. **How many other ways can we find to partition the number 7841? How can we prove that what we have made still equals 7841?**
- Repeat the above activity for place value counters. For example, 7841 can be 7000 + 800 + 40 + 1, or 6000 + 1800 + 40 + 1, or 7000 + 700 + 140 + 1. **How many other ways can we find to partition the number 7841 using place value counters? How can we check our solutions?**

Unit 2.4 Number sequences

- Revise and discuss the patterns in our place value system. Focus on bridging 10, 100 and 1000 in sequences involving 4-digit numbers. For example:

  3457, 3458, 3459, 3460, 3461, 3462.......

  7198, 7199, 7200, 7201, 7202, 7203.......

  5996, 5997, 5998, 5999, 6000, 6001.......
- Extend to include 5-digit numbers. **What is the number after 9999? What about the number after that? And the next number?**

- Practise reciting sequences of 5-digit numbers. Focus on bridging the hundreds, thousands and tens of thousands. For example:

  Start at 57 396 and count forwards in ones up to 57 406.

  Start at 23 993 and count forwards in ones up to 24 003.

  Start at 49 995 and count forwards in ones up to 50 005.
- Repeat the above activity but count backwards.
- Practise counting forwards and backward in 10s, 100s, 1000s, 10 000s from a range of 5-digit numbers. **Which sequences do you find most challenging? Why is that? Would base 10 blocks or place value counters help? What about a calculator?** Link counting forwards/backwards in 10s, 100s, 1000s and 10 000s to adding/subtracting 10, 100, 1000 and 10 000.

Unit 2.5 Comparing and ordering whole numbers

- Display a set of 5-digit numbers and work together to compare and order them. **Which number will come first/last? Why? How many numbers are there in-between ______ and ______? What effect will changing the first digit of the number ______ to a ______ have on the order?**
- Repeat the above activity for a mixed set of 3-, 4- and 5-digit numbers.
- Ask each pupil to write a 5-digit number on a piece of paper or card. They must work together to peg their cards in order, from smallest to largest or from largest to smallest, on a 'washing line'. **Are the numbers in the correct order? If not, what do we need to change?** Challenge each pupil to write another card that fits between their card and the next so that the order remains correct. **What number could go between, for example, 34 707 and 34 708?**
- Ask each pupil to write a number greater than 900 but less than 20 000 on their mini whiteboard. They should then work together to compare numbers and arrange themselves in a given order around the room. **How did you work out the correct order? What strategy did you use?**

Unit 2.6 Identifying and placing positive and negative numbers

- Practice counting forwards and backwards on both horizontal and vertical number lines showing positive and negative numbers. Vary the starting number.
- Give each pupil a card with a positive or negative number written on it. Pupils should work together to form a 'human number line' around the room. **Who has a positive number? Who has a negative number? Who has the number that comes in-between –3 and –5? How many numbers come in-between minus four and plus six? Which of these are positive numbers? Which are negative numbers? What about zero?**

- Challenge pupils to imagine jumping forwards and backwards on a number line showing positive and negative numbers. **You are on 3. What number will you be on if you jump backwards five places? Etc.**
- Examine the packaging on a range of frozen foods. **Can you find all the negative numbers? Which is the lowest temperature? How do you know?** Ask pupils to arrange the temperatures in order from lowest to highest.

Unit 2.7 Reading and writing decimal fractions

- Discuss the 'Let's learn' section on page 24. Practise reading *different* decimal fractions written in numerals. **What does this decimal fraction say? What does it mean?**
- Demonstrate how to write the same set of decimal fractions as mixed numbers. **Why do you think they are called mixed numbers? How would you write, for example, six point one nine as a mixed number? What does six point one nine actually mean?**
- Extend the above activity to include examples involving zeros; for example, three point zero seven. **What does three point zero seven mean? Why is the zero needed when we write three point zero seven as a decimal number? Is the zero needed when we write three point zero seven as a mixed number? Why not?**
- Compare three point zero seven and three point seven zero. **What is the same and what is different? How will we write three point seven zero as a decimal number/mixed number? Which zeros are needed this time? Why is that?**
- Discuss where decimal fractions with hundredths are used in real life; for example, in money and measurement. **Why do we need a decimal point when writing amounts such as £2·49, or lengths such as 10·25 m?** Focus the discussion on hundredths of a pound/metre. **Why would it be wrong to write £2·49p? What if we were to remove the decimal points? What do £249 and 1025 m say now?**

Unit 2.8 Representing and describing decimal fractions

- Compare pairs of price labels such as £11·08 and £11·80. **What does the 8 mean in each label? Which is dearer/cheaper? Why?**
- Display a decimal fraction written in words; for example, two point zero five. Ask each pupil to represent this number on squared paper. Compare results. **Has everyone represented the decimal fraction correctly? Has everyone shown two point zero five in the same way? If not, how can everyone be correct?** Possible solutions are:

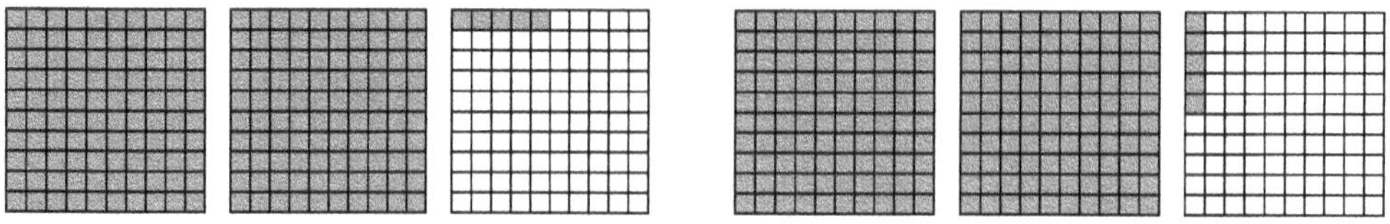

Ask pupils to write a mixed number to match their diagram. **Is there more than one correct answer this time? Justify your thinking.**

- Repeat the above activity for different decimal fractions. **How many different ways are there to represent the decimal fraction 0·99 on squared paper? Prove it!**
- Introduce decimal place value arrow cards. Ask pupils to predict the resulting decimal fractions when three cards (showing ones, tenths and hundredths) are 'layered' and the arrows aligned. **Which cards would we need to make….?** Challenge pupils to make a decimal fraction with three digits using only two place value arrow cards. **Read the decimal fraction you have made out loud. Write it as a mixed number. How many hundredths does your decimal fraction have?**

Unit 2.9 Comparing and ordering decimal fractions

- Write down a decimal fraction with two decimal places within a given range (for example, between 1 and 2) but don't let pupils see it. Pupils must work together to try to guess your number (in ten guesses or fewer) by asking questions to which you may only reply yes or no. **Does it matter which question you start with? How can we ensure we don't 'waste' questions?**
- Ask pupils to work in pairs. Each pair will need a set of place value arrow cards sorted into four piles, i.e. tens in one pile, ones in another, tenths in another and hundredths in another. Shuffle the cards and place them face down on the table. Each player picks one card from the top of each pile, makes the number and says it out loud. The person with the smaller number/larger number/number nearest to a chosen whole number, etc. wins that round and keeps the cards. Keep playing until all of the cards have been picked up. The person with the most cards is the overall winner.
- Ask pupils to start at 0 and keep adding 0·25 on a calculator until they reach 2. Record the sequence on a number line. Repeat for 0·5. Talk with pupils about the patterns they see. **Which numbers are in both sequences? Why does this happen? Can you extend each pattern to 4 *without* using a calculator? Which number will be next to appear in both sequences?**

Number - order and place value 2C

Section overview

Textbook 2C units	Review units	Prior learning	Benchmarks/Es&Os
Unit 2.1 Reading and writing whole numbers	**2B** Unit 2.1	• I can read 5-digit numbers written in both words and numerals. • I can write 5-digit numbers in both words and numerals. • I know that zero is a place holder and can use it correctly when writing 5-digit numbers.	**MNU 2-02a** Reads, writes and orders whole numbers to 1 000 000, starting from any number in the sequence.
Unit 2.2 Representing and describing whole numbers	**2B** Unit 2.2	• I can represent 5-digit numbers in different ways. For example, with place value houses, place value arrow cards or place value counters. • I can describe the position of each digit in a 5-digit number using the words tens of thousands, thousands, hundreds, tens and ones. • I can explain the value of each digit in a 5-digit number using 10 000s, 1000s, 100s, 10s and 1s.	**MNU 2-02a** Explains the link between a digit, its place and its value for whole numbers to 1 000 000.
Unit 2.3 Place value partitioning of whole numbers	**2B** Unit 2.3	• I know that our place value system is based on tens: 10 ones = 1 ten, 10 tens = 1 hundred, 10 hundreds = 1 thousand, 10 thousands = 10 thousand, 10 tens of thousands = 100 thousand. • I can partition any 5-digit number into 10 000s, 1000s, 100s, 10s and 1s. For example, 18 634 = 10 000 + 8000 + 600 + 30 + 4. • I can partition 4-digit numbers into thousands, hundreds, tens and ones in different ways.	**MNU 2-02a** Partitions a wide range of whole numbers and decimal fractions to three decimal places; for example, 3·6 = 3 ones and 6 tenths = 36 tenths.
Unit 2.4 Number sequences	**2B** Unit 2.4	• I can count forwards and backwards in ones from any 5-digit number, including those that bridge a ten, hundred, thousand or ten thousand. • I can say and write the number that comes before or after any 5-digit number. • I can count forwards and backwards in tens, hundreds, thousands and tens of thousands. • I can say and write the number that is 10 more/less, 100 more/less, 1000 more/less or 10 000 more/less than any 5-digit number.	**MNU 2-02a** Reads, writes and orders whole numbers to 1 000 000, starting from any number in the sequence.

Unit 2.5 Comparing and ordering whole numbers	**2B** Unit 2.5	• I can compare two 5-digit numbers by looking at the position of the digits and thinking about their value. • I can use place value to order a set of 5-digit numbers from smallest to largest. • I can use place value to order a set of 5-digit numbers from largest to smallest. • I can use the symbols =, ≠, < and > to compare numbers.	**MNU 2-02a** Reads, writes and orders whole numbers to 1 000 000, starting from any number in the sequence.
Unit 2.6 Negative numbers	**2B** Unit 2.6	• I know that negative numbers are whole numbers below zero and positive numbers are whole numbers above zero. • I can give examples of where negative numbers are used in everyday life. • I can use the words 'plus' and 'minus', and the + and – symbols, to talk and write about positive and negative numbers. • I can identify and place negative numbers on a number line.	**MNU 2-04a** Identifies familiar contexts in which negative numbers are used. Orders numbers less than zero and locates them on a number line.
Unit 2.7 Reading and writing decimal fractions	**2B** Units 2.7 and 6.4	• I know that a decimal point separates a whole number and a fraction part. • I can read decimal fractions with one or two decimal places. • I can write decimal fractions with one or two decimal places.	**MNU 2-02a** Reads, writes and orders sets of decimal fractions to three decimal places.
Unit 2.8 Representing and describing decimal fractions	**2B** Unit 2.8	• I can represent decimal fractions with two decimal places in different ways. • I can describe decimal fractions with two decimal places in words and numerals. • I can read and write decimal fractions with three decimal places. • I can write mixed numbers by combining whole numbers and hundredths.	**MNU 2-02a** Explains the link between a digit, its place and its value for numbers to three decimal places.
Unit 2.9 Zero as a place holder in decimal fractions	**2A** Unit 2.1	• I understand zero as a place holder in whole numbers. • I can write any whole number with up to six digits correctly, including those that have zero in more than one place. • I can read and write decimal fractions with up to three decimal places. • I can interpret 3rd place decimals as made up of wholes and thousandths.	**MNU 2-02a** Explains the link between a digit, its place and its value for numbers to three decimal places.

Unit 2.10 Partitioning decimal fractions	**2B** Unit 2.3	• I understand the '10 × relationship' between the digits in a whole number and so can find the value of each digit by multiplying it by its place value. • I can partition any 6-digit number into 100 000s, 10 000s, 1000s, 100s, 10s and 1s. For example: 734 945 = 700 000 + 30 000 + 4000 + 900 + 40 + 5. • I can read and write decimal fractions with up to three decimal places.	**MNU 2-02a** Partitions a wide range of whole numbers and decimal fractions to three decimal places, for example, 3·6 = 3 ones and 6 tenths = 36 tenths.
Unit 2.11 Comparing and ordering decimal fractions	**2B** Units 2.9 and 6.8	• I can position decimal fractions with two decimal places on a number line. • I can compare decimal fractions with one decimal place and arrange them in order. • I can compare decimal fractions with two decimal places and arrange them in order. • I can read, write, build and describe decimal fractions with three decimal places.	**MNU 2-02a** Reads, writes and orders sets of decimal fractions to three decimal places.

Conceptual context

Appreciating the multiplicative relationship between the places is an important aspect of place value. Opportunities to manipulate and discuss a variety of base 10 materials will help pupils understand that by multiplying the value of a digit by its place we can find the amount it represents. For example, in the number 4712 the 7 represents 700 because it is 7 lots of 100 or 7 × 100.

Pupils' understanding of the multiplicative relationship between the places can be further enhanced by linking the above idea to the repeating pattern of ones, tens and hundreds in the way we read and write whole numbers, i.e. we have the ones of ones, the tens of ones and the hundreds of ones, the ones of thousands, the tens of thousands and the hundreds of thousands.

Whilst the expectation of the national benchmarks is for pupils to be able to read and write numbers *up to* one million, pupils who understand this repeating pattern will be able to use their knowledge to read and write numbers in the millions, or indeed the billions, and should be challenged to do so.

Research tells us that, for many pupils, understanding *decreasing* values in powers of ten is considerably more challenging. It is therefore possible that pupils who can read and write decimal fractions (i.e. 'name' the places) may do so without fully understanding the link between each digit, its place and its value.

As for whole numbers, it is important that we provide pupils with lots of opportunities to represent decimal fractions with concrete materials and pictorially. This will support their understanding of the meaning of the digits and so enable them to compare, order and partition decimal fractions with up to three decimal places.

Resources

- Digits cards 0-9
- Graduated number line for positive and negative numbers
- Place value arrow cards
- Place value houses
- Base 10 blocks
- Place value counters
- Mini whiteboards
- Calculators
- Squared paper
- Four 25 cm × 10 cm grids
- Post-it notes
- Washing line and pegs
- 0-9 dice/spinner per pair/group

Key vocabulary

- **Celsius**
- **Decimal fraction**
- **Decimal point**
- **Digits**
- **Multiple**
- **Negative number**
- **Numerals**
- **Partition**
- **Place holder**
- **Positive number**
- **Sequence**
- **Symbol**
- **Value**

Misconceptions

Potential misconception	Addressing the misconception
Units 2.2, 2.3, 2.4 and 2.5 The pupil may be able to name the places in whole numbers without fully understanding the multiplicative relationship between them.	Pupils who do not yet fully appreciate the multiplicative relationship between the places will tend to think additively in activities such as those described in the Teaching Ideas for Unit 2.4 (page 48 final bullet point) and will attempt to change the number into 100 000 by counting on. These pupils need practice in building and describing smaller numbers, using concrete materials that clearly illustrate the ×10 relationship (for example base 10 blocks), to help them move away from an over-reliance on 'named places'.
Units 2.7, 2.8, 2.9, 2.10 and 2.11 The pupil may try to create 'symmetry' around the decimal point, i.e. may try to create a 'oneths' column.	Provide lots of opportunities to read, write and illustrate decimal fractions on place value houses. 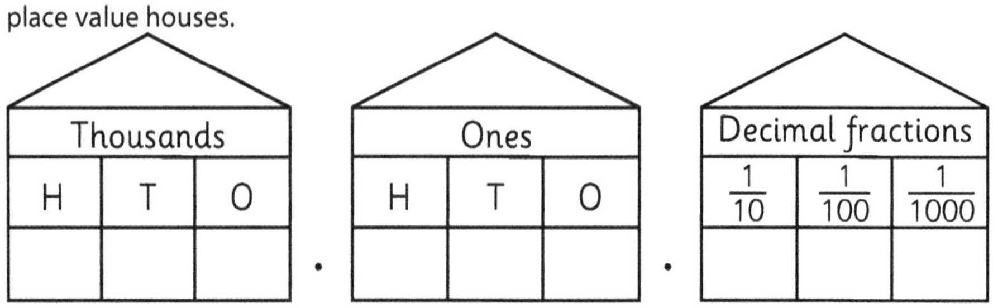

Potential misconception	Addressing the misconception
Unit 2.11 The pupil may think that the more digits a number has, the larger it is. For example, the pupil may believe that 1·625 is greater than 3·1. Conversely, they may think that the more digits there are after the decimal point, the smaller the number is. For example, the pupil may believe that 1·408 is smaller than 1·08.	Teachers must think carefully about their choice of language to avoid inadvertently exacerbating misconceptions, and should use a variety of tools aimed at helping pupils get a feel for the order and relative magnitude of numbers. Such tools include the number line and base 10 materials.
Unit 2.10 The pupil may be able to name the places to the right of the decimal point as tenths, hundredths and thousandths, and partition a decimal fraction into ones, tenths, hundredths and thousandths, but may be unable to equate this with other ways to partition it. For example, may identify 1·234 as $1 + \frac{2}{10} + \frac{3}{100} + \frac{4}{1000}$ but does not equate this with $\frac{1234}{1000}$. Conversely, they may think that decimals with three places are always thousandths and write 1234/1000 but do not equate this with 1 + 2/10 + 3/100 + 4/1000.	Provide pupils with lots of opportunities to combine and partition place value arrow cards and talk about what they notice.

Teaching ideas

Unit 2.1 Reading and writing whole numbers

- Show the number 99 999 on place value houses. **Using what we know about the patterns in the way we say, read and write whole numbers, can you predict the number after 99 999? How could you use a calculator to check your prediction?**
- Show the number 100 000 on place value houses. One hundred thousand is a 6-digit number. It is made up of hundreds of thousands, tens of thousands, thousands, hundreds, tens and ones.

- Represent a series of 6-digit numbers on place value houses. Ask the pupils to say the numbers represented out loud. Emphasise the slight pause between the 'thousands' number and the 'ones' number. For example, 141 850 is one hundred and forty-one thousand (pause) eight hundred and fifty. **How many 'hundreds of thousands'? How many 'tens of thousands'? How many 'ones of thousands'? How many thousands is this altogether?** Emphasise, for example, that 3 hundreds of thousands plus 9 tens of thousands plus 5 ones of thousands makes 395 thousands altogether.
- Revise how we write 5-digit numbers, i.e. with a small space between the thousands and the rest of the number. Call out a range of 6-digit numbers and ask pupils to 'show you' each number in numerals on a mini whiteboard. Compare answers with a partner. **Are they the same? If not, why not? Who is correct?** Discuss correct use of spacing. Emphasise zero as a place holder, where appropriate.
- Repeat the above activity for 6-digit numbers written in words. **How will we show the slight pause between the thousands and the rest of the number?** (small space).
- Display a whole number that has been incorrectly written in numerals and ask pupils to spot the error. For example, display 100 000 274. **Amman thinks he has written one hundred thousand, two hundred and seventy-four. What has he done wrong?** Challenge pupils to come up with as many reasons as they can to explain why 100 000 274 cannot say one hundred thousand, two hundred and seventy-four. **How could we use place value houses to work out how to say Amman's number correctly?**
- Challenge more able pupils to use place value houses like those above to read 7-, 8- and 9-digit numbers and say/write them in words.

Unit 2.2 Representing and describing whole numbers

- Show a range of 6-digit numbers on place value houses and discuss the value of the digits. For example: **What is the value of the 8? What about the 5? Which digit has the greatest value? What is it worth? Which digit is worth least? What is its value? What would happen to the value of the digits if we were to swap the position of the seven and the zero? What if we swapped the one and the zero?**

Thousands			Ones		
H	T	O	H	T	O
5	1	8	4	7	0

- If available, introduce 'hundreds of thousands' place value counters. Ask the pupils to work in pairs to create the 6-digit number shown by each place value house. **Can you write this number in words and in numerals?**
- Repeat the above activity for 'hundreds of thousands' place value arrow cards. **Is it possible to represent a 6-digit number using only one, two, three, four or five place value arrow cards? Why is that?**

Unit 2.3 Place value partitioning of whole numbers

- Ask pupils to work in pairs to investigate 'ten times greater' than given numbers with two, three, four or five digits. For example, 45, 67, 208, 712, 560, 9331, 6815, 30 277, 61 818... **Can you explain what is happening to the numbers? Why does it happen? What will ten times 100 000 be? Test it and see.** Focus discussion on the value of the digits. Challenge answers such as, 'Add a zero' by asking pupils to find, for example, 3·14 × 10.
- Extension of the above: ask pupils to key a number into their calculator, for example 6. Ask questions such as: **Which keys would you need to press to change this number into 600? What would you need to do to change this new number into 600 000?**
- Present pupils with a range of 6-digit numbers written on place value houses. Ask them to work in pairs to write each number in numerals, then partition them into thousands, hundreds, tens and ones in three different ways. For example:

  231 143 = 231 000 + 100 + 40 + 3

  231 143 = 200 000 + 31 000 + 100 + 40 + 3

  231 143 = 200 000 + 30 000 + 1000 + 100 + 40 + 3
- Represent the same set of 6-digit numbers with place value counters.

  How many thousands?

  2 × 100 000 = 200 000, 3 × 10 000 = 30 000, 1 × 1000 = 1000 so 231 thousands.

  How many ones?

  1 × 100 = 100, 1 × 40 = 40, 1 × 3 = 3 so 143 ones.

  231 143 = 231 thousands and 143 ones.

  How many other ways can we find to partition the number 231 143 using place value counters? How can we check our solutions?

Unit 2.4 Number sequences

- Present pupils with a 6-digit number. Ask them to work in pairs and count forwards by ones to say and write the next twenty numbers in the sequence. Compare answers with another pair. **Did both pairs say and write the same set of numbers? If not, what do you think has happened?** Focus discussion on the multiplicative relationship between the places and the repeating pattern in the way we say, read and write whole numbers. Ask pairs to check their answers using a calculator.
- Repeat the above activity for different 6-digit numbers. Choose numbers so that the resulting sequence bridges a thousand, ten thousand or one hundred thousand.
- Repeat the above activities but count backwards.

- Practise counting forwards and backward in 10s, 100s, 1000s, 10 000s and 100 000s from a range of 6-digit numbers. **Which sequences do you find most challenging? Why is that? Would a calculator help?** Link counting forwards/backwards in 10s, 100s, 1000s, 10 0000s and 100 000s to adding/subtracting 10, 100, 1000, 10 000 and 100 000.
- Extension activity: Present pupils with a 3-, 4- or 5-digit number. **How can you use your calculator to change this number into a 6-digit number?** Pupils who understand the multiplicative relationship between the places will quickly realise that the simplest way to do this is to multiply. For example, we can change 345 into a 6-digit number by entering 345 × 1000 or 345 × 10 × 10 × 10. These pupils could be challenged to investigate ten/one hundred/one thousand times smaller.

Unit 2.5 Comparing and ordering whole numbers

- Display a set of 6-digit numbers and work together to compare and order them. **Which number will come first/last? Why? How many numbers are there in-between ______ and ______? What effect will changing the first digit of the number ______ to a ______ have on the order?** Repeat this activity for a mixed set of 4-, 5- and 6-digit numbers.
- Ask each pupil to each write a 6-digit number on a piece of paper or card. They must work together to peg their cards in order, from smallest to largest or from largest to smallest, on a 'washing line'. **Are the numbers in the correct order? If not, what do we need to change?** Challenge each pupil to write another card that fits between their card and the next so that the order remains correct. **What number could go in-between ______and______?**
- This game can be played by pairs or small groups. Each pair/group will need a 0-9 dice or spinner. Each pupil will need a sheet of blank place value houses, or one place value house with a wipe-clean surface. Pupils take turns to roll/spin the dice/spinner and choose a column in which to write the number shown. After six rounds each pupil will have created a 6-digit number. The winner is the person with the largest number. Before the game: **What other rules could we have for deciding the winner?** (smallest number, number closest to 500 000, largest even number, etc.) During/after the game: **Why did you decide to put 9 in the ones column? Why isn't the hundreds of thousands column a good place for 1? What are your chances of winning the game if you put zero in the hundreds of thousands column? Why? How might a change of rule affect your strategy?**

Unit 2.6 Negative numbers

- Discuss the 'Let's learn' section on page 22. **Have you seen a number line like the one in the photograph before? If so, where? Where else might we see number lines showing both positive and negative numbers? What is meant by the phrase, 'average daily temperature'?**

- Ask pupils to work in pairs to research the maximum and minimum winter temperatures in places around the world over a period of one week. Record on a number line. Ask questions such as: **What was the difference between the maximum and minimum temperature in Warsaw on Tuesday? Which place recorded the highest temperature? Which place recorded the lowest temperature? Which is colder, 26°C or –6°C? How much colder?**
- Draw a simple function machine on the board. Choose a function, for example +7. Provide pupils with an input number and ask them to show you the output number on their mini whiteboard. For example, teacher calls out 'negative twelve', pupil writes –5. This activity can be made more challenging by providing pupils with the output number and asking them to show you the input number.

Unit 2.7 Reading and writing decimal fractions

- Discuss the 'Let's learn' section on page 24. Practise reading different numbers with three decimal places. Include numbers within the range 100-1000. Stress the difference in how we read the numbers either side of the decimal point, for example 675·675. **What does this decimal fraction say?** *Six hundred and seventy-five point six, seven, five.* **What does it mean?** Six hundred and seventy-five wholes and 675 thousandths.
- Show pupils a range of decimal fractions represented by place value houses. **How would you read the decimal fraction shown by these place value houses? What do they mean?**
- Discuss examples such as 5·006. 5·060 and 5·600. **What is the same and what is different? How many thousandths in each example? How would we write each decimal fraction as a mixed number?**
- Create a 10 × 100 grid on centimetre squared paper by taping four 10 × 25 grids together. Use to illustrate the difference between, for example, 0·006, 0·060 and 0·600
- Discuss decimal fractions in measurement: 1000 m = 1 km so 1 m = 0·001 km; 1000 g = 1 kg so 1 g = 0·001 kg; 1000 ml = 1 l so 1 ml = 0·001 l. Look for examples in daily life and discuss what they mean, e.g. 2·272 litres of milk. If a millisecond is one thousandth of a second, what do these times mean: 5·001 seconds; 0·930 seconds; 11·064 seconds, etc.

Unit 2.8 Representing and describing decimal fractions

- Present pupils with a 3rd place decimal fraction written in numerals, for example, 0·543. **What does this decimal fraction say? What does it mean? How would you write this as a fraction? Show me.** Repeat for different decimal fractions less than one.
- Discuss numbers greater than one, for example 2·139. **What does this decimal fraction say? What does it mean? How would you write this as a mixed number? Show me.** Repeat for different decimal fractions greater than one.
- Introduce base 10 blocks. **How many small cubes equal one large cube?** One thousand thousandths = 1 whole so each small cube is *one thousandth* of the large cube. **How many 'flats' equal one large cube? What fraction of the large cube is each 'flat'? How many 'rods' equal one large cube? What fraction of the large cube is each 'rod'?**

- Discuss the 'Let's learn' section on page 26.
- Draw a decimal point on a large sheet of paper. Call out a decimal number; for example, three point six two five. Ask a pupil/pair of pupils to lay base 10 blocks out on the paper to represent the number called. Ask the rest of the class to write the number in numerals on their mini whiteboards. Compare and discuss solutions. **Has everyone written 3·625? Is the base 10 model an accurate representation of this number? What makes you say that? Which part of the model represents the 2/3/5/6? How many thousandths can you see?** Draw out that whilst 3·625 means 3 ones and 625 thousandths it can also mean 3 ones, 6 tenths, 2 hundredths and 5 thousandths. Show this on place value houses, e.g.

				Decimal fractions		
H	T	O		$\frac{1}{10}$	$\frac{1}{100}$	$\frac{1}{1000}$
		3	•	6	2	5

- Write different decimal fractions on the board in numerals. Ask a pupil/pair of pupils to lay base 10 blocks out on the paper to represent the number shown. The rest of the class should record the number of ones, tenths, hundredths and thousandths needed on place value houses.

Unit 2.9 Zero as a place holder in decimal fractions

- Discuss the 'Let's learn' section on page 28. Show pupils a range of decimal numbers, each containing at least one zero. For example: 0·112, 9·708, 6·220, 7·005, 3·094, 5·030, 0·107. **What does the number say? What does it mean? What is the purpose of the zero(s)? What would happen if we removed the zero(s)? Would the value of the number change? How?** If appropriate, challenge pupils by asking: **Is every zero a place holder? What makes you say that? Can you use base 10 blocks to prove your idea?**
- Ask pupils to represent the same set of decimals as fractions or mixed numbers. **Which zeros are needed this time? Can you explain why zeros are needed when we write 7·005 but not when we write $7\frac{5}{1000}$?**
- Use base 10 blocks to illustrate the multiplicative relationship between tenths, hundredths and thousandths. Show pupils the large cube. Ask them to predict how many of each of the other base 10 blocks (i.e. flats, rods, small cubes) would be needed to create another large cube. Challenge pupils to work in groups to build a large cube with flats, rods and small cubes to check their predictions. **If the large cube is 'one' what is the value of each flat/rod/small block? How do you know? How many thousandths equal one hundredth?** Record ten thousandths = one hundredth; 0·010 = 0·01. **How many hundredths equal one tenth?** Record ten hundredths = one tenth; 0·10 = 0·1. **How many thousandths equal one tenth?** Record 100 thousandths = one tenth. 0·100 = 0·1.
- Discuss the above statements. **What do you notice?** Draw out that 0·1 = 0·10 = 0·100. **Which zeros are place holders and which are not? What makes you say that?**

Unit 2.10 Partitioning decimal fractions

- Display a set of decimal place value arrow cards showing ones, tenths, hundredths and thousandths. Ask pupils to predict the resulting decimal fraction when the four cards are 'layered' and the arrows aligned. Repeat for different sets of arrow cards.
- **Which place value arrow cards would we need to make the number _____? What makes you say that?** Ask pupils to check their predictions by building the number with place value arrow cards.
- Challenge pupils to make a 4-digit decimal fraction within a given range using only two or three place value arrow cards. **What number have you made? In what position(s) is/are the place holder(s)? Can you make a number with 2/3 decimal places that has the digit 0/5/7, etc. in the tenths/hundredths/thousandths place?** Read your number to a partner. Explain what your number means.
- Ask pupils to make the same set of numbers using place value counters. **Which type of model do you prefer? Why?**
- Write some mixed numbers on the board. For example: two and thirty-four hundredths, six and twenty-nine thousandths. Ask pupils to represent each number using place value arrow cards or place value counters. **Which cards/counters did you choose? Why?** Ask pupils to compare models with a partner. **What is the same and what is different? Are you both correct?** Ask pupils to use digit cards to represent each decimal fraction created on place value houses.
- Ask pupils to enter a decimal fraction with one, two or three decimal places into a calculator. **What will happen to the digits if we press ×10? How will the number change? What will it say?** Draw a decimal point on a large sheet of paper. Use digit cards to model the action of each digit moving one place to the left. **What number did we start with? What does the number say now?** Discuss the effect of multiplying a decimal number by 10 on the need to include zero. For example: 0·126 × 10 = 1·26 (zero no longer required), 5·04 × 10 = 50·4 (zero required).

Unit 2.11 Comparing and ordering decimal fractions

- Hang up a 'washing line'. Peg '1' on the far left and '2' on the far right. **Where should 1·5 go? Why? Should 1·436 come before or after 1·5? Why?** Give each pupil a piece of paper and ask them to write a number with three decimal places that comes in-between 1 and 2 on the number line. Point to a position on the washing line and invite anyone with a number that could take that position to come to the front. If more than one pupil volunteers their number they should work together to agree the order. Point to a different position on the number line and repeat the process. **Why did you place your number there? Which number is greater/smaller?** When all cards are placed, have pupils read the numbers aloud to check the order. **Is the order correct? If not, what needs to be changed?** Increase the level of challenge by asking pupils to communicate in silence.
- Once pupils are confident, the above activity can be extended to include a mixture of numbers with one, two or three decimal places.
- Ask pupils to enter a decimal fraction with one or two decimal places into a calculator. **What will happen to the digits if we press ÷10? How will the number change? What will it say?** Draw a decimal point on a large sheet of paper. Use digit cards to model the action of each digit moving one place to the right. **What number did we start with? What does the number say now?** Discuss the effect of dividing a decimal number by 10 on the need to include zero. For example: 10·1 ÷ 10 = 1·01, 7·02 ÷ 10 = 0·702.

Addition and subtraction 2A

Section overview

Textbook 2A units	Review units	Prior learning	Benchmarks/Es&Os
Unit 3.1 Mental addition and subtraction of 2- and 3- digit numbers	**1C** Units 1.1, 3.1, 3.6 and 3.7	• I can round 2- and 3-digit numbers to the nearest 10. • I can use known facts and place value to add or subtract a 1-digit number to or from a 2- or 3-digit number. • I can use doubles and near doubles to add two 2-digit numbers. • I can use round and adjust to add or subtract 8 or 9 to/from a 2- or 3-digit number. • I understand that addition is commutative but subtraction is not.	**MNU 2-03a** Adds and subtracts whole numbers and decimal fractions to two decimal places, within the number range 0 to 1 000 000.
Unit 3.2 Adding and subtracting 1, 10, 100 and 1000	**1B** Units 2.4, 2.5, 2.6 and 2.7 and 2.8 **1C** Unit 2.3	• I can count forwards and backwards in 1s, 10s and 100s from any 3-digit number. • I can say and write the number before or after any 3-digit number. • I can say and write the number that is 10 more/less or 100 more/less than any 3-digit number.	**MNU 2-03a** Adds and subtracts multiples of 10, 100 and 1000 to and from whole numbers and decimal fractions to two decimal places.
Unit 3.3 Adding and subtracting multiples of 100	**1C** Units 3.2 and 3.3 **2A** Unit 3.2	• I can add or subtract a multiple of ten to or from a 3-digit number. • I can recognise and use the patterns in our number system to add or subtract 10, 100 or 1000 to/from a 4-digit number.	**MNU 2-03a** Adds and subtracts multiples of 10, 100 and 1000 to and from whole numbers and decimal fractions to two decimal places.
Unit 3.4 Adding and subtracting by making 10s or 100s	**1C** Units 3.4, 3.5, 3.8, 3.10, 3.11, 3.12 and 3.13	• I can identify pairs of 2-digit numbers that add to make 100. • I understand the inverse relationship between addition and subtraction. • I can use an empty number line to add 3-digit numbers together. • I can subtract a 3-digit number from another 3-digit number by partitioning the second number into hundreds, tens and ones.	**MNU 2-03a** Adds and subtracts whole numbers and decimal fractions to two decimal places, within the number range 0 to 1 000 000.

Unit 3.5 Adding and subtracting multiples of 1000	**1C** Units 3.3, 3.4 and 3.8	• I can work out number bonds to 100. • I can use place value to mentally add 3-digit multiples of 10 and 100. • I can choose between counting on and counting back to solve a subtraction problem.	**MNU 2-03a** Adds and subtracts multiples of 10, 100 and 1000 to and from whole numbers and decimal fractions to two decimal places.
Unit 3.6 Adding and subtracting multiples of 10 and 100	**1C** Units 3.3 and 3.8	• I can use my understanding of place value to mentally add and subtract multiples of 10 and 100 to/from 3-digit numbers. • I understand that the equals sign symbolises balance in a number sentence. • I can choose between counting on and counting back to solve a subtraction problem.	**MNU 2-03a** Adds and subtracts multiples of 10, 100 and 1000 to and from whole numbers and decimal fractions to two decimal places.
Unit 3.7 Solving word problems	**1C** Units 3.4, 3.8, 3.9, 3.14 and 3.15	• I can use a bar model to help me solve addition and subtraction problems. • I can choose the correct operation to solve a wide range of addition and subtraction problems with 2- and 3-digit numbers. • I can solve two-step problems with 2- and 3-digit numbers.	**MNU 2-03a** Adds and subtracts whole numbers and decimal fractions to two decimal places, within the number range 0 to 1 000 000.
Unit 3.8 Representing word problems in different ways	**1C** Units 3.4, 3.8, 3.9, 3.14 and 3.15	• I can solve addition and subtraction word problems with 2-digit and 3-digit numbers. • I can represent the same word problem in different ways; for example, with concrete materials, by drawing an empty number line or bar model, or by using words and symbols. • I can write a number sentence to match how I solved the problem.	**MNU 2-03a** Adds and subtracts whole numbers and decimal fractions to two decimal places, within the number range 0 to 1 000 000.
Unit 3.9 Using non-standard place value partitioning	**1C** Units 2.6, 2.7 and 2.8	• I can partition 3-digit numbers into hundreds, tens and ones. For example 485 = 400 + 80 + 5. • I can partition 2-digit numbers in different ways. For example, I know that 42 is 40 + 2 but it can also be 30 + 12, 20 + 22, 10 + 32. • I can count forwards in hundreds, tens and ones.	**MNU 2-02a** Partitions a wide range of whole numbers and decimal fractions to three decimal places, for example, 3·6 = 3 ones and 6 tenths = 36 tenths.

Unit 3.10 Adding using a semi-formal written method	**1C** Units 3.1, 3.10 and 3.12 **2A** Unit 2.3	• I can instantly recall all addition facts to 20. • I can use my knowledge of place value to partition 3-digit numbers in different ways. • I know that calculations can be set down horizontally or vertically. • I can add 3-digit numbers by partitioning both numbers and jotting down each step in the calculation.	**MNU 2-03a** Adds and subtracts whole numbers and decimal fractions to two decimal places, within the number range 0 to 1 000 000.
Unit 3.11 Adding 3-digit numbers using standard algorithms	**1C** Units 3.1, 3.6 and 3.7 **2A** Units 3.1, 3.9 and 3.10	• I can instantly recall all number bonds to 20. • I can add 2-digit numbers mentally using strategies such as doubles and near-doubles, round and adjust, give and take. • I can use my knowledge of place value to partition 3-digit numbers in different ways. For example, I know that 140 = 1 hundred and 4 tens = 14 tens. • I can add 3-digit numbers using a semi-formal written method.	**MNU 2-03a** Adds and subtracts whole numbers and decimal fractions to two decimal places, within the number range 0 to 1 000 000.
Unit 3.12 Representing and solving word problems	**1C** Units 3.9, 3.14 and 3.15	• I can use mental strategies, models and jottings to solve different types of addition and subtraction problems. • I can represent the same problem in different ways; for example, with concrete materials, by drawing an empty number line or bar model, or by using words and symbols. • I can write a number sentence to match how I solved the problem.	**MNU 2-03a** Adds and subtracts whole numbers and decimal fractions to two decimal places, within the number range 0 to 1 000 000.

Conceptual context

Pupils need lots of opportunities to represent addition and subtraction situations in multiple ways, for example with concrete materials, pictures and diagrams, words and symbols. Representations such as the bar model, which highlight the part-part-whole relationship between addition and subtraction, are key to helping pupils understand how these operations are connected, with subtraction being the inverse of addition. Part-part-whole is also key to helping pupils appreciate why addition is commutative but subtraction is not.

Underpinning effective mental and written calculation is a deep *conceptual* understanding of place value. Pupils who understand the value of the places – and can partition whole numbers in different ways according to their place values – are able to think flexibly and can choose the most efficient strategy to solve a number problem.

Mental and written calculation strategies introduced at first level (for example, compensation, partitioning, known facts, number bonds to 100 and counting in 10s and 100s on an empty number line) are extended to include larger numbers. Book 2A sees the introduction of semi-formal and formal written methods (column addition and the addition algorithm) as a more efficient way to solve calculations that are too complex to deal with mentally. It is important that pupils understand why, and how, standard written methods work so that they do not become confused by trying to follow a process with little or no understanding. Teachers should therefore use their discretion in deciding when to introduce these methods, i.e. when pupils demonstrate sound *conceptual* understanding of place value. Pupils should also continue to develop mental agility supported by informal jottings.

It is important that pupils experience a wide range of problem types, not simply missing sum or missing difference problems. Pupils who understand the inverse relationship will be able to solve missing addend problems by subtracting, and missing minuend problems by adding. They will also appreciate that missing subtrahend problems can be solved by subtracting or by counting on and will choose the most efficient method given the numbers involved. Part-whole diagrams such as the bar model will help pupils compare and equalise two quantities and articulate their understanding of two-step problems.

Resources

- Triangles template for Challenge question, Unit 3.5
- Think Board template
- Place value counters
- Bar model template
- Mini whiteboard per pupil
- Calculator per pupil
- 10-sided dice showing multiples of 100
- 10-sided dice showing the digits 0-9 per pair, or a pack of 0–9 digit cards per pupil
- Flipchart paper and pens

Key vocabulary

- Calculate
- Compensation
- Decreasing
- Difference
- Exchange
- Increasing
- Inverse
- Jottings
- Mental agility
- Method
- Number bonds
- Partition
- Place value
- Sequence
- Strategy
- Sum
- Think Board

Misconceptions

Potential misconception	Addressing the misconception
Unit 3.1 Pupils who do not have a *conceptual* understanding of place value may become confused when attempting to use the compensation strategy, being uncertain about when to add and when to subtract. The premature introduction of rules such as 'add 10 then take 1 away', before the pupil has acquired full conceptual place value understanding, may exacerbate this misconception.	Pupils need time to explore *why* strategies such as round and adjust work. Initially, they should use concrete materials such as base 10 blocks or place value counters to prove, for example, why the answer to 146 – 19 must be a bigger number than the answer to 146 – 20. Once pupils are confident with the above, ask them to predict whether the words 'bigger than', 'smaller than' or 'equal to' is required to make number sentences such as those below true: 28 + 49 is ________ 51 + 30 85 + 36 is ________ 83 + 38 95 – 19 is ________ 94 – 20 71 – 58 is ________ 73 – 60 **What makes you say that? How could you check? Were you correct? If not, where did you go wrong?** Extend the above ideas to include a 3-digit number +/– a 2-digit number.

<table>
<tr><th>Potential misconception</th><th>Addressing the misconception</th></tr>
<tr><td>Unit 3.4
Pupils may generate an answer that is 100 out by using a split strategy to solve missing addend problems. For example: 460 + ___ = 900 think, '400 + 500 = 900' and 60 + 40 = 100 so arrives at the incorrect answer of 540.</td><td>Ask pupils to jump along an empty number line to solve the same problem. Which answer do you think is the correct answer? Why? Ask pupils to use base 10 blocks to check. Why can't the answer be 540?</td></tr>
<tr><td>Units 3.4, 3.5 and 3.8
Pupils may not fully understand the inverse relationship and so add to solve calculations such as ___ + 189 = 500 and subtract to solve problems such as ___ − 1305 = 4000</td><td>Present pupils with a completed bar model, for example:
<table><tr><td>311</td><td>189</td></tr><tr><td colspan="2">500</td></tr></table>Which number represents the whole? Which numbers represent the parts?

What add 189 makes 500? What must be added to 311 to make 500? Which two numbers total 500? What must be subtracted from 500 to give an answer of 311? What number minus 189 gives an answer of 311, etc.

If pupils are still having difficulty, use smaller quantities that can be more easily visualised.</td></tr>
<tr><td>Units 3.7 and 3.12
Pupils may automatically think they should add every time they read the word 'more' and subtract every time they see the word 'fewer'. Providing pupils with 'rules', in the form of lists of key words linked to specific mathematical operations/symbols, may cause or exacerbate this misconception.</td><td>Present the pupils with word problems of a similar type to those in the relevant units but using smaller quantities that can be easily visualised or 'acted out'. Instead of focusing on the words 'more' and 'fewer', focus on the 'action' inferred by the problem.</td></tr>
</table>

Teaching ideas

Unit 3.1 Mental addition and subtraction of 2- and 3-digit numbers

- Write 24 + 49 on the board. Ask pupils to calculate the answer mentally and 'show you' on their mini whiteboards. **Do we all have the same answer? How did you work it out?** Discuss the various strategies used. **Did anyone think of the calculation as 23 + 50? Why did you change the numbers in this way? What did you need to do to change them?** Introduce new vocabulary: compensate, compensation, adjust.
- Ask pupils to use compensation to calculate the following: 48 + 47, 69 + 21, 82 + 34, 99 + 52, 17 + 76. Agree and discuss answers. **Why did you add/subtract 1/2/3? How did adding/subtracting 1/2/3 change the calculation?**

- Write the 'new' calculation alongside the original calculation on the board. For example:

  48 + 47 gives the same answer as 50 + 45
  82 + 34 gives the same answer as 80 + 36
  17 + 76 gives the same answer as 20 + 73
  69 + 21 gives the same answer as 70 + 20
  99 + 52 gives the same answer as 100 + 51

  Ask pupils to check the answers on a calculator. **Can you explain why this strategy works?**
- Write 56 – 19 on the board. Ask pupils to calculate the answer mentally and 'show you' on their mini whiteboards. **Does everyone agree? How did you work it out?** Repeat for 73 – 29, 65 – 18, 192 – 59, 217 – 48.
- Compare compensation strategies for addition and subtraction. **What is the same and what is different?**
- Challenge more able pupils to investigate whether or not 'give and take' works for subtraction. **Does 'give and take' work for subtraction? Why is that?**

Unit 3.2 Adding and subtracting 1, 10, 100 and 1000

- Ask pupils to key a 4-digit number into their calculator and to predict the number they will get when they add 1. Repeat to create a sequence. Focus on numbers that bridge a thousand. For example: 2998 + 1 = 2999, 2999 + 1 = 3000, 3000 + 1 = 3001, 3001 + 1 = 3002....
- Repeat the above activity for adding 10, 100, 1000. **Which sequence did you find most challenging to predict? Why is that?**
- Repeat the above activities for subtracting 1, 10, 100, 1000. **What happens to the digit in the ones, tens, hundreds, thousands place each time we subtract 1, 10, 100, 1000?**
- Screen a row of place value counters with a strip of paper or card. Present pupils with a four-digit number and have them add 1, 10, 100 or 1000 as the counters are revealed, one at a time. Pupils keep a running total. For example:

1849
starting number

(counters hidden and revealed one at a time)

1849	1000	1000	100	10	1000	100	1	100	1000
starting number	2849	3849	3949	3959	4959	5059	5060	5160	6160

(counters hidden and revealed one at a time)

Which digit(s) changed each time? Why does that happen?

- Present pupils with a collection of place value counters. Have them count in 1000s, 100s, 10s and 1s to find the total. Remove the counters one at a time. Ask pupils to write down the value of the remaining counters each time, without counting them. **How do you know the value of the remaining counters is ______? How did you work it out? How can we check?**
- Ask pupils to make jumps of 1, 10, 100 and 1000 along an empty number line to solve additions and subtractions such as: 1345 + 3516 = ______, 8438 − 5174 = ______, 2912 + ______ = 4735, 9078 − ______ = 6129 and check their solutions on a calculator.

Unit 3.3 Adding and subtracting multiples of 100

- Play 'Multiples of 100 dice game'. Pupils play in pairs/groups. They decide on a target number that is a 4-digit multiple of 100; for example, 4700. Pairs/groups take turns to roll a 10-sided dice showing the multiples of 100, from 0–900 or from 10–1000, and write down the number they roll. They each add the next number they roll to their first roll and record the total. Play continues with players recording the new total each time. The first player to reach the *exact* target number is the winner. Should the number rolled exceed the target number, players should subtract the next number rolled. **Is it always possible to reach the exact target number first time? If you were on 4200 and rolled 900, what number would you have reached? How much more than the target number is this? So, what will you need to do in the next round to try to reach the target number?**
- Adapt the above game for subtraction. Pupils subtract multiples of 100 from an agreed starting number. The winner is the first player to reach zero. The rules of the game can be adapted to suit the pupil's ability. For example, pupils could either miss a turn if the number rolled is greater than the number required, or add to a negative number in the next round. **Is it always possible to reach zero exactly? If you were on 500 and you rolled 700, what number would you have reached? How much less than the target number is this? So, what will you need to do in the next round to try to reach the target number?**
- Increase challenge in the above activity by asking pupils to roll two or more 10-sided dice and subtract the total. Ask: **How else could we adapt this game?**
- Play 'The answer is______, what is the question?' Display a 4-digit multiple of 100; for example, 1600. Using only multiples of 100, challenge pupils to work in small groups to write as many additions and subtractions as they can that have the answer 1600, within a given timeframe. **How effectively did you work together as a group? What was your strategy? Could it be improved? If so, how?**

Unit 3.4 Adding and subtracting by making 10s or 100s

- Display a 10 × 10 grid that has been partitioned into two parts, for example 71 and 29. Challenge pupils to say how many squares there are in each part of the grid, without counting. **How did you know there were 71 and 29? What strategy did you use?**

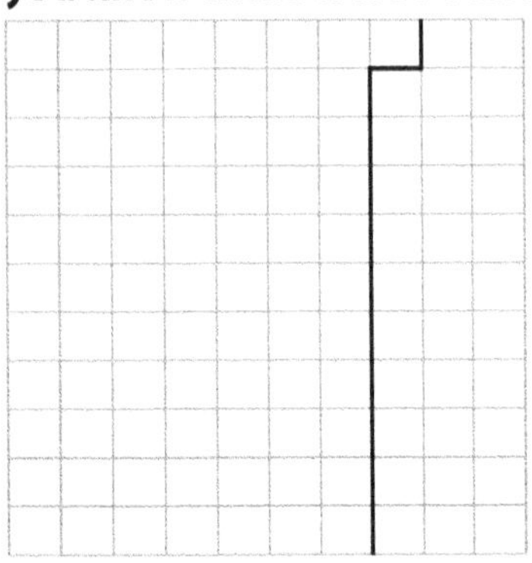

Repeat for other complements to 100.

- Extend the above activity by taping ten 10 × 10 grids together to make a 1000 square grid. Draw a line to partition the grid into two parts. **How many squares on the left? How many on the right? How many altogether?** Compare the partitioned grids. **What do you notice?**

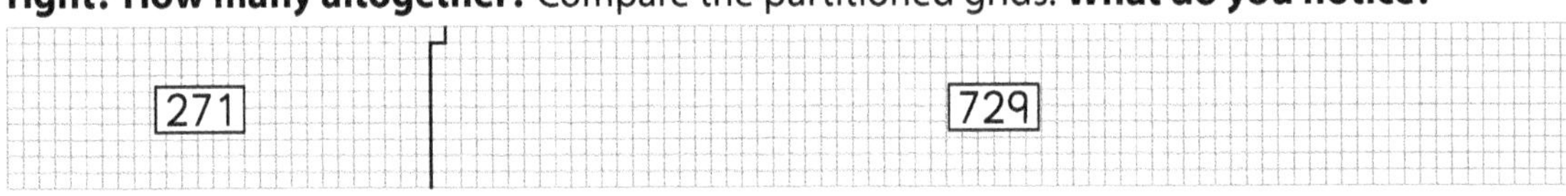

Repeat for other complements to 1000.

- Pose questions that are similar to those in the 'Let's learn' section on page 38. **Which numbers will you add together first? Why? What number fits with ____ to make the next multiple of 10/100?** Pupils should be encouraged to predict their answers by visualising a hundred grid/thousand grid, and by making jottings to help them keep track of their thinking. They should then be invited to check each answer by drawing an empty number line.
- Challenge pupils to write their own number sentences that can be solved by a partner by making multiples of 10 or 100.

Unit 3.5 Adding and subtracting multiples of 1000

- Refer to the 'Let's learn' section on page 40. **Why do you think Isla decided to count on? How has she thought about the problem? Why has Nuria decided to count back? What might she have been thinking? Why do both counting on and counting back work for this problem? How would you solve it?**

- Write two 4-digit numbers on the board. One should be a multiple of 1000. Ask the pupils to draw an empty number line on their mini whiteboards and to use a jump strategy to find the difference between the two numbers given. Compare answers and strategies. **Did everyone get the same answer? If not, what might have happened? How can we check our answers? Is it easier to add/subtract the ones or the thousands first? Always?** Ask pupils to write an addition number sentence and a subtraction number sentence for each example.
- Present pupils with a range of missing sum, missing addend, missing difference, missing subtrahend and missing minuend calculations. Discuss whether it is better to count on or count back for each example. **Which number will you count on/back from? Which number will you count on or back to? Is it possible to solve a problem such as ______ + 2167 = 7000 by subtraction? Why is that? Is it possible to solve a problem such as ______ – 5689 = 1000 by addition? Why is that?** Ask pupils to write an addition number sentence to partner each subtraction and vice-versa.

Unit 3.6 Adding and subtracting multiples of 10 and 100

- Present a 4-digit addition where the total number of each of the thousands, hundreds, tens or ones is 9 or less, for example 5300 + 1248. Ask pupils to predict the answer without writing anything down. **How did you do it? How easy was it to keep track of your thinking? Would jottings have helped?** Demonstrate the preferred strategy, jotting down each step on a place value house, for example: '*5 thousands plus 1 thousand equals 6 thousands*' (record 6 in the thousands column), *3 hundreds plus 2 hundreds equals 5 hundreds* (record 5 in the hundreds column), *0 tens plus 4 tens equals 4 tens* (record 4 in the tens column), *0 ones plus 8 ones equals 8 ones* (record 8 in the ones column). *So, 5300 + 1248 = 6548.*
- Repeat the above activity for further examples, ensuring that either addend is a multiple of 10 or 100. **How did you work it out?** Encourage pupils to name the places when explaining their thinking. For example, '*5 thousands add 1 thousand makes' 6 thousands'* (note: <u>not</u> 5 add 1 makes 6).
- Repeat for 4-digit subtractions where the subtrahend only is a multiple of 10 (for example, 9496 – 2420) or where both the minuend and subtrahend are multiples of 10, for example, 3670 – 410. Care should be taken to ensure that no negative digits are created (this is addressed in Book 2B).
- Present a 4-digit addition where the total number of hundreds, tens or ones is greater than 9; for example, 4620 + 3755. **What makes this question more challenging? Would our previous strategy of recording the number of thousands, hundreds, tens and ones work for this example? Why not? How else could we work it out?** Discuss pupils' suggestions. Lead them towards the use of place value partitioning. For example, 4620 + 3755:

  4000 + 3000 = 7000; 600 + 700 = 1300; 20 + 50 = 70; 0 + 5 = 5 → 7000 + 1300 + 70 + 5 → 8375. So, 4620 + 3755 = 8375
- Ask pupils to key a 4-digit number into a calculator; for example, 9480. **Which keys will you need to press to change the thousands digit into a 6? The hundreds digit into a 1? The tens digit into a 4? The ones digit into a 9, etc.? Why?**
- In all of the above, less confident pupils should be encouraged to check their solutions using base 10 blocks or place value counters, counting in thousands, hundreds, tens and ones and keeping a running total (jottings) as they go.

Unit 3.7 Solving word problems

- Present pupils with a range of word problems of different types. Ask them to represent each problem as a bar model on a mini whiteboard. **Is this an addition problem or a subtraction problem? How do you know?** Focus the discussion on 'part-part-whole'. Pupils who fully understand the inverse relationship will appreciate that if the whole is unknown, then addition is the correct choice of operation. If one of the parts is unknown, they need to subtract.
- Ask pupils to solve the problems presented using a strategy of their choice. **What calculation(s) did you do? What strategy did you choose? Why did you choose it? Is there another way?** Pupils who fully understand the inverse relationship will be able to choose between counting on and counting back (mentally or on an empty number line) to solve missing addend, difference and subtrahend tasks depending on the relative sizes of the numbers involved. For example, in 312 – 16 = ______ it is better to count back, whereas in 312 – 296 = ______ it is better to count on.

Unit 3.8 Representing word problems in different ways

- Ask pupils to work in pairs or small groups to sort a range of addition and subtraction word problems into those that can be solved by addition, those that can be solved by subtraction, and those that can be solved using either addition or subtraction. **How do you know? Show me.** Focus the discussion on 'part-part-whole'.
- Pupils work in pairs. Allocate a different word problem from the above set to each pair. Give each pair a 'Think Board' and ask them to represent their problem as a bar model, a number sentence and on an empty number line. **What is missing from the bar model? Is it the whole or one of the parts? Did you put the 'missing number box' before or after the equals sign? Why? Which numbers will you write on the number line? Will you need to jump forwards or backwards to find the solution? Why?** Pairs should solve the problem(s) allocated to them then share their solution and explain their Think Board to their peers.
- Ask pupils to write a word problem for their partner to solve, stating which operation they believe is required and why. **Do you agree with your partner's judgement? Is there more than one way to solve this problem? What number sentence will you write? Why?**

Unit 3.9 Using non-standard place value partitioning

- Present pupils with a 3-digit whole number, for example 632. Ask them to sketch base 10 blocks to represent the chosen number on their mini whiteboard and 'show you'. **Have we all drawn the same thing?** Most will likely have drawn 6 hundreds, 3 tens and 2 ones. Introduce a tray of base 10 blocks in which there are only 5 hundreds but at least 33 tens. **Can we make the number 632 using some or all of the materials in this tray?** Challenge pupils to justify their answer. **How can we create the extra hundred block(s) needed?** Build the number 632 by combining 10 tens to make 1 hundred. **How many hundreds? How many tens?** Record on a flipchart or whiteboard:

  632 = 6 hundreds, 3 tens and 2 ones

632 = 5 hundreds, 13 tens and 2 ones

Replace one of the hundreds with 10 tens. **How many hundreds now? How many tens?** Record:

632 = 4 hundreds, 23 tens and 2 ones

Replace a further hundred with 10 tens. **How many hundreds now? How many tens?** Record:

632 = 3 hundreds, 33 tens and 2 ones

What do you notice? Challenge pupils to write the next three lines of the pattern by imagining the movement of the blocks.

- Make the same 3-digit number with place value counters. Write '= 600 + 30 + 2' alongside '6 hundreds, 3 tens and 2 ones'. We know that 10 × 10 = 100. We can *exchange* a 100 counter for 10 tens counters. **We have 5 × 100 counters. What is their value (500)? How many tens counters do we have now? What is their value (130)? The value of the ones counters is still 2 so we have 500 + 130 + 2.** Write '= 500 + 130 + 2' alongside '5 hundreds, 13 tens and 2 ones'. Ask pupils to predict the next five additions in the pattern and check using place value counters.
- **What if we exchange 1 ten for 10 ones in each example? How many hundreds, tens and ones will we have? What will the addition number sentence say?**
- Pupils work in small groups. Give each group a piece of flipchart paper with a 3-digit number written on it. Challenge them to record as many different ways as they can to make the number with place value counters. For example, 547 could be 500 + 40 + 7, 400 + 140 + 7, etc. It could also be 500 + 30 + 17, 500 + 20 + 27, etc. **Have you found all possibilities? How can you be sure?**

Unit 3.10 Adding using a semi-formal written method

- Write a 3-digit addition on a flipchart or whiteboard; for example, 485 + 339. Ask pupils to find the answer by partitioning both numbers into hundreds, tens and ones. **Which numbers will you add first?** Stress that they should start by adding the hundreds, i.e.

  400 + 300 = 700; 80 + 30 = 110; 5 + 9 = 14 → 700 + 110 + 14 → 824 so 485 + 339 = 824.

  Did everyone get the same answer? If not, where did you go wrong? Why do you think that happened? Demonstrate how the same calculation can be presented vertically and solved by partitioning, i.e.

```
  485
+ 339
  ---
  700
  110
   14
  ---
  824
```

Which layout do you prefer? Why? Why do you think the 7 is written underneath the 4 and the 3? Ask similar questions about the other digits.

- Pupils work in pairs. Each pupil rolls a 0–9 dice three times, or chooses three digit cards from a 0–9 pack, to create a 3-digit number. The pairs write down the two numbers they have created. Each pupil then works independently to find the sum of the two numbers created using column addition. Finally, they should compare answers and layout. **Is each digit in the correct column? Did you both get the same answer? If not, whose answer is correct? How do you know? Is any answer greater than 1000? Where will you write the thousands digit? Why?**
- Write two 4-digit numbers on a flipchart/whiteboard. **What must be added to/subtracted from each digit to change the first number into the second number?** For example, to change 8721 into 3645 it is necessary to subtract 5 thousands (5000), subtract 1 hundred (100), add 2 tens (20) and add 4 ones (4). So 8721 – 5000 – 100 + 20 + 4 = 3645.

Unit 3.11 Adding 3-digit numbers using standard algorithms

- Write the two column additions below on a flipchart/whiteboard and ask pupils to compare them. **What is the same? What is different?**

```
    369        369
  + 487      + 487
    ---        ---
    700         16
    140        140
     16        700
    ---        ---
    856        856
```

- Discuss the 'Let's learn' section on page 52. Which digits has Amman added first? Tell pupils that when we are using the standard written algorithm we must always add the ones first.
- Ask the pupils to check Amman's calculation using column addition. **Can you explain how the addition algorithm works? Why do you think the first digit of the number 11 is carried over into the tens column? Why has Amman recorded 2 tens in his answer? Why has he written 6 in the hundreds column instead of 5?** Encourage responses such as, 'There are 11 ones. 11 ones is the same as 1 ten and 1 one or 10 + 1', or 'There are 11 tens. 11 tens is like 10 tens and 1 ten. 10 tens make 100, add the other ten makes 110, so 11 tens = 110.' **Why are the 'carried' digits necessary? Why are they written in these places? What would happen if we missed them out or wrote them in the wrong place?**
- Write some 3-digit additions that are similar to those below on a flipchart/whiteboard. Ask pupils to predict whether or not they will need to carry a ten and/or a hundred in each example. **How do you know?**

```
    174        236        468        541
  + 732      + 538      + 174      + 359
    ---        ---        ---        ---

    ---        ---        ---        ---
```

- Write the algorithms below on a flipchart/whiteboard. **Can they be correct? Why not?** Stress that, prior to using an algorithm, it is important to estimate the approximate answer so that we can judge if our answer is reasonable.

```
    1                          1 4         1 1
  1 2 6        3 9 1          4 2 7       5 8 7
+ 6 5 4      + 2 6 5        + 2 5 7     + 1 9 5
  7 7         515 6          7 1 1       6 7 2
```

- Ask pupils to estimate the answer to each algorithm displayed, work out the exact answers on their mini-whiteboards and compare them with a partner.

Unit 3.12 Representing and solving word problems

- Present pupils with the following word problem: *There are 354 cars, some motorbikes and 106 vans in a multi-storey car park. If all 500 spaces are taken, how many motorbikes are there?* Ask pupils to paraphrase the problem, then work in pairs to represent the problem as a bar model. **How does your bar model help you decide which calculations to do? Which strategy/strategies will you use?** Pairs should agree a strategy, work together to solve the problem, then compare methods and solutions with at least one other pair.
- Ask pupils to represent the following word problems as bar models and/or number sentences.

  There are 354 cars in the car park. 126 of the cars are white. How many cars are not white?

  Pat delivered 368 letters on Tuesday. This is 147 fewer letters than she delivered on Monday. How many letters did Pat deliver on Monday?

  Jack delivered 416 letters on Monday. Jan delivered 96 fewer letters than Jack that day. How many letters did Jan deliver?

  Sally sells 371 ice-creams on Saturday and 432 ice-creams on Sunday. How many more ice-creams did she sell on Sunday?

  107 of the ice-creams sold were chocolate chip flavour. 213 more vanilla than chocolate chip were sold. How many vanilla ice-creams did Sally sell?

  Do we need to find the whole or one of the parts? Which operation will you use? Which strategy will you choose? (For example, the pupils may use mental with jottings, number line, column method, algorithm.) Compare methods and solutions.

Addition and subtraction 2B

Section overview

Textbook 2B units	Review units	Prior learning	Benchmarks/ Es&Os
Unit 3.1 Mental addition and subtraction	**2A** Units 2.3 and 3.1	• I can use known facts and place value to mentally add or subtract a 2-digit number to/from a 3-digit number. • I can partition 2- and 3-digit numbers in different ways. • I can 'round and adjust' to add or subtract a 2-digit number to/from a 3-digit number. • I can use the strategy of 'give and take' to make a calculation easier.	**MNU 2-03a** Adds and subtracts whole numbers and decimal fractions to two decimal places, within the number range 0 to 1 000 000.
Unit 3.2 Adding and subtracting a string of numbers	**1C** Units 3.4 and 3.5 **2A** Units 3.3, 3.4, 3.5 and 3.6	• I can identify multiples of ten that total 100. • I can add and subtract 3-digit multiples of 100. • I can add and subtract 3-digit numbers by making 10s or 100s. • I can add and subtract to and from 1000 and multiples of 1000. • I can add and subtract multiples of 10 and 100 to and from a 4-digit number.	**MNU 2-03a** Adds and subtracts whole numbers and decimal fractions to two decimal places, within the number range 0 to 1 000 000.
Unit 3.3 Adding and subtracting multiples of 10, 100 and 1000	**1C** Unit 3.8 **2A** Units 3.3, 3.5 and 3.6	• I can use my understanding of place value to mentally add or subtract multiples of 10, 100 or 1000 to/from 4-digit whole numbers. • I understand that the equals sign symbolises balance in a number sentence. • I understand the inverse relationship between addition and subtraction.	**MNU 2-03a** Adds and subtracts multiples of 10, 100 and 1000 to and from whole numbers and decimal fractions to two decimal places.
Unit 3.4 Using place value partitioning	**1C** Units 3.12 and 3.13 **2A** Units 2.3, 3.9 and 3.10 **2B** Unit 2.3	• I know that calculations can be set down horizontally or vertically. • I can use my knowledge of place value to partition 4-digit numbers in different ways. • I can add 3-digit numbers using a semi-formal written method.	**MNU 2-02a** Partitions a wide range of whole numbers and decimal fractions to three decimal places; for example, 3·6 = 3 ones and 6 tenths = 36 tenths.

Unit 3.5 Adding 4-digit numbers using standard algorithms	**2A** Units 3.1 and 3.11 **2B** Units 2.3 and 3.4	• I can use a range of strategies to mentally add 2-digit numbers. • I can add 4-digit numbers using a semi-formal written method. • I can add 3-digit numbers using a standard algorithm. • I can partition 4-digit numbers in different ways. For example, I know that 3700 can be 3 thousands and 7 hundreds or 2 thousands and 17 hundreds.	**MNU 2-03a** Adds and subtracts whole numbers and decimal fractions to two decimal places, within the number range 0 to 1 000 000.
Unit 3.6 Column subtraction: 3-digit numbers	**1C** Units 3.1, 3.8, 3.12 and 3.13 **2A** Units 2.8, 3.4 and 3.6 **2B** Unit 2.6	• I can instantly recall all subtraction facts within 20. • I understand that subtraction is the inverse of addition. • I know that calculations can be set down horizontally or vertically. • I can count back in tens and ones on a number line displaying both positive and negative numbers. • I can subtract a 3-digit number from a larger 3-digit number by partitioning and jotting down each step in the calculation.	**MNU 2-03a** Adds and subtracts whole numbers and decimal fractions to two decimal places, within the number range 0 to 1 000 000.
Unit 3.7 Column subtraction: 4-digit numbers	**1C** Units 3.1 and 3.8 **2A** Units 3.4 and 3.6 **2B** Unit 3.6	• I can instantly recall all subtraction facts within 20. • I understand that subtraction is the inverse of addition. • I can subtract a 3-digit number from a larger 3-digit number using a semi-formal written method.	**MNU 2-03a** Adds and subtracts whole numbers and decimal fractions to two decimal places, within the number range 0 to 1 000 000.
Unit 3.8 Subtracting 3-digit numbers using standard algorithms	**Book 2A** Units 2.3, 3.1 and 3.9 **Book 2B** Units 2.3 and 3.6	• I can subtract 2-digit numbers mentally using strategies such as halving, round and adjust and give and take. • I can subtract 3-digit numbers using a semi-formal written method. • I can use my knowledge of place value to partition 3-digit numbers in different ways. For example, I know that 278 can be 2 hundreds, 7 tens and 8 ones; 1 hundred, 17 tens and 8 ones; 2 hundreds, 6 tens and 18 ones.	**MNU 2-03a** Adds and subtracts whole numbers and decimal fractions to two decimal places, within the number range 0 to 1 000 000.

Unit 3.9 Subtracting 4-digit numbers using standard algorithms	**2A** Unit 2.3 **2B** Units 2.3 and 3.7	• I can subtract a 4-digit number from a larger 4-digit number using a semi-formal written method. • I can use my knowledge of place value to partition 4-digit numbers in different ways. For example, I know that 1724 can be: 1 thousand, 7 hundreds, 2 tens and 4 ones; 1 thousand, 72 tens and 4 ones; 1 thousand, 7 hundreds and 24 ones; 17 hundreds, 2 tens and 4 ones.	**MNU 2-03a** Adds and subtracts whole numbers and decimal fractions to two decimal places, within the number range 0 to 1 000 000.
Unit 3.10 Mental and written strategies	**2B** Units 3.1, 3.2, 3.3, 3.4, 3.5, 3.6, 3.7, 3.8 and 3.9	• I can use a range of mental strategies to solve addition and subtraction number problems. • I can use an empty number line, or informal jottings, to help me keep track of my thinking. • I can add and subtract 4-digit numbers using a semi-formal written method. • I can use standard algorithms to add and subtract 4-digit numbers.	**MNU 2-03a** Adds and subtracts whole numbers and decimal fractions to two decimal places, within the number range 0 to 1 000 000.
Unit 3.11 Representing and solving word problems	**2A** Units 3.7, 3.8 and 3.12 **2B** Unit 3.10	• I can solve addition and subtraction word problems involving 2- and 3-digit numbers. • I can represent the same problem in different ways; for example, with concrete materials, by drawing an empty number line or bar model, or by using words and symbols. • I can choose between addition and subtraction to make a direct calculation possible. • I understand how algorithms work and can use them appropriately to solve problems involving 'tricky' numbers.	**MNU 2-03a** Adds and subtracts whole numbers and decimal fractions to two decimal places, within the number range 0 to 1 000 000.
Unit 3.12 Solving multi-step word problems	**2A** Units 3.7, 3.8 and 3.12 **2B** Units 3.10 and 3.11	• I can use mental strategies, models and jottings to solve different types of addition and subtraction problems. • I can represent and solve two-step word problems in different ways. • I understand how algorithms work and can use them appropriately to solve problems involving 'tricky' numbers.	**MNU 2-03a** Adds and subtracts whole numbers and decimal fractions to two decimal places, within the number range 0 to 1 000 000.

Unit 3.13 Adding decimal fractions and whole numbers	**2A** Units 2.5, 2.6, 2.7 and 3.1 **2B** Units 2.3, 2.4 and 3.1	• I can use a range of mental strategies, including 'round and adjust' and partitioning, to add whole numbers together. • I understand the patterns in our place value system, i.e. 10 tenths = 1 one, 10 ones = 1 ten, 10 tens = 1 hundred, 10 hundreds = 1 thousand, 10 thousands = 10 thousand. • I can read, write, build and describe decimal fractions with tenths. • I can use a number line to compare and order decimal fractions with tenths.	**MNU 2-03a** Adds and subtracts whole numbers and decimal fractions to two decimal places, within the number range 0 to 1 000 000.
Unit 3.14 Adding and subtracting decimal fractions	**2B** Units 3.2 and 3.13	• I can round and adjust to add a whole number and a decimal fraction with one decimal place. • I can use partitioning to add a whole number and a decimal fraction with one decimal place. • I can recall number bonds to 100. • I understand the inverse relationship between addition and subtraction.	**MNU 2-03a** Adds and subtracts whole numbers and decimal fractions to two decimal places, within the number range 0 to 1 000 000.

Conceptual context

Pupils who understand the patterns in our place value system and the multiplicative relationship between the places can use this knowledge to calculate efficiently. Mental calculation strategies previously introduced are extended to include transformation. All are highlighted as efficient ways to add and subtract some 4- and 5-digit whole numbers, and decimal fractions with one decimal place. Semi-formal and formal written methods are extended to include column subtraction and the subtraction algorithm for 4-digit whole numbers. As stated in section 2A, algorithms are advocated as an effective way to solve calculations that are too complex to deal with mentally and a focus on mental calculation, supported by jottings, should be maintained. Adding and subtracting decimal fractions is introduced through the use of empty number lines and bar models, which highlight the inverse relationship between the operations.

As also stated in section 2A, it is important that pupils experience the full range of addition and subtraction problem types: join (missing sum and missing addend); separate (missing minuend, missing subtrahend or missing difference); combine (whole unknown or one part unknown); equalise; and compare. They should continue to be provided with ample opportunities to represent the same word problem in different ways (see 'Think Board'), and write their own word problems for others to solve, thereby enhancing their ability to understand and solve word problems of increasing complexity.

Resources

- Think Board templates
- Bar model template
- Mini whiteboards
- Base 10 blocks
- Place value counters

Key vocabulary

- **Calculate**
- **Compensate**
- **Difference**
- **Empty number line**
- **Exchange**
- **Jottings**
- **Method**
- **Multi-step word problem**
- **Number bonds**
- **Partition**
- **Place value**
- **Solution**
- **Strategy**
- **Think Board**
- **Transform**
- **Value**

Misconceptions

Potential misconception	Addressing the misconception
Units 3.1 and 3.10 The pupil is uncertain about whether to add or subtract to adjust the numbers when using a transformation strategy.	Display pairs of calculations such as those below. Ask pupils to calculate the answers using a strategy they are comfortable with then insert = or ≠ in each box to make the number statements true. Discuss the true statements. **What do you notice? How have the numbers been transformed?** Provide further examples as necessary. 128 + 167 _____ 130 + 165 762 – 129 _____ 763 – 130 128 + 167 _____ 130 + 169 762 – 129 _____ 761 – 130 114 + 149 _____ 115 + 150 891 – 132 _____ 889 – 230 114 + 149 _____ 113 + 150 891 – 132 _____ 893 – 230

Potential misconception	Addressing the misconception
Units 3.2, 3.3 and 3.10 The pupil arrives at an answer that is 100 or 1000 out when adding numbers to make a multiple of 100 or 1000.	Have pupils practise adding complements of 100 and 1000 by mentally partitioning both numbers, supported by jottings where necessary. **Which parts did you add first? How did you keep track of your thinking?** Record pupils' explanations, for example: **83 + 17** 80 + 10 = **90**, 3 + 7 = **10** ⟶ 90 + 10 = **100** **254 + 746** 200 + 700 = **900**, 50 + 40 = **90**, 4 + 6 = **10** ⟶ 900 + 90 + 10 ⟶ **900 + 100 = 1000** Ask pupils to work in pairs to generate lists of numbers that add to 100 or 1000. Once confident, extend to multiples of 100 or 1000.
Units 3.6 and 3.7 The pupil may know that 9135 – 5672 ≠ 5672 – 9135 without fully appreciating the impact of 'flipping' some parts of the subtraction on the final answer. For example, in the calculation 9135 – 5672 pupils may think they can turn 30 – 70 around to say 70 – 30 or 100 – 600 around to say 600 – 100.	Display a calculation vertically on a flipchart or whiteboard, for example: 9 1 3 5 – 5 6 7 2 _______ _______ Ask pupils to draw a number line graduated in thousands to show that 9000 – 5000 = 4000. Ask pupils to draw a number line graduated in hundreds to show 100 – 600. Demonstrate how the number line needs to be extended to the left to show negative numbers. Ask pupils to draw a number line graduated in tens to show 30 – 70. Demonstrate how the number line again needs to be extended to the left to show negative numbers. Practise counting backwards in 1s, 10s, 100s, 1000s from a positive number to a negative number by visualising a number line. Link this with subtraction. For example: Start on 200. Count back in hundreds to –500. How many hundreds? 200 – 700 = –500.
Units 3.8 and 3.9 The pupil is unable to relate decomposition with place value partitioning and so has difficulty seeing why the subtraction algorithm works and/or makes decomposition errors.	Display a calculation vertically on a flipchart or whiteboard, for example: 8 1 3 – 4 5 7 _____ _____ Ask pupils to make the minuend (i.e. 813) with base 10 blocks or place value counters. Remind them that when we are using standard algorithms, we start with the ones. **Can we subtract 7? Why not? How can we get enough ones without changing the overall value of the number?** Lead discussion towards exchanging the ten for 10 ones. **How can we subtract 5 tens?** Lead discussion towards exchanging a hundred for 10 tens.

Potential misconception	Addressing the misconception
	We had 8 hundreds, 1 ten and 3 ones. What does 8 hundreds, plus 1 ten, plus 3 ones make? What do we have now? (7 hundreds, 10 tens, 13 ones). **How much is that altogether?** Link each step of the process with the written algorithm, i.e. First we exchanged a ten for 10 ones. We added the 10 ones in with the 3 ones. That gave us 13 ones. 13 ones – 7ones = 6 ones 0 1 8 ~~1~~ 3 – 4 5 7 6 Then we exchanged a hundred for 10 tens. That meant we had 7 hundreds and 10 tens. 10 tens – 5 tens = 5 tens 7 10 1 ~~8~~ ~~1~~ 3 – 4 5 7 5 6 7 hundreds – 4 hundreads = 3 hundreads. 7 10 1 ~~8~~ ~~1~~ 3 – 4 5 7 3 5 6 Ask pupils to check the answer by subtracting 457 from the concrete representation of 813 created. Provide further examples, including 4-digit subtractions. Encourage pupils to round to find an approximate answer before calculating.
Units 3.13 and 3.14 The pupil may think, for example, that 3·9 + 1·9 = 4·18.	Represent each addend with concrete materials in which the individual tenths that make up one whole are visible, for example base 10 blocks or ten frames. Focus discussion on the value of the digits and on the multiplicative relationship between the places, i.e. 10 tenths = 1 whole so 18 tenths = 1 whole and 8 tenths.

Teaching ideas

Unit 3.1 Mental addition and subtraction

- Display a calculation such as 432 + 128. Ask pupils to find the answer using a strategy of their choice and 'show you' on their mini whiteboards. **Does everyone have the same answer? How did you work it out?** Possible strategies include: mental place value partitioning with jottings, i.e. 500, 50, 10 → 560; jump along an empty number line or count on mentally from 432, i.e. 532, 552, 560; column addition; standard algorithm; and round and adjust. Discuss the advantages and disadvantages of each. **Which is the most efficient strategy? Always?**

- Repeat the above activity for a calculation such as 765 – 359. **Do the same strategies work for subtraction? Always? Sometimes? Never? Is there another way?**
- Discuss Isla's addition example from the 'Let's learn' section on page 30. **Which strategy does this remind you of?**
- Discuss Nuria's subtraction example from the 'Let's learn' section on page 30. **What is the same and what is different?**
- Ask pupils to work in pairs to transform a set of additions, for example 169 + 276, 141 + 658, 377 + 135. **How did you transform each number sentence? Why did you choose to do that? How do you transform the numbers for addition?** Challenge pupils to write a simple rule for transforming additions to explain to their parents or a friend in another class.
- Ask pupils to work in pairs to transform a set of subtractions, for example 1083 – 499, 967 – 318, 751 – 247. **How did you transform each number sentence? Why did you choose to do that? How do you transform the numbers for subtraction?** Challenge pupils to write a simple rule for transforming subtractions to explain to their parents or a friend in another class.
- Present pupils with written statements such as, 'Amman thinks that 284 – 119 is the same as 283 – 120' and have them work in pairs to prove whether or not Amman is correct.

Unit 3.2 Adding and subtracting a string of numbers

- Display word problems on a flipchart or whiteboard and ask pupils to solve them by using combinations to 100 or 1000; for example, *There are 764 spectators in the North stand and 1150 in the South stand. How many more spectators are in the South stand?* **Sixty-four and *what* makes one hundred? Eight hundred and *what* makes one thousand?**
- Pose questions similar to those in the 'Let's learn' section on page 32, for example 2560 + 1240 + 3096. Ask pupils to calculate the answer and 'show you' on their mini whiteboard. They should be encouraged to make jottings to help them keep track of their thinking. Compare and discuss answers with a partner. **Which numbers did you add together first? Why did you do that?**
- Repeat the above activity for subtractions, for example 8700 – 299 – 301. **How did you work it out? Why did you choose to do it in that way? Can you explain why your strategy works?**
- Pose missing addend and missing subtrahend problems where the missing number is a multiple of 10, 100 or 1000; for example, 3300 + ______ + 2700 = 9640 or 4675 – 1050 – ______ = 3175. **Will the missing number be a multiple of 10, 100 or 1000? How do you know?** Ask pupils to work in pairs to solve the problems presented.
- Challenge pupils to write their own number sentences that can be solved by making multiples of 10, 100 or 1000 and ask a partner to solve them.

Unit 3.3 Adding and subtracting multiples of 10, 100 and 1000

- Present a 5-digit addition where one or both addends is a multiple of 10, 100 or 1000; for example, 67 850 + 37 498. Ask pupils to solve the problem by partitioning. Discuss and compare methods and answers. **Which numbers did you add first? Why?** Repeat for different additions. **Which examples did you find most challenging? Why?**
- Repeat the above activity for 5-digit subtractions. Care should be taken to ensure that each digit of the minuend is larger or the same as the equivalent digit of the subtrahend; for example, 18 679 – 14 325 but not 18 325 – 14 679. Examples such as the latter are addressed in units 3.6 and 3.7.
- Present a 5-digit addition where the total number of each of the tens of thousands, hundreds, tens or ones is 9 or less; for example, 24 210 + 12 754. Ask pupils to predict the answer without writing anything down. **How did you do it? How easy was it to keep track of your thinking? Would jottings have helped?** Demonstrate the use of number facts, place value and jottings on place value houses, i.e. *24 thousands + 12 thousands = 36 thousands* (write 36 in the appropriate columns in the thousands house), *2 hundreds + 7 hundreds = 9 hundreds* (write 9 in in the hundreds column), *0 ones + 4 ones = 4 ones* (write 4 in in the ones column).
- Repeat the above activity for 5-digit subtractions; for example, 57 894 – 23 540.
- Ask pupils to key a 5-digit number into a calculator; for example, 19 311. **Which keys will you need to press to change the tens of thousands digit into an 8? The thousands digit into a 6? The hundreds digit into a 1? The tens digit into a 4? The ones digit into a 9, etc.? Why?**

Unit 3.4 Using place value partitioning

- Display a 5-digit addition on a flipchart/whiteboard. Ask pupils to find the answer by partitioning both numbers into tens of thousands, thousands, hundreds, tens and ones. Ask one half of the class to record the partitions horizontally and the other half to record the partitions vertically in a column, for example:

```
  36 451
+ 23 728
  50 000
   9 000
   1 100
      70
       9
  60 179
```

36 451 + 23 728 = 30 000 + 20 000 + 6000 + 3000 + 400 + 700 + 50 + 20 + 1 + 8
= 50 000 + 9000 + 1100 + 70 + 9
= 60 179

Discuss the position of the digits in the vertical layout. **Why is it important to write the digits in the correct column? What makes 'eleven hundred' different to the other numbers?**

- Write different 5-digit additions on a flipchart/whiteboard and ask pupils to use the column method to find the answers. **How many ten thousands/thousands/hundreds/tens/ones?** Swap with a partner and check each other's layout. **Are the digits in the correct columns?**

- Repeat for 5-digit subtractions. Care should be taken to ensure that no negative numbers are created (this is addressed in units 3.6 and 3.7).
- Write two 5-digit numbers on a flipchart/whiteboard. **What must be added to/subtracted from each digit to change the first number into the second number?** For example, to change 12 905 into 26 450 it is necessary to add 10 thousands (10 000), add 4 thousands (4000), subtract 5 hundreds (500), add 5 tens (50) and subtract 5 ones (5).

Unit 3.5 Adding 4-digit numbers using standard algorithms

- Display a 4-digit addition vertically on a flipchart or whiteboard. Ask pupils to find the answer using the column method, i.e. by partitioning both numbers into thousands, hundreds, tens and ones. Ask one half of the class to start with the tens of thousands and the other half to start with the ones. **Does it matter which numbers we add together first?** Tell pupils that when we use the standard algorithm, we always begin by adding the ones first.
- Discuss the 'Let's learn' section on page 38. Display a different four-digit addition (for example 4575 + 3087) vertically. Ask pupils to predict the number of times it will be necessary to carry a digit over into the next column. **How do you know?** Ask pupils to use the addition algorithm to calculate the answer and compare their solution with a partner. **Did you get the same answer? If yes, how could you check that your answer is correct? If no, whose answer is correct? How do you know?** Provide further practice as appropriate, including examples involving adding three 4-digit numbers together.
- Tell pupils that Amman and Isla added 1656 + 7804 like this.

$$\begin{array}{ccccc} & {}^{1} & & {}^{1} & \\ & 1 & 6 & 5 & 6 \\ + & 7 & 8 & 0 & 4 \\ \hline & 8 & 4 & 6 & 0 \\ \hline \end{array} \qquad \begin{array}{cccccc} & & & & {}^{1} & \\ & & 1 & 6 & 5 & 6 \\ + & & 7 & 8 & 0 & 4 \\ \hline & 8 & 1 & 4 & 6 & 0 \\ \hline \end{array}$$

  Is either child correct? Can you spot Amman and Isla's mistakes? Tell pupils that it is useful to estimate the answer before performing a calculation so that we can judge whether or not our answer is reasonable. 1656 + 7804 is approximately 1660 + 7800 so we should expect the answer to be around 9460. Had Amman estimated first, he might have remembered to add the extra one thousand. Isla would have realised that an answer of 81 460 was unreasonable.
- Provide a range of additions, some of which have been answered incorrectly. Pupils should use estimation and rounding to judge whether or not the answer is reasonable, then check each answer by performing the calculation.

Unit 3.6 Column subtraction: 3-digit numbers

- Write a calculation such as 976 – 523 on a flipchart or whiteboard. Invite pupils to calculate the answer using a method of their choice. Discuss solutions. **Why did you do it that way? How might we use the column method to find the answer?** Invite pupils to 'show you' on their mini whiteboards. Model and encourage appropriate mathematical language, i.e. '*nine hundred minus five hundred is four hundred, seventy minus twenty leaves fifty, six minus three equals three. The answer is four hundred and forty-three*'.
- Ask pupils to articulate a calculation such as 683 – 126 in the same way. **What makes this calculation more challenging? What is 3 minus 6?** Invite pupils to suggest a solution to this 'challenge'.
- Demonstrate how the answer to the above calculation can be found using the column method.

```
  6 8 3
– 1 2 6
  5 0 0
    6 0
    – 3
  5 5 7
```

  Can you explain why this method works? Why do we subtract three? Ask pupils to check the solution by drawing an empty number line.
- Repeat the above activity for a subtraction involving a negative number of tens, for example 529 – 283. **What is 20 minus 80? How will we record this as a column subtraction?** Ask pupils to investigate whether or not the answer is affected by the order in which they combine the numbers. **Does the answer to 300 – 60 and 6 ones give the same answer or a different answer to 306 – 60? Which do you find easier? Can you explain why this method works?**
- Provide further practice as appropriate. Challenge pupils to prove their answers by drawing an empty number line.

Unit 3.7 Column subtraction: four-digit numbers

- Set the calculation 5731 – 1268 down vertically on a flipchart or whiteboard. Drawing on what pupils already know about the column method for 3-digit subtractions, invite suggestions about how column subtraction can be used to solve this calculation. **Which number will we start with? What is 5000 – 1000? Where should I write 4000? What is 700 – 200? Who can show us where to write 500? What about 30 – 60? What do I need to write down? Why? What is 1 – 8? Where will I write minus 7? So, we have 4000, 500, –30 and –7. How will we subtract 37? Which is easier, 4500 – 37 or 4500 – 30 – 7?** Ask pupils to check the solution by drawing an empty number line on their mini whiteboards.

- Present pupils with a range of 4-digit subtractions where there is a negative number of ones, tens, hundreds or a combination of these. For example, 8587 – 6713 (results in –200), 7182 – 4953 (results in –800 and –1), 9319 – 7867 (results in –500 and –50). **What is the most efficient way to subtract the negative numbers?** Encourage pupils to predict how many negative numbers the column will contain prior to subtracting. Encourage pupils to check their solutions by drawing an empty number line.
- Ask pupils to predict whether or not they will get the same answer or a different answer if they begin by subtracting the ones first, then the tens, then the hundreds, then the thousands. Ask them to check their prediction by performing the same set of calculations in this way. **Which setting makes it easier to subtract the negative numbers?**

Unit 3.8 Subtracting 3-digit numbers using standard algorithms

- Pupils work in small groups. Allocate a different method to each group, for example: use base 10 blocks; use place value counters; draw an empty number line; use transformation; use column subtraction. Present all groups with the same calculation 782 – 346 = 436. Groups must use their method to prove to the rest of the class why the number sentence is true. **What did you do first? Why did you do that? What would have happened if…?**
- Present the same calculation as a standard algorithm.

$$\begin{array}{r} 7\,\overset{7}{\not8}\,{}^{1}2 \\ -\ 3\,4\,6 \\ \hline 4\,3\,6 \\ \hline \end{array}$$

  Challenge pupils to work as a group to figure out and explain the method used. Encourage responses such as: *'7 hundreds, 8 tens and 2 ones is the same as 7 hundreds, 7 tens and 12 ones', '80 + 2 = 82 and 70 + 12 = 82', 'It's just like when we used the place value counters. We didn't have enough ones so we had to exchange a ten for 10 ones.'*
- Silently model solving 738 – 486 = 252 using the subtraction algorithm. Ask pupils to observe the procedure carefully, then discuss as a group. **Can you explain what I did? Why did I need to do that?** Repeat for 524 – 278 = 246. Ask pupils to compare their explanations with those in the 'Let's learn' section on page 45.

Unit 3.9 Subtracting 4-digit numbers using standard algorithms

- Write a set of 3-digit subtractions, which are best solved using a standard written algorithm, on a flipchart or whiteboard, i.e. calculations that are too tricky to solve mentally by partitioning, counting on or using a compensation strategy. Ask pupils to estimate the answer then 'show you' their solutions on a mini whiteboard. Include examples that have zero in the answer, for example 872 – 564, 453 – 247, 721 – 613. **Which zeros are needed and which are not? Why?**
- Write a set of 4-digit subtractions, which are best solved using a standard written algorithm, on a flipchart/whiteboard; for example, 6633 – 2537, 3409 – 1631, 8216 – 4342, 5250 – 4576. Ask pupils to estimate the answer by rounding. Invite pupils to take turns to 'be the teacher' and talk the class through the calculation. **Is that correct? How can we check? What is different about the last example? Do any of the answers contain a zero? Is it necessary? Why?**
- Introduce examples where it is first necessary to exchange a hundred/thousand before exchanging a ten/hundred, for example 7404 – 1677 and 9033 – 3656. **What makes these examples more challenging?** Ask pupils to 'think, pair and share' suggestions for how this challenge might be overcome.
- Write the following calculations on a flipchart/whiteboard. Ask pupils to work in pairs to identify and correct the mistakes.

7 1	5 1 16 1	2 14 1
8 4 0 7	6 0 7 1	3 5 1 2
– 3 5 6 1	– 4 3 8 5	– 7 6 3
4 9 6 6	1 7 8 6	2 7 8 2

- Ask pupils to work in pairs or small groups to write a set of instructions for the class next door, explaining how to use the subtraction algorithm. Their instructions should work for any pair of 4-digit numbers.

Unit 3.10 Mental and written strategies

- Write a 4-digit number on a flipchart or whiteboard. Ask pupils to work in small groups to brainstorm a list of ten additions and ten subtractions that have the chosen number as the answer. For example, the number 7654 could be written as 4601 + 3053 or 8000 – 346. Compare solutions with other groups. **How many different possibilities can the class generate between them?** This activity can be easily differentiated by setting 'rules'; for example, all chosen numbers must be odd/even, all numbers must have four digits, one number can be a single-digit number, one number must be a multiple of 10, etc.

- Increase the level of challenge in the above activity by telling pupils that each statement should include two addition symbols, two subtraction symbols or one addition symbol and one subtraction symbol; for example, 3000 + 1650 + 3004, 9654 – 1000 – 1000, 2600 + 5100 – 46
- Discuss the 'Let's learn' section on page 50. **Whose strategy do you think is the most efficient? Justify your answer.** Write a subtraction on a flipchart/whiteboard. Ask pupils to find the answer using their preferred strategy. Invite some pupils to explain their strategy and solution to the rest of the class and convince them that theirs is the most efficient strategy. **Who do you agree with? Why?** Repeat for different additions and subtractions.

Unit 3.11 Representing and solving word problems

- Display a list of word problems that can be solved using addition or subtraction. Provide pupils with a set of cards with number sentences written on them and have them work together in pairs or small groups to match a number sentence to each problem. **Why did you choose that number sentence? Can the same word problem be matched to more than one number sentence? Why?**
- Provide pupils with a word problem that can be solved by either counting on or counting back. For example: *The first combine harvester was invented in 1834. The first tractor was invented in 1892. How many years before the tractor was the combine harvester invented?* Ask pupils to imagine jumping along a number line and write down the answer only. **What did you say to yourself? Why? Who worked it out a different way? Is it easier to count forwards or backwards in your head? How might jottings have helped you?**
- Ask pupils to write a word problem for a partner to solve. They should tell their partner which operation to use (i.e. addition or subtraction) but leave the choice of strategy to the individual. **Does your partner agree that the problem can be solved by addition/subtraction? If not, why not? What strategy did they use? How would you have solved the problem?**

Unit 3.12 Solving multi-step word problems

- Ask each pupil to write a word problem that can be solved using either addition *or* subtraction, then swap problems with a partner. To solve their partner's problem they should draw a bar model to represent the problem, identify the operation required and write a number sentence to show how they thought about it. **Which operation do we use if the whole is unknown? Which operation(s) can we use if one of the parts is unknown?**
- Write a two-step word problem on a flipchart or whiteboard. For example: *The Edinburgh Playhouse has 3059 seats. 860 seats are in the Balcony, 680 seats are in the Circle and the rest are in the Stalls. How many seats are in the Stalls?* Ask pupils to work in pairs to identify the questions that need to be answered and write a number sentence to match each. **What calculations will you need to do? Why? What is/are the unknown(s)? Which strategies will you use?**

- Repeat the above activity for multi-step problems. **What steps will you take to solve this problem? What do you know already that can help you? Is there more than one way to solve this problem? How important is the order in which you do the calculations?**

Unit 3.13 Adding decimal fractions and whole numbers

- Randomly display a set of ten decimal fractions with one decimal place on a flipchart or whiteboard; for example, 12·6, 6·1, 10·5, 0·8, 7·7, 20·9, 15·2, 4·5, 9·3, 8·4. Ask pupils to work with a partner to predict which pairs of decimal fractions will add to give a whole number. **Which decimal fraction has 12·6 as its partner? Does everyone agree? Why did you choose that decimal fraction? What about 20·9, etc?** Ask pupils to write an addition number sentence for each pair and work together to find the answers. **How did you do it?**
- Extend the above idea to include using basic facts to add two decimal fractions. **How could we use the fact that 4 + 2 = 6 to add 0·4 and 0·2? How about 3 + 5 = 8 to add 0·3 and 0·5, etc? What about 0·9 + 0·7?** Draw out that 9 tenths add 7 tenths equals 16 tenths, which is the same as 1 whole and 6 tenths, so the answer is 1·6.
- Discuss Finlay's strategy for finding 4·5 + 199 (see 'Let's learn', page 56). **Why did Finlay change the 4 but not the 5?** Provide similar examples (i.e. examples that involve rounding the whole number and adjusting the decimal fraction) and ask pupils to 'show you' the solution on their mini whiteboards.
- Discuss Isla's strategy for finding 13·9 + 5·3 (see 'Let's learn', page 56). **Can you explain how Isla's strategy works?** Provide similar examples (i.e. examples that involve rounding one addend to make it a whole number, then adjusting) and ask pupils to 'show you' on their mini whiteboards.
- Display a calculation such as 12·6 + 11·7 on a flipchart or whiteboard. **How could we use partitioning to make this calculation easier?** Lead pupils towards the suggestion of adding the whole numbers first, then the decimal fractions, i.e. 12 + 11 = 23 and 0·6 + 0·7 = 1·3 so the answer is 24·3. Repeat for different additions.

Unit 3.14 Adding and subtracting decimal fractions

- Display a missing addend calculation on a flipchart or whiteboard; for example, 41·5 + _____ = 50. Ask pupils to calculate the answer using a strategy of their choice. Discuss the strategies used. **Did everyone get the same answer (8·5)? Who counted on/back? Who drew/visualised a number line?** Discuss the merits of the strategies used. Repeat for different missing addend questions; for example _____ + 16·8 = 30, 60 = 11·9 + _____, 100 = 80·6 + _____.
- Display a set of subtractions on a flipchart or whiteboard; for example, 70·7 – 26·3, 52·9 – 2·2, 14·8 – 6·8, 25·3 – 0·6, 88·1 – 88. Ask pupils to calculate the answers using a strategy of their choice and agree on the answers. **Did you use the same strategy for each example? Why/why not? Which answers contain a zero? Is the zero needed? Why/why not? How might we use addition to check our answers?**
- Display a set of calculations, each containing an error; for example, 13·7 + 14·6 = 27·13, 12·5 – 9·9 = 3·1, 57 – 7·2 = 50·2. Ask pupils to discuss each example with a partner and decide whether or not the answer in reasonable. **Why can't 57 – 7·2 = 50·2, etc.?**
- Ask pupils to work in pairs to solve this decimal number puzzle:

  Think of a decimal number with one decimal place and write it down.

  Add 3·4 to your chosen number.

  Double the answer.

  Subtract your first number.

  Subtract 4·8.

  Subtract your first number.

  Your answer should be 2.

  Repeat for five different decimal numbers with one decimal place. **Does the puzzle work each time? Can you explain how it works?**
- Challenge pupils to write their own decimal fraction puzzle for another pair to solve.

Addition and subtraction 2C

Section overview

Textbook 2C units	Review units	Prior learning	Benchmarks/Es&Os
Unit 3.1 Mental addition and subtraction	**2A** Unit 3.1 **2B** Units 3.1, 3.2 and 3.3	• I can use known facts and place value to mentally add or subtract 2- and 3-digit numbers. • I can use near doubles and near halves to simplify a calculation. • I can use 'round and adjust' to simplify a calculation. • I can simplify a calculation by 'transforming' both numbers, e.g. by changing 553 − 119 into 554 − 120. • I understand the inverse relationship between addition and subtraction and can use this to simplify a calculation.	**MNU 2-03a** Adds and subtracts whole numbers and decimal fractions to two decimal places, within the number range 0 to 1 000 000.
Unit 3.2 Adding and subtracting a string of numbers	**2A** Units 3.2, 3.3, 3.4, 3.5 and 3.6 **2B** Units 3.2 and 3.3	• I can add and subtract 4-digit multiples of 10, 100 and 1000. • I can add and subtract 4-digit whole numbers by making multiples of 10, 100 or 1000. • I can add and subtract multiples of 10, 100 or 1000 to and from whole numbers with up to six digits.	**MNU 2-03a** Adds and subtracts whole numbers and decimal fractions to two decimal places, within the number range 0 to 1 000 000.
Unit 3.3 Using place value partitioning to add and subtract	**2B** Unit 2.3 **2C** Units 2.2 and 2.3	• I can represent 6-digit numbers in different ways; for example, with place value houses, place value arrow cards or place value counters. • I understand that the value of a digit can be found by multiplying it by its place; for example, in the number 5431 the value of the five is 5 × 1000 = 5000. • I can partition any five-digit number into 10 000s, 1000s, 100s, 10s and 1; for example, 31 681 = 30 000 + 1000 + 600 + 80 + 1. • I can partition four-digit numbers in different ways; for example, I know that 4352 = 4000 + 300 +50 + 2 but it can also be 4000 + 200 + 150 + 2.	**MNU 2-02a** Partitions a wide range of whole numbers and decimal fractions to three decimal places; for example, 3·6 = 3 ones and 6 tenths = 36 tenths. **MNU 2-03a** Adds and subtracts whole numbers and decimal fractions to two decimal places, within the number range 0 to 1 000 000.

Unit 3.4 Adding whole numbers using standard algorithms	**2B** Units 3.4, 3.5 and 3.10 **2C** Unit 2.3	• I can use a range of mental strategies to add 3- and 4-digit whole numbers. • I can add 5-digit whole numbers using a semi-formal written method. • I can add 4-digit whole numbers using a standard algorithm. • I can partition 5- and 6-digit numbers in different ways; for example, I know that 54 900 can be 54 thousands and 9 hundreds or 53 thousands and 19 hundreds.	**MNU 2-03a** Adds and subtracts whole numbers and decimal fractions to two decimal places, within the number range 0 to 1 000 000.
Unit 3.5 Subtracting whole numbers using standard algorithms	**2B** Units 3.1, 3.2, 3.3, 3.4, 3.7 and 3.10 **2C** Units 2.3, 3.1, 3.2 and 3.3	• I can use a range of mental strategies to subtract whole numbers mentally. • I can subtract 4- and 5-digit whole numbers using a semi-formal written method. • I can subtract 4-digit whole numbers using a standard algorithm. • I can partition 5- and 6-digit numbers in different ways; for example, I know that 54 900 can be 54 thousands and 9 hundreds or 53 thousands and 19 hundreds.	**MNU 2-03a** Adds and subtracts whole numbers and decimal fractions to two decimal places, within the number range 0 to 1 000 000.
Unit 3.6 Mental and written calculation strategies	**2B** Units 3.1, 3.2 and 3.3 **2C** Units 2.3, 3.1, 3.2, 3.3, 3.4 and 3.5	• I can use a wide range of mental strategies to solve addition and subtraction problems. • I can use jottings to help me keep track of my thinking. • I can partition whole numbers with up to six digits in different ways. • I can use standard algorithms to add and subtract whole numbers with up to six digits.	**MNU 2-03a** Adds and subtracts whole numbers and decimal fractions to two decimal places, within the number range 0 to 1 000 000.
Unit 3.7 Representing word problems	**2B** Units 3.10 and 3.11 **2C** Unit 3.6	• I can choose the most efficient mental or written strategy to calculate the answer to a number problem. • I understand that the same word problem can be represented and solved in different ways. • I can represent and solve word problems involving whole numbers with up to five digits.	**MNU 2-03a** Adds and subtracts whole numbers and decimal fractions to two decimal places, within the number range 0 to 1 000 000.

Unit 3.8 Solving multi-step word problems	**2B** Units 3.10 and 3.12 **2C** Units 3.6 and 3.7	• I can use a range of mental and written strategies and can choose the most efficient strategy to calculate the answer to a number problem. • I can represent the same word problem in different ways; for example, with concrete materials, diagrams, words and symbols. • I understand the relationships between the four operations and can choose the correct operations to solve a multi-step word problem.	**MNU 2-03a** Adds and subtracts whole numbers and decimal fractions to two decimal places, within the number range 0 to 1 000 000.
Unit 3.9 Adding whole numbers and decimal fractions	**2B** Units 2.3, 2.4, 2.7, 2.8, 2.9, 3.13 and 3.14	• I can use a range of mental strategies, including 'round and adjust' and partitioning, to add whole numbers and decimal fractions with one decimal place. • I understand the patterns in our place value system, i.e. 10 hundredths = 1 tenth, 10 tenths = 1 one, 10 ones = 1 ten, 10 tens = 1 hundred, 10 hundreds = 1 thousand, 10 thousands = 10 thousand. • I can read, write, build and describe decimal fractions with hundredths. • I can use a number line to compare and order decimal fractions with hundredths.	**MNU 2-03a** Adds and subtracts whole numbers and decimal fractions to two decimal places, within the number range 0 to 1 000 000.
Unit 3.10 Adding decimal fractions using standard written algorithms	**2B** Units 2.3, 2.4 and 3.5 **2C** Units 3.4 and 3.9	• I understand the patterns in our place value system, i.e. 10 hundredths = 1 tenth, 10 tenths = 1 one, 10 ones = 1 ten, 10 tens = 1 hundred, 10 hundreds = 1 thousand, 10 thousands = 10 thousand. • I can use known facts and mental strategies to add decimal fractions with two decimal places. • I can use a written algorithm to add whole numbers that are too tricky to add mentally.	**MNU 2-03a** Adds and subtracts whole numbers and decimal fractions to two decimal places, within the number range 0 to 1 000 000.
Unit 3.11 Subtracting decimal fractions	**2B** Units 2.3, 2.4, 2.7, 2.8, 2.9 and 3.14	• I can use known facts, partitioning and the inverse relationship between addition and subtraction to subtract decimal fractions with one decimal place. • I understand the patterns in our place value system, i.e. 10 hundredths = 1 tenth, 10 tenths = 1 one, 10 ones = 1 ten, 10 tens = 1 hundred, 10 hundreds = 1 thousand, 10 thousands = 10 thousand. • I can read, write, build and describe decimal fractions with hundredths. • I can use a number line to compare and order decimal fractions with hundredths.	**MNU 2-03a** Adds and subtracts whole numbers and decimal fractions to two decimal places, within the number range 0 to 1 000 000.

Unit 3.12 Subtracting decimal fractions using standard algorithms	**2B** Units 2.3, 2.4 and 3.9 **2B** Units 3.5 and 3.11	• I understand the patterns in our place value system, i.e. 10 hundredths = 1 tenth, 10 tenths = 1 one, 10 ones = 1 ten, 10 tens = 1 hundred, 10 hundreds = 1 thousand. • I understand the inverse relationship between addition and subtraction. • I can use known facts and mental strategies to subtract decimal fractions with two decimal places. • I can use a written algorithm to subtract whole numbers that are too tricky to subtract mentally.	**MNU 2-03a** Adds and subtracts whole numbers and decimal fractions to two decimal places, within the number range 0 to 1 000 000.
Unit 3.13 Adding and subtracting decimal fractions	**2B** Units 2.3, 2.4, 3.13 and 3.14 **2C** Units 2.9, 3.9, 3.10, 3.11 and 3.12	• I understand the patterns in our place value system, i.e. 10 hundredths = 1 tenth, 10 tenths = 1 one, 10 ones = 1 ten, 10 tens = 1 hundred, 10 hundreds = 1 thousand. • I make correct use of zero as a place holder in decimal fractions. • I can use a range of mental strategies to add and subtract decimal fractions with one or two decimal places. • I can use written algorithms to add and subtract decimal fractions with one or two decimal places where the numbers are too tricky to deal with mentally.	**MNU 2-03a** Adds and subtracts whole numbers and decimal fractions to two decimal places, within the number range 0 to 1 000 000.
Unit 3.14 Representing word problems involving decimal fractions	**2C** Units 3.7, 3.9, 3.10, 3.11, 3.12 and 3.13	• I understand that the same word problem can be represented and solved in different ways. • I can represent and solve word problems involving whole numbers with at least six digits. • I can choose the most efficient mental or written strategy to calculate the answer to a number problem involving decimal fractions.	**MNU 2-03a** Adds and subtracts whole numbers and decimal fractions to two decimal places, within the number range 0 to 1 000 000.
Unit 3.15 Multi-step word problems	**2C** Units 3.8, 3.9, 3.10, 3.11, 3.12, 3.13 and 3.14	• I understand the relationships between the four operations and can choose the correct operations to solve a multi-step word problem involving whole numbers. • I can use a range of mental and written strategies and can choose the most efficient strategy to calculate the answer to a number problem involving decimal fractions. • I can represent the same word problem in different ways; for example, with concrete materials, diagrams, words and symbols.	**MNU 2-03a** Adds and subtracts whole numbers and decimal fractions to two decimal places, within the number range 0 to 1 000 000.

Conceptual context

By this stage, pupils should be able to instantly recall basic addition and subtraction facts. This, coupled with a sound *conceptual* understanding of place value, will enable them to access, and accurately use, a wide range of mental calculation strategies (for example, doubling and halving, bonds to 100, compensation, standard and non-standard partitioning). These strategies can be supported by informal jottings where appropriate. Standard algorithms are endorsed as an effective method for calculations that are too complex to work out mentally. The focus turns to helping pupils choose the most *efficient* method (mental, written or a combination of both) to solve a given problem. It is important that pupils are encouraged to routinely estimate and sense-check their solutions prior to performing a calculation so that they can judge whether or not their answer is reasonable.

A sound understanding of the relationships between the four operations enables pupils to solve increasingly complex word problems where there are no obvious verbal or written clues to indicate the operation(s) they should use. It is important that pupils experience the full range of addition and subtraction problem types: join (missing sum and missing addend); separate (missing minuend, missing subtrahend or missing difference); combine (whole unknown or one part unknown); equalise; and compare. They should continue to be provided with ample opportunities to represent the same word problem in different ways, and write their own word problems for others to solve. This will help to equip them with the knowledge and skills to tackle word problems that require multiple steps to reach a solution.

In Book 2C, all of the above ideas are applied to 5- and 6-digit whole numbers and decimal fractions to two decimal places. As highlighted in the place value section, a small space separates the thousands from the hundreds digit when writing 5- and 6-digit numbers.

Resources

- Think Board template
- Bar model template

Key vocabulary

- **Algorithm**
- **Bar model**
- **Calculate**
- **Compensate**
- **Difference**
- **Empty number line**
- **Exchange**
- **Jottings**
- **Method**
- **Number sentence**
- **Palindrome**
- **Partition**
- **Place value**
- **Strategy**
- **Think Board**
- **Value**

Misconceptions

Potential misconception	Addressing the misconception
Units 3.1, 3.2 and 3.6, The pupil considers the individual digits but does not consider the values of the digits or the number as a whole. For example, when calculating 143 + ______ = 400 thinks '*1 +* ***3*** *= 4; 4 +* ***6*** *= 10; 3 +* ***7*** *= 10*' and so arrives at an incorrect answer of 367.	Provide four numbers, e.g. 354, 156, 243, 146 and tell pupils that we want to find two numbers that when added together make a multiple of 100. **Which number can't it be? Why?** (eliminate the number 243) Look at the remaining three numbers. **Which *pair* can't it be? Why?** (146 + 156 because 6 + 6 = 12). So, we are left with 354 and 146, or 354 and 156. Ask pupils to represent the numbers 354, 146 and 156 with place value counters, i.e. 354 = 300 + 50 + 4 146 = 100 + 40 + 6 156 = 100 + 50 + 6 Ask pupils to combine 354 and 146 to find the total. Repeat for 354 and 156. **Which pair add to make a multiple of 100? Can you explain why it can't be 354 + 156?**
Units 3.9 and 3.11 The pupil may be able to name the places to the right of the decimal point as tenths, hundredths and thousandths, and partition a decimal fraction into ones, tenths, hundredths and thousandths, but is unable to equate this with other ways to partition it. For example, may identify 1·234 as $1 + \frac{2}{10} + \frac{3}{100} + \frac{4}{1000}$ but does not equate this with $\frac{1234}{1000}$. Conversely, pupils may think that decimal fractions with three places are always thousandths and write $\frac{1234}{1000}$ but do not equate this with $1 + \frac{2}{10} + \frac{3}{100} + \frac{4}{1000}$. **NB The premature introduction of standard algorithms (i.e. before the pupil has acquired conceptual place value understanding) may exacerbate this problem.*	Provide pupils with lots of opportunities to combine and partition place value arrow cards and talk about what they notice.

Teaching ideas

Unit 3.1 Mental addition and subtraction

- Display a set of calculations with 5- and 6-digit numbers that can be solved mentally, using number facts, place value, partitioning, compensation, doubling and halving. Ask pupils to work out the answers on a mini whiteboard. They should be encouraged to make jottings to help them keep track of their thinking. Invite pupils to take turns to 'be the teacher' and explain to the rest of the class how they worked the answers out. **How many different ways are there? Which method to you prefer? Why? Which method is the most efficient? Always?**
- Display a set of calculations, and sets of jottings, on a flipchart or whiteboard. Challenge pupils to work out which set of jottings matches which calculation. For example:

  Does everyone agree on the pairings? What strategy has been used in each example? Is this the most efficient strategy for this calculation?
- Discuss the 'Let's learn' section on page 36. Invite pupils to share their thoughts about the strategies used. **Has each character used the most efficient method for their calculation? How would you have worked it out?** Draw out that whilst efficiency is our goal, it is important to choose a strategy you are comfortable with.

Unit 3.2 Adding and subtracting a string of numbers

- Display a string of numbers on a flipchart or whiteboard, for example:

  1000 + 143 + 5000 + 260 + 340 + 57

  How can we rearrange the order of the numbers to make this calculation easier? Draw out suggestions such as add the thousands first, look for pairs of numbers that combine to make a multiple of 10, 100 or 1000.

  5000 + 1000 + 260 + 340 + 143 + 57

  6000 + 600 + 200

  Model rearranging and grouping the numbers to solve the problem. Provide a different example for pupils to try themselves and 'show you' on a mini whiteboard. Discuss the ways that the numbers were rearranged. **Did everyone rearrange and group the numbers in the same way? Which arrangement is easiest to work with?** Repeat the activity for more complex numbers; for example, 1399 + 3500 + 1209 + 6500 + 1601 + 321.

- Write the numbers 800, 600, 1412 and 88 on a flipchart or whiteboard. Tell pupils that Amman has been asked to subtract these numbers from 53 900. Write the number 53 900 on the flipchart or board. Ask pupils to brainstorm with a partner the most efficient way to do this. **Is it easier to subtract the numbers one at time or group them together? Which numbers go together to make a number that is easier to subtract? Does it matter if we subtract 1500 (the total of 1412 and 88) or 1400 (the total of 800 and 600) first? Is it easier to subtract 1500 and 1400 from 53 900 separately or combine them?** Repeat for different sets of numbers.
- Display missing addend and missing minuend examples such as 2720 + ______ + 718 = 3938 and ______ – 116 – 534 = 9050. Ask pupils to work in pairs to calculate the answers. Discuss strategies and solutions.

Unit 3.3 Using place value to add and subtract

- Display a calculation such as 319 436 + 185 287. Remind pupils that when we add or subtract on an empty number line we keep the larger number whole and partition the smaller number according to the values of the places. **How can we partition 185 287?** Ask pupils to draw an empty number line and use place value partitioning to find the answer to 319 436 + 185 287. **Which number was trickiest to add? Why? Did anyone partition 80 into 70 and 10? Why did you do that?**
- Display the same addition vertically and ask pupils to calculate the answer using the column method. Remind pupils that this method requires us to partition both numbers according to their place values; for example, we need to think 300 000 + 100 000; 10 000 + 80 000, etc. **Which method do you prefer? Why?**
- Repeat the above activities for the calculation 306 537 – 61 328. **Which parts of the calculation did you find most tricky this time? Why? Which is your preferred method? Why?**
- Ask pupils to investigate whether or not the answer is affected by the order in which they combine the numbers in the column method; for example, **Does 300 000 – 60 000 + 5000 + 209 give the same answer or a different answer than calculating 305 210 – 60 000 – 1? Which do you find easier? Is there another way in which we can combine the numbers? Can you explain why this method works?**

Unit 3.4 Adding whole numbers using standard algorithms

- Write two 5-digit numbers on a flipchart or whiteboard (for example, 54 326 and 22 978) and ask pupils to round each to the nearest hundred. Remind pupils that before performing a written algorithm it is helpful to work out the approximate answer so that we can judge whether or not our answer is reasonable. 54 326 + 22 978 is approximately 54 300 + 23 000 so we should expect our answer to be around 77 300.
- Ask pupils to find the exact answer to the above calculation using the standard written algorithm for addition. **How will you write the numbers down? What do you need to remember?** (digits aligned correctly, start by adding the ones, record the carrying figure at the top of the appropriate column where necessary). **How accurate was your estimate/calculation?** Discuss any discrepancies. **Why has that happened? Where did you go wrong?**
- Repeat the above for different 5-digit and 6-digit additions. Ask pupils to predict the number of times it will be necessary to carry a digit over into the next column. **How do you know?**
- Display a calculation such as 254 906 + 117 586. Ask one half of the class to round each addend to the nearest 100 and the other half to round each addend to the nearest 1000. Ask pupils to work out the exact answer using the addition algorithm. **Which way of rounding gives the best estimate? Which calculation is easier to do mentally, 254 900 + 117 600 or 255 000 + 118 000?** Discuss the relative merits of each method.

Unit 3.5 Subtracting whole numbers using standard algorithms

- Write two 5-digit numbers on a flipchart or whiteboard (for example, 84 831 and 39 067) and ask pupils to round each to the nearest hundred. Remind pupils that before performing a written algorithm it is helpful to work out the approximate answer so that we can judge whether or not our answer is reasonable. 84 831 – 39 067 is approximately 84 800 – 39 000 so we should expect our answer to be around 45 800.
- Ask pupils to find the exact answer to the above calculation using the standard written algorithm for subtraction. **How will you write the numbers down? What do you need to remember?** (digits aligned correctly, start by subtracting the ones, exchange where there is an insufficient number of ones, tens, hundreds or thousands) **How accurate was your estimate/calculation?** Discuss any discrepancies. **Why has that happened? Where did you go wrong?**
- Repeat the above for different 5-digit and 6-digit subtractions. Ask pupils to predict the number of times it will be necessary to exchange in each example. **How do you know?**
- Tell the pupils that Finlay used the standard algorithm to calculate 170 201 – 85 465 and got the answer 115 264. **Is Finlay's answer reasonable? Explain.** Ask pupils to work in pairs to calculate the correct answer. **Can you work out what Finlay has done wrong?**

Unit 3.6 Mental and written calculation strategies

- Tell the pupils that Nuria and Isla each decide to work out the answer to 21 738 + 1046 mentally.

  Nuria jots down: 22 738, 22 778, 22 780, 22 784

  Isla jots down: 22 000, 22 700, 22 770, 22 784

  Can you work out what each girl was thinking? What is the same and what is different? Whose strategy do you prefer? Why? Ask pupils to perform the following calculations mentally, making jottings to help them keep track of their thinking:

  8138 + 10 458

  120 504 + 53 235

  14 000 – 7310

  33 855 – 16 222

  Compare answers with a partner. Do you agree? Now show your jottings to you partner. Can they work out how you solved each problem? How does your partner's strategy compare with the way in which you solved the problem?
- Display a list of additions and a list of subtractions on a flipchart or whiteboard. Ask pupils to work in pairs or small groups to decide on the most efficient strategy for each example, then use their chosen strategies to find the answers.

Unit 3.7 Representing word problems

- Display a range of word problems that can be solved by adding and/or subtracting large whole numbers. Ask pupils to work in pairs to represent each problem as a bar model. **Did each pair draw the same bar model? What is the same and what is different? Why did you put that number there? How does your bar model help you to work out the number sentence for this problem? What is the most efficient way to solve this problem? Will you calculate the answer mentally with jottings, or will you use a written algorithm?** Ask pupils to calculate the solution to the problem, then check their answer by drawing an empty number line.
- Provide pupils with a completed number line and have them write an addition/subtraction word problem to match it.
- Provide pupils with a completed bar model and have them write an addition/subtraction word problem to match it.
- Challenge pupils to work in pairs to write addition and subtraction problems for each other to solve. Discuss strategies and solutions. **Why did you choose to use that strategy? Is there a more efficient way? Which problems were best solved using the standard written algorithm? Which problems were best solved mentally, with jottings?**

Unit 3.8 Solving multi-step word problems

- Copy the word problem from the 'Let's learn' box on page 50 onto a flipchart or whiteboard. Read the problem aloud. **What do we know already? What are we trying to find out?** Model entering the known information into a bar model. Ask pupils to perform the necessary calculations in order to reach a solution. Compare methods and solutions with the example on page 50. Ask pupils to suggest ways in which they could check the accuracy of their calculations/solution.
- Write multi-step word problems on a flipchart or whiteboard. Ask pupils to work in pairs to identify the questions that need to be answered and write number sentences to match them. **What steps will you take to solve this problem? What do you know already that can help you? Is there more than one way to solve this problem? How important is the order in which you do the calculations?**

  What calculations will you need to do? Which strategies will you use? How can you check the accuracy of your answers?
- Challenge pupils to write their own multi-step word problem for others to solve.

Unit 3.9 Adding whole numbers and decimal fractions

- Discuss Amman and Isla's strategies in the 'Let's learn' section on page 52. **Can you explain how the pupil's strategy works?** Display similar calculations. Ask pupils to use 'round and adjust' to find the answers and 'show you' on their mini whiteboards.
- Discuss Nuria's strategy in the 'Let's learn' section on page 52. Challenge pupils to partition the same decimal fraction in different ways by playing, 'The answer is… What is the question?' **The answer is 12·87, what could the addition be? Can you partition the same decimal fraction into more than two parts? Can one addend be a whole number? Prove it! Can one addend be a decimal fraction with one decimal place? Prove it! How many different ways can you find to partition the number 12·87?**
- Display a set of additions on a flipchart or whiteboard; for example, 14·25 + 3·6, 7·03 + 15·82, etc. Ask pupils to round each addend to the nearest whole number to find the approximate total, then use a mental strategy of their choice to find the exact answer. Have them compare answers with a partner. **Is your answer reasonable? If not, where have you gone wrong?**

Unit 3.10 Adding decimal fractions using standard algorithms

- Discuss Finlay's method in the 'Let's learn' section on page 54. **What do we need to remember when carrying out standard written algorithms?** (digits aligned in columns, work from right to left) **Has Finlay followed these rules correctly?**

- Tell pupils that Isla did the same calculation and got the answer 82 051. **What do you think has happened?** Stress that as well as aligning the digits, we also need to remember to align the decimal points. Had Isla remembered to round and estimate first she would have known that her answer was unreasonable.
- Display a set of addition algorithms similar to those in this unit on a flipchart or whiteboard. Ask pupils to use rounding to estimate each answer. For example, 756·77 + 89·89 is roughly 757 + 90 so we should expect our answer to be about 847. Ensure that pupils agree on the approximate answer, then ask them to find the exact answer and 'show you' on their mini whiteboards. Discuss any errors as they arise and encourage pupils to suggest possible reasons for these.
- Challenge more able pupils by displaying a calculation such as 416·51 + 203·5 and asking them to use rounding to find the approximate answer. **Who rounded 203·5 up to 204? Who rounded 203·5 down to 203? Why did you do that? Which gives an estimate that is closest to the actual answer, 417 + 204 or 417 + 203? Can you explain why?** Ask pupils to work in pairs to identify the upper and lower limits for similar calculations.

Unit 3.11 Subtracting decimal fractions

- Discuss the merits of Nuria and Isla's strategies on page 56. **How would you use counting on to check Nuria's solution? Why doesn't partitioning work for Isla's calculation? What would you need to do to the number 15 to make partitioning work?**
- Display a set of subtractions on a flipchart or whiteboard. **Which subtractions are best solved by partitioning? Which are best solved by counting on?** Ask pupils to solve each calculation one at a time, recording their jottings and solution on a mini whiteboard. **How can we use the inverse relationship to check our answers?**
- Pupils work in pairs or small groups and agree a starting number; for example, 50. Each pupil takes it in turns to spin two spinners, combine the numbers spun and subtract the total from the starting number using either partitioning or counting on. Players check each other's calculations on a calculator. Play continues with each player subtracting from their previous total. The player closest to zero (without going below zero) after four rounds is the winner. **The pupils play the same game. After three rounds Finlay is left with the number 7·32. Which numbers will he need to spin in the next round to have the best possible chance of winning the game? Which numbers won't he be hoping for? Why?**

Unit 3.12 Subtracting decimal fractions using standard algorithms

- Discuss Finlay's method in the 'Let's learn' section on page 58. **What do we need to remember when carrying out standard written algorithms?** (digits aligned in columns, work from right to left, subtract the bottom number from the top number exchanging where necessary) **Has Finlay followed these rules correctly?**
- Tell pupils that Amman and Isla each did the same calculation. They rounded first to find the approximate answer. Amman got an approximate answer of 17. Isla got 13. **Whose answer is the most reasonable? Why? What has Isla done wrong?**
- Ask pupils to use addition to check the accuracy of Finlay's answer. **Who used place value partitioning? Who counted on? Why did you choose that method?**
- Label four trays 0-10, 10-30, 30-50, 50-70. Ask pupils to use rounding to work out the approximate answers to each example in question 1. They should write each *calculation* on a piece of paper or card and place it in the appropriate tray. Discuss the results. **Is each calculation in the correct tray? What makes you say that? In which tray should the incorrectly placed calculations go? Why?**
- Ask pupils to verify their decisions in the previous activity by completing question 1.

Unit 3.13 Adding and subtracting decimal fractions

- Discuss the 'Let's learn' section on page 60. Focus the discussion on the values of the places and zero as a place holder. **How should Isla set the calculation 126·2 – 78·87 out as a standard algorithm? How many digits will change when Nuria adds 0·7 to 134·15? Which digit will that be? Why?**
- Display a range of calculations involving whole numbers and decimal fractions with one or two decimal places, some that can be solved mentally and others for which a standard algorithm would be appropriate. Ask pupils to work in groups to compare mental and written strategies and decide on the most efficient strategy in each case. Compare group responses. **Does everyone agree? Why did you choose round and adjust/transformation/number facts and place value/partitioning/written algorithm? Are there equally efficient ways to solve the same problem?**
- Tell pupils that the children are having a long jump competition. Amman jumps 3·75 m in the first round, 3·3 m in the second round and 3·59 m in the third round. Ask pupils to find Amman's total score using a strategy or strategies of their choice. Repeat for the other pupil's jumps:

  Finlay: 2·9 m, 3·12 m, 2·9 m

  Isla: 2·96 m, 3·04 m, 3·28 m

  Nuria: 3·84 m, 4·2m, 4·11 m

  How did you work each answer out? Who jumped furthest? How much further than each of the other children did he/she jump?

Unit 3.14 Representing word problems involving decimal fractions

- Ask pupils to survey their families about the strategies they use to add and subtract. **What is the same and what is different about the way your parents/grandparents add and subtract and the way that you add and subtract? How do the strategies used by you parents compare with those used by your grandparents? Are there any strategies that are new to you? Can you explain how they work?**
- Display word problems involving decimal fractions with up to two decimal places on a flipchart or whiteboard. For example:

  Finlay's dad cycled 67·5 km. His Uncle Jamie cycled 9·75 km further. How far did Finlay's Uncle Jamie cycle?

  Nuria walked 16 km. Her sister walked 9·5 km. How much further did Nuria walk?

  Ask pupils to draw a bar model to represent each problem. **Is it the whole or is it one of the parts that is unknown? How does this help us decide which operation to use? How will you calculate the unknown? Which strategy is best for this particular problem?** Ask pupils to find the solution either mentally, supported by jottings, or by performing a written algorithm. Finally, they should check their solution by drawing an empty number line.

Unit 3.15 Multi-step word problems

- Present pupils with a multi-step word problem involving decimal fractions and two or more of the four operations. For example, *P7 have a budget of £20 for their party. They decide to buy 27 bottles of fizzy drink at 40p per bottle, 2 family bags of savoury snacks at £1·29 each and 2 bags of sweet treats at £1·99 each. Do they have enough money left to buy a bag of party blowers costing £3·99?* **How many steps are needed to solve this problem? Which operations will you need to use? How do you know? How important is the order in which you do the calculations?**
- Ask pupils to partition amounts of money to avoid small change. For example, Nuria's mum bakes some cakes for the P7 party. The ingredients cost £5·37. She gives the shopkeeper a £10 note and thirty-seven pence. **Why do you think she gave the shopkeeper the extra 37p?**
- Ask pupils to work in pairs or small groups. Give each pair/group a partially completed bar model and have them create and solve a word problem to fit it. Swap bar models and word problems with another pair/group. **Work together to solve the problem given to you. Does it work? If not, what advice do you have for the writers of this word problem?**
- Ask pupils to use everyday events as themes for their own word problems. Swap problems with a partner. **Which operations will you need to use to solve your partner's word problems? How do you know?**

4 Number – multiplication and division

Multiplication and division 2A

Section overview

Textbook 2A units	Review units	Prior learning	Benchmarks/ Es&Os
Unit 4.1 Recalling multiplication and division facts for 2, 5 and 10	**1A** Units 4.3, 4.4 and 4.5 **1B** Units 4.2, 4.3 and 4.5 **1C** Unit 4.7	• I can skip count in 2s, 5s and 10s. • I know double and halving facts. • I know how many tens are in a given decade. • I can recognise multiplication problems and represent them with materials. • I can read and write number sentences involving multiplication and division.	**MTH 2-03c** • Uses multiplication and division facts to the 10th multiplication table. • Multiplies and divides whole numbers by multiples of 10, 100 and 1000. • Multiplies and divides decimal fractions to two decimal places by 10, 100 and 1000. • Multiplies whole numbers by two digit numbers. • Multiplies decimal fractions to two decimal places by a single digit.
Unit 4.2 Recalling multiplication and division facts for 3	**1B** Units 4.1, 4.2 and 4.3 **1C** Unit 4.7	• I can skip count in 3s. • I can recall 2, 5 and 10 times-table facts. • I can recognise multiplication and division problems and represent them with materials. • I can read and write number sentences involving multiplication and division.	
Unit 4.3 Using 2, 5 and 10 multiplication facts to solve multiplication problems	**1B** Units 4.2, 4.3, 4.6 and 4.7 **1C** Unit 4.7	• I know 2, 5 and 10 times-table facts. • I can recognise multiplication problems and represent them with materials. • I can read and write number sentences involving multiplication and division.	
Unit 4.4 Multiplying numbers by 10 and 100	**1B** Units 4.2, 4.3, 4.6 and 4.7 **1C** Units 2.1, 2.2 and 2.6 **2A** Units 2.1 and 2.2	• I can recognise multiplication problems and represent them with materials. • I can read and write number sentences involving multiplication and division. • I know the meaning of each digit in the ones, tens, hundreds and thousands columns.	

Unit 4.5 Dividing by 10 and 100	**1B** Units 4.2, 4.3, 4.6 and 4.7 **1C** Units 2.1, 2.2 and 2.6 **2A** Units 2.1 and 2.2	• I can recognise division problems and represent them with materials. • I can read and write number sentences involving multiplication and division. • I know the meaning of each digit in the ones, tens, hundreds and thousands columns.	
Unit 4.6 Partitioning to multiply a 2-digit number by a 1-digit number	**1C** Units 2.1, 2.2 and 4.7 **2A** Units 4.1 and 4.2	• I can recall 2, 3, 5 and 10 times-table facts. • I know how many tens and ones are in 2-digit numbers. • I can recognise multiplication problems and represent them with materials. • I can read and write number sentences involving multiplication and division.	
Unit 4.7 Using knowledge of doubles and halves to solve multiplication and division problems	**1B** Units 4.2, 4.3, 4.6 and 4.7 **1C** Unit 4.7	• I can recognise multiplication and division problems and represent them with materials. • I can read and write number sentences involving multiplication and division. • I know double facts to 20. • I know halving facts to 20.	
Unit 4.8 Recalling multiplication and division facts for 4	**1B** Units 4.2, 4.3, 4.6 and 4.7 **1C** Unit 4.7	• I can skip count in 2s. • I can recall doubling and halving facts to 20. • I can recall multiplication and division facts for 2. • I can recognise multiplication and division problems and represent them with materials. • I can read and write number sentences involving multiplication and division.	
Unit 4.9 Solving division problems using repeated addition or subtraction	**1B** Units 4.1, 4.2, 4.3, 4.6 and 4.7 **1C** Unit 4.7	• I can recognise multiplication and division problems and represent them with materials. • I can read and write number sentences involving multiplication and division. • I can skip count in 2s, 5s, 3s and 10s. • I know doubles and halves to 20.	

Unit 4.10 Multiplying by changing the order of the factors	**1B** Units 4.1, 4.2, 4.3, 4.6 and 4.7 **1C** Unit 4.7 **2A** Unit 5.1	• I can recognise multiplication and division problems and represent them with materials. • I can read and write number sentences involving multiplication and division. • I can recall multiplication and division facts for 2, 3, 5 and 10 times tables. • I know that factors are numbers we multiply together to get another number.	
Unit 4.11 Solving multiplication problems using known facts and doubling	**1B** Units 4.2, 4.3, 4.6 and 4.7 **1C** Unit 4.7 **2A** Unit 4.2	• I can recognise multiplication and division problems and represent them with materials. • I can read and write number sentences involving multiplication and division. • I know doubles and halves to 20. • I can recall multiplication and division facts for 2, 3, 5 and 10 times tables.	
Unit 4.12 Solving division problems using known facts and halving	**1B** Units 4.2, 4.3, 4.6 and 4.7 **1C** Unit 4.7 **2A** Unit 4.2	• I can recognise multiplication and division problems and represent them with materials. • I can read and write number sentences involving multiplication and division. • I know doubles and halves to 20. • I can recall multiplication and division facts for 2, 3, 5 and 10 times tables.	

Conceptual context

At the beginning of second level, pupils need to have a secure understanding of multiplication as being the joining of equal groups. Through experience of exploring a wide range of concrete materials (e.g. arrays) pupils should see the connection between repeated addition and multiplication (e.g. understanding that $2 + 2 + 2$ is the same as 3×2). Pupils need to be able to represent multiplication problems using materials or pictures, and, vice-versa, to record arrays as multiplication sentences. Also, they should have developed the ability to skip count in a range of multiples, but particularly in 2s, 5s and 10s, and be able to use skip counting to solve multiplication problems.

At this level, pupils also require a strong understanding of equal sharing and understand that division problems can be solved by equal sharing or by forming equal groups and counting the number of groups. Pupils need to have been exposed to division problems in both of these contexts. For example, the problem, 'share 12 marbles between 3 friends,' would involve equal sharing, while the problem, 'Nuria has 12 marbles and puts 3 in each bag – how many bags does she have?' involves forming equal groups. Pupils should also understand that division problems may have remainders. They should have explored a range of problems using real-life contexts where remainders may occur.

It is important to check pupils' understanding of the symbols used in multiplication and division and for them to be able to show a strong understanding of how to read and write multiplication and division sentences.

Moving on to develop knowledge and strategies within multiplication and division, it is vital that concrete materials are continued to be used when introducing key ideas and strategies to promote conceptual understanding of the processes they are developing.

This unit develops knowledge of multiplication and division facts, initially focusing on revision of facts for 2, 5 and 10 and then extending recall of facts for 3 and 4. Pupils should be encouraged to develop this knowledge through games, practice and making connections with existing knowledge, rather than by disconnected rote memorisation. Concrete materials, such as arrays, bead strings, dot patterns and abacuses, are useful to promote this recall and develop understanding of connections, e.g. using an array to show 5 rows of 2 then adding one more row to show 6 rows of 2.

Pupils now need to understand how this knowledge of facts can be used to solve a range of problems in different ways. If pupils have strong knowledge of 2, 5 and 10 multiplication facts, we can begin to partition problems by splitting the factors additively. For example, we can use materials such as squared paper or arrays to show that 7×6 can be split into $(5 \times 6) + (2 \times 6)$, allowing pupils to use their recall of facts to solve the problem. Notation of brackets is important to introduce at this stage to allow pupils to be able to clearly record their thinking and strategies.

This can then be extended to partitioning by place value; materials such as ten frames or bundles of tens and ones can be used to promote the understanding that 2-digit numbers can be split into tens and ones to solve the problem. Again, pupils should be encouraged to use bracket notation to record the thinking used to solve problems.

As pupils explore multiplying and dividing by 10 and 100, it is essential to check understanding of place value for 3- and 4-digit numbers. Materials such as place value sticks, cubes or money can be used to show that when we multiply by 10 the number gets 10 times bigger, and when we divide by 10 the number gets 10 times smaller. Place value houses are very useful to promote the understanding that the digits move to the left or right depending on the operation, and that we use zero as a placeholder.

Pupils need to further extend their understanding of the connection between repeated addition and subtraction to solve division problems. Number lines are a useful way to show how we can use skip counting forwards or backwards to work out division problems, e.g. to solve $25 \div 5$ we could skip count in 5s backwards using repeated subtraction and count the number of jumps to zero:

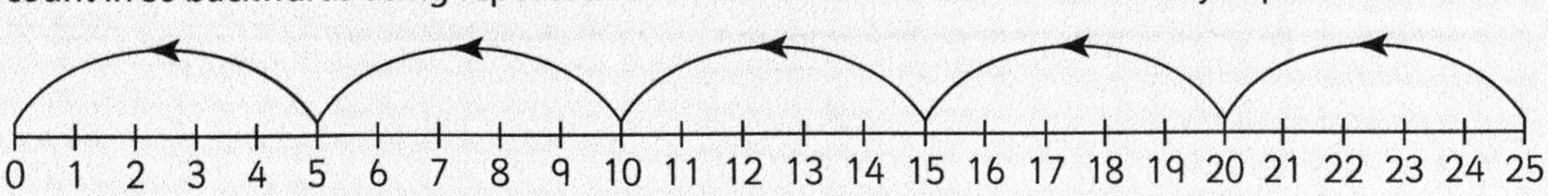

Or alternatively skip count forwards from zero and count the number of jumps to 25.

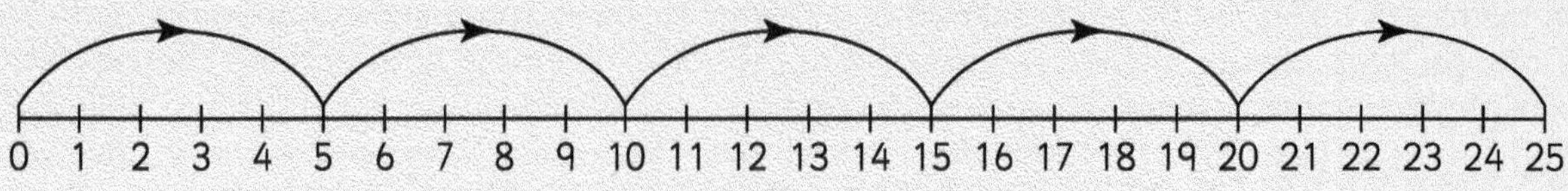

Once pupils have experience using number lines, these could then be turned over or covered with a screen and pupils encouraged to use a mental image of the number line to work out the answer.

If this is difficult, remove the screens so the number line is visible again and allow more experiences with the visible number line before again promoting a mental image by screening.

Pupils should understand the commutative properties of multiplication. Arrays are a very useful way to show that the order we multiply the factors does not matter. They can be introduced to the term *commutative property*, so they understand what this means, although would not necessarily be expected to use this term themselves at this stage. Pupils should be encouraged to think how understanding this property of multiplication can help them when solving problems; for example, they might find it easier to solve 2 × 8 than 8 × 2 because they more easily connect their knowledge of the double fact for 8 rather than skip counting 8 groups of 2.

Pupils are also encouraged to make connections between their knowledge of doubles and halves and multiplication and division problems, for instance understanding that halving a number is the same as dividing by 2 and vice versa. They can then use this knowledge to solve problems; for instance, they can use their knowledge of double 5 to work out 3 × 5 as double 5 is 10 then add 5 more. They can also be encouraged to double and double again and see the connection between multiplying by 4 and vice versa for halving and halving again and dividing by 4.

Resources

- Bead strings
- Arrays
- Dot cards
- Money – mock £10 notes and £100 notes for multiplying by 10 and 100
- Number lines
- Screens, such as large pieces of card
- Abacuses
- Place value materials, such as sticks, counters or cubes
- Place value houses
- Ten frames
- Hundred squares

Key vocabulary

- Array
- Brackets
- Commutative property
- Dividend
- Divisor
- Factor
- Grouping
- Multiplicand
- Multiplier
- Product
- Quotient
- Repeated addition
- Repeated subtraction
- Skip counting
- Times

Misconceptions

Potential misconception	Addressing the misconception
Units 4.1, 4.2 and 4.8 The pupil shows only rote understanding of 'times tables' always recited in order, without an understanding of each fact as a number of equal groups.	• Explore visual representations of each fact using arrays, dot cards or squared paper. Link with repeated addition so pupil is clear what each fact represents.
Units 4.3 and 4.6 The pupil uses partitioning incorrectly, e.g. for 7×6 partitioning into $(5 \times 6) + 2 = 32$, rather than $(5 \times 6) + (2 \times 6) = 42$.	• Use concrete materials to show problem and partitioning, e.g. squared paper, and colour each step of partitioning or bundles of tens and ones.
Units 4.4 and 4.5 The pupil uses rules such as, 'add a zero' or 'take away a zero' rather than understanding that the number gets 10 times bigger or 10 times smaller.	• Use concrete materials to show ten bundles of a number to demonstrate how number has got ten times bigger. Similarly for division, bundle loose objects into tens to show how number has got ten times smaller. Connect with place value house to show how digits have moved 1 place to the left or right. • Explain that these rules will not apply when we move on to decimal fractions; it is important we can show understanding by explaining what has happened to the numbers.
Units 4.9 and 4.12 Difficulty with division as repeated addition or subtraction due to lack of understanding of division as forming equal groups as well as sharing.	• Ensure pupils have experience of a range of different contexts for division including grouping as well as sharing. Discuss both ways of dividing and share and record different ways of solving division problems. • Use concrete materials to allow pupils to physically form them into equal groups and model skip counting, as well as opportunities to share one-by-one.

Teaching ideas

Unit 4.1 Recalling multiplication and division facts for 2, 5 and 10

- Use an abacus to show double facts, e.g. 2 rows of 6. Ask pupils questions such as: **What fact does this show? What multiplication fact of 2 does this also show? What do you notice?** Link pupils' understanding of doubles and multiplication facts for 2. You could also pair-wise ten frames as a visual representation.
- Use an abacus. Move 8 rows of ten to the left-hand side. **How many rows of beads have I moved? How many beads altogether? How do you know?** Ask pupils how they could record this as a multiplication fact. Link knowledge of tens in decade numbers to multiplication facts for 10.

  If you know that 8 × 10 = 80 what division fact do you also know? Link understanding of 80 as 8 rows of ten. You could also use 8 ten frames or 8 rows of 10 cubes as visual representations.
- Explain that we are going to work out 6 × 5. With the abacus now move 1 group of five to the left-hand side. Move another group of five leaving a slight gap between the two groups of five. Repeat with another 4 groups of five on the following 2 rows (i.e. 6 groups of five over 3 rows). **How many altogether? What else do you notice?** Look for understanding that 6 × 5 is the same as 3 × 10. Explain that we can use our knowledge of tens to help us with multiplication facts for 5. If we know that there are 3 tens in 30 and we also know each ten is double 5, then we know that there are 6 fives in 30 too.
- Encourage the pupils to play a range of games using materials such as dice, playing cards or using digital games or apps.

Unit 4.2 Recalling multiplication and division facts for 3

- Use a hundred square and ask pupils to colour multiples of 3 or to put a counter on each multiple of 3. **What patterns can you see?**
- Use a bead string and move 6 multiples of 3 leaving a small gap between each group of 3. **What is 6 × 3? How do you know?** Encourage pupils to look for patterns in the bead string, in this case they can see 10 and 8 to quickly get the answer of 18:

  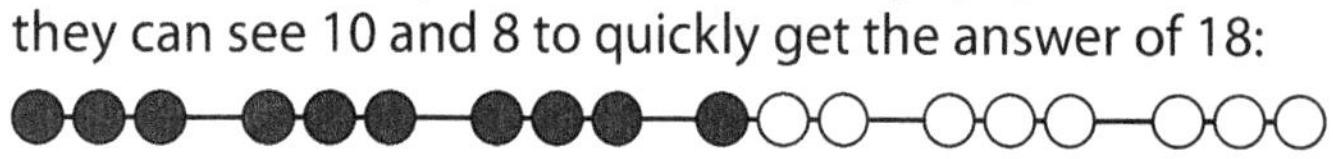
- Ask pupils to close their eyes, or screen the bead string and move multiples of 3 on the bead string. **I have moved 4 groups of 3. How many is 4 × 3?** Encourage pupils to visualise the beads moved, then show the bead string.
- Play games with playing cards such as taking each card in turn and multiplying it by 3, e.g. turning over a 10 and giving the answer 10 × 3 = 30. Encourage pupils to work with a partner, placing all the facts they know in one pile and all the facts they need to practise in another.

Unit 4.3 Solving multiplication problems

- Make an array showing 7 rows of 8 using squared paper. Ask the pupils to write a multiplication sentence for this array. **This shows 7 × 8. What strategies could we use to work out the answer?** Encourage the pupils to share different strategies they may have, such as using doubles, skip counting or using other facts they know. Explain that we are going to use the facts we have been learning for 2, 5 and 10 to help us. With a crayon quickly shade 5 rows of 8. **I have shaded 5 rows of 8. How many have I shaded altogether? How do you know?** (Make sure pupils know 5 × 8 = 40; if not, review unit 4.1 and ensure pupils have more experience working on recall of facts for 2, 5 and 10.) Record this on a whiteboard. Now shade the remaining 2 rows of 8 using a different colour. **I have shaded 2 rows of 8. How many have I shaded altogether? How do you know?** (Make sure pupils know 2 × 8 = 16; if not, review unit 4.1 and ensure pupils have more experience working on recall of facts for 2, 5 and 10.) Record this on a whiteboard. **Now we add these calculations together to work out the answer for 7 × 8.** (Record (5 × 8) + (2 × 8) = 40 + 16 = 56.) Explain that we used the brackets to show how we grouped our calculations. Explain that the brackets are important because they show which calculations we did first. Show how we would get a different answer by removing the brackets.
- Repeat with a range of different examples using multiplication sentences involving 6, 7, 8 or 9.
- You could also cut the squared paper to show how you split the factors or you could use an array and show you split the rows up by moving the rows into two groups.

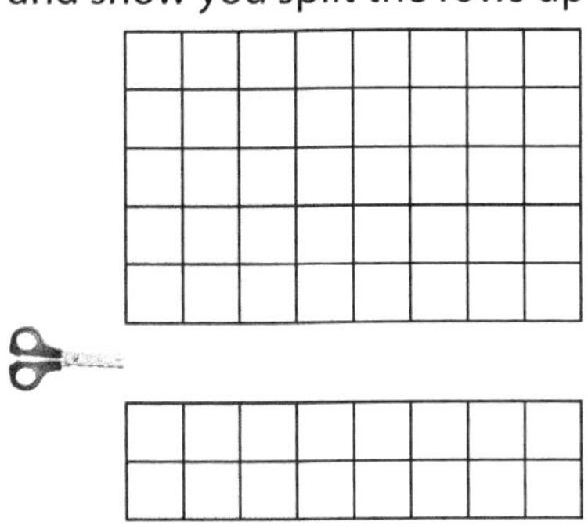

5 × 8

2 × 8

7 × 8 = (5 × 8) + (2 × 8)

Unit 4.4 Multiplying whole numbers by 10 and 100

- Give a problem in context, such as: **Six children have £10 each. How much do they have altogether?** Ask the pupils to write a multiplication sentence for this problem. Make an array, or use rows of cubes to model the problem and ask pupils to work out the answer. Check they are able to skip count or use recall to get the answer, 6 × £10 = £60.
- Use a place value grid to show how the 6 has moved one place to the left into the tens column and explain that we use zero as a placeholder. Explain that when we multiply by 10 the number gets ten times bigger. If a pupil says that they can just add zero explain that it is important they understand that the number gets 10 times bigger and can explain what happens to each digit rather than just following this rule, because when we move on to decimal fractions this rule will not work.

- Repeat example for 6 × £100. Use mock £100 notes to model 6 × 100. **How much money do they have altogether? What has happened to the digits this time?** Ensure pupils now understand that the digits have moved 2 places this time and use a place value grid to model what has happened. Repeat with a wide range of examples.

Unit 4.5 Dividing whole numbers by 10 and 100

- Review learning from unit 4.4. Now give a problem in context, such as: **Amman wants to buy a game that costs £70. He earns £10 a week in his paper round. How many weeks will it take him to save up?** Ask pupils to write a number sentence to show how they would work out this problem. Check understanding that this is a division problem. Use mock £10 notes to represent £70. Ask pupils to work out the answer and check that they are able to skip count or use recall to get the answer, 70 ÷ 10 = 7, so 7 weeks.
- Use a place value grid to show how the 7 has moved one place to the right into the ones column. Explain that when we divide by 10 the number gets ten times smaller. If a pupil says that they can just take away zero explain that it is important they understand that the number gets 10 times smaller and can explain what happens to each digit rather than just following this rule, because when we move on to decimal fractions this rule will not work.
- Repeat example for 700 ÷ 100. **What has happened to the digits this time?** Ensure pupils now understand that the digits have moved 2 places this time and use a place value grid to model what has happened. Repeat with a wide range of examples.

Unit 4.6 Solving multiplication problems by partitioning

- Check pupils' recall of multiplication facts for 2, 3, 5 and 10.
- Use ten frames to represent 12 × 3.

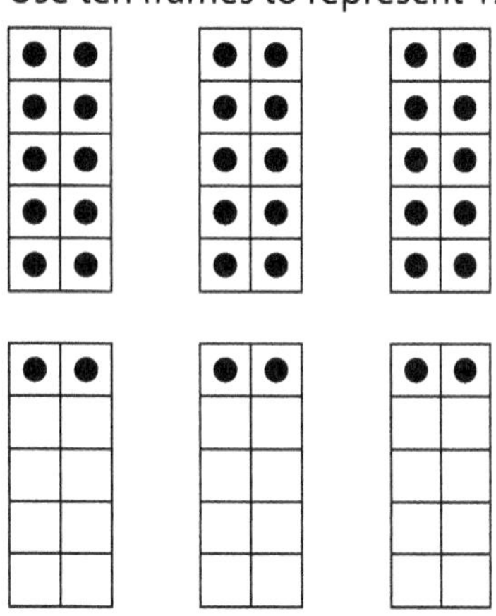

- (You could also use sticks or straws bundled into tens and ones.) Ask pupils for ideas how we could work out this problem. Model how we could split the ten frames into tens and ones to split the factors. **What is 10 × 3?** Record on a whiteboard. **What do we need to work out next?** Check pupils understand that next we need to work out 2 × 3. Record on whiteboard. Model how we would record this using brackets: (10 × 3) + (2 × 3) = 30 + 6 = 36. Explain that the brackets are important because they show how we have partitioned the problem (review unit 4.3).
- Repeat for a wide range of examples, such as 22 × 4, 31 × 2, 18 × 5.

Unit 4.7 Solving multiplication and division problems

- Check pupils' understanding of doubles and halves. Give problems such as: **I am thinking of a number. I'm going to double it and now I have 22. What number did I start with?**
- Use a pack of playing cards. Turn over each card and give a double for that number, e.g. if I turn over a 6 then I need to say double 6 = 12. Encourage the pupils to link it to multiplication facts for 2, so double 6 is the same is 2 × 6.
- Encourage pupils to link double and half facts. Ask questions such as: **What is double 14? What halving fact do you know now too?**
- Give pupils True and False cards. Make up statements such as: **Half of 16 is 9. True or False?** Pupils hold up their True or False cards and justify their answers, e.g. I know it's not 9 because double 9 is 18. Link to multiplication and division, with statements such as, **I know that 50 × 2 is 100 because I know half of 100 is 50.**

Unit 4.8 Recalling multiplication and division facts for 4

- Ask pupils, **What knowledge do we already have that might help us recall multiplication and division facts for 4?** Look for answers such as, recall of doubles and halves, recall of multiplication facts for 2. Model how we can use skip counting in 2s to help us skip count in 4s by missing out every other number.
- Use a bead string or hundred square to count in 4s and explore patterns pupils find.
- Discuss facts we already know for 4 because we know multiplication facts for 2, 5 and 10
- Model what happens when we double a number, then double it again. **What do you notice?** Link halving and halving again to dividing by 4.

Unit 4.9 Solving division problems

- Give a problem in context, such as: **I have 24 pupils in my class and I want to put them into groups of 3. How many groups will I have?**
- Draw a large number line and explain that we are going to start at 24 and explore jumping back in different multiples. Model skip counting backwards in multiples of 3, and draw the jumps on the number line.

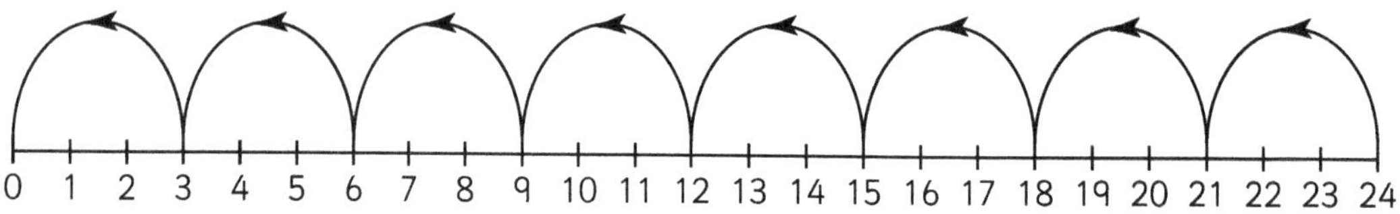

- **How many jumps did I make?** Check pupils can see there are 8 jumps back to zero. **Now try jumping back in fours. Draw a number line and start at 24. Count how many jumps backward in multiples of 4.** Record as a division sentence, $24 \div 3 = 8$.
- Repeat for other multiples such as 6 or 8.

Unit 4.10 Multiplying by changing the order of the factors

- Make an array showing 6 rows of 10 on a piece of card.
- **Write this as a multiplication sentence. How many altogether?** Check understanding that this represents $6 \times 10 = 60$. Now rotate the card by 90 degrees so they can now see the array as 10 rows of 6.
- **Write this as a multiplication sentence. How many altogether? What do you notice?** Check for understanding that the answer has not changed, even though we have changed the order of the factors. Explain that this is called the commutative property; it doesn't matter what order the factors are in.

Unit 4.11 Solving multiplication problems

- Ask pupils to share examples of multiplication facts that they know already. Explain that we can use these to solve other multiplication problems. Give a problem in context, such as: **I have 4 trays of biscuits. Each has 7 biscuits. How many biscuits do I have altogether?**

  If we know 2×7 we can use that to help us.

  $2 \times 7 = 14$

  We could double the answer to give us 4×7. Add 2 more rows of 7 to the array.

  $4 \times 7 = 28$

Could this help us work out 8 × 7?

We could double the answer again. Add 4 more rows of 7 to the array.

$8 \times 7 = 56$

- Repeat for a range of different examples. Encourage pupils to draw their own arrays to help them, or screen arrays using a large piece of card to encourage them to form mental images. Once confident, remove arrays altogether so pupils are using their knowledge of multiplication facts and doubling.

Unit 4.12 Using known facts and halving to solve division problems

- Check understanding of linking halving facts to division facts for 2 by asking questions such as: **What is half of 18? How could you write that as a division fact?**
- Explain that we can use our knowledge of halves to help us work out division problems. Give a problem in context, such as: **I have 24 stickers to share between 8 children. How many stickers will they each get?** Put out 24 cubes or counters in 2 rows of 12. To divide by 2 we can half the 24. **What is half of 24?** Record division sentence:

  $24 \div 2 = 12$

  To divide by four we half, then half again. Rearrange the 24 counters into 4 rows of 6 by halving each row of 12 and creating a new row. Now we have halved them again. **What is half of 12?** Record division sentence:

  $24 \div 4 = 6$

  To divide by eight we half, half and then half again. Now half each of the rows of 6 and create new rows of 3, so array shows 8 rows of 3. Now we have halved them one more time. **What is half of 8?** Record division sentence:

  $24 \div 8 = 3$

- Repeat with a range of examples to show how we can use halving and known facts to solve division problems.

Multiplication and division 2B

Section overview

Textbook 2B units	Review units	Prior learning	Benchmarks/ Es&Os
Unit 4.1 Recalling multiplication and division facts for 6	**2A** Units 4.1, 4.2, 4.8, 4.11 and 4.12	• I know doubling and halving facts for numbers up to 20. • I can recall multiplication and division facts for 2, 3, 4, 5 and 10. • I can use facts I know to work out new multiplication facts. • I can read and write number sentences involving multiplication and division.	**MTH 2-03c** • Uses multiplication and division facts to the 10th multiplication table. • Multiplies and divides whole numbers by multiples of 10, 100 and 1000. • Multiplies and divides decimal fractions to two decimal places by 10, 100 and 1000. • Multiplies whole numbers by two digit numbers. • Multiplies decimal fractions to two decimal places by a single digit.
Unit 4.2 Recalling multiplication and division facts for 9	**2A** Units 4.1, 4.2, 4.8, 4.11 and 4.12	• I know doubling and halving facts for numbers up to 20. • I can recall multiplication and division facts for 2, 3, 4, 5 and 10. • I can use facts I know to work out new multiplication facts. • I can read and write number sentences involving multiplication and division.	
Unit 4.3 Multiplying multiples of 10, 100 or 1000	**2A** Units 2.1, 2.2, 4.1, 4.2, 4.8, 4.11 and 4.12	• I can recall multiplication and division facts for 2, 3, 4, 5 and 10. • I can use facts I know to work out multiplication and division problems. • I know the meaning of each digit in the ones, tens, hundreds and thousands columns.	
Unit 4.4 Dividing by multiples of 10, 100 or 1000	**2A** Units 2.1, 2.2, 4.1, 4.2, 4.8, 4.11 and 4.12	• I can recall multiplication and division facts for 2, 3, 4, 5 and 10. • I can use facts I know to solve multiplication and division problems. • I know the meaning of each digit in the ones, tens, hundreds and thousands columns.	
Unit 4.5 Solving multiplication problems	**2A** Units 2.1, 2.2, 4.1, 4.2, 4.8, 4.11 and 4.12	• I can recall multiplication and division facts for 2, 3, 4, 5 and 10. • I can use facts I know to work out multiplication and division problems. • I know the meaning of each digit in the ones, tens, hundreds and thousands columns.	

Unit 4.6 Solving division problems	**2A** Units 2.1, 2.2, 4.1, 4.2, 4.8, 4.11 and 4.12	• I can recall multiplication and division facts for 2, 3, 4, 5 and 10. • I can use facts I know to work out multiplication and division problems. • I know the meaning of each digit in the ones, tens, hundreds and thousands columns.	
Unit 4.7 Solving multiplication problems	**2A** Units 2.1, 2.2, 4.1, 4.2, 4.8, 4.11 and 4.12	• I can recall multiplication and division facts for 2, 3, 4, 5 and 10. • I can use facts I know to work out multiplication and division problems. • I know the meaning of each digit in the ones, tens, hundreds and thousands columns.	
Unit 4.8 Solving multiplication problems	**2A** Units 1.1, 4.1, 4.2, 4.8, 4.11 and 4.12	• I can recall multiplication and division facts for 2, 3, 4, 5 and 10. • I can use facts I know to work out multiplication and division problems. • I can round numbers to the nearest 10.	
Unit 4.9 Solving problems involving addition, subtraction, multiplication and division	**2A** Units 3.7, 3.8, 4.11 and 4.12	• I can solve addition and subtraction problems using a range of strategies. • I can solve multiplication and division problems using a range of strategies.	
Unit 4.10 Multiplying decimal fractions	**2A** Units 4.4, 4.5, 4.11 and 4.12	• I can use facts I know to work out multiplication and division problems. • I know the meaning of each digit in the tenths, ones, tens, hundreds and thousands columns. • I know that when we multiply whole numbers by 10 the digits move one place to the left. • I know that when we multiply whole numbers by 100 the digits move two places to the left.	
Unit 4.11 Dividing whole numbers by 10, 100 and 1000	**2A** Units 4.4, 4.5, 4.11 and 4.12	• I can use facts I know to work out multiplication and division problems. • I know the meaning of each digit in the tenths, ones, tens, hundreds and thousands columns. • I know that when we divide whole numbers by 10 the digits move one place to the right. • I know that when we divide whole numbers by 100 the digits move two places to the right.	

Unit 4.12 Solving division problems with remainders	**2A** Units 2.5, 2.6, 2.7, 4.1, 4.2, 4.8, 4.11 and 4.12	• I can recall multiplication and division facts for 2, 3, 4, 5 and 10. • I can solve multiplication and division problems using different strategies. • I can read and write decimal numbers to 2 places. • I can read and write simple fractions.	
Unit 4.13 Solving multiplication and division problems	**2A** Units 4.1, 4.2, 4.8, 4.11 and 4.12 **2B** Units 4.1, 4.2, 4.6 and 4.8	• I can use known facts to solve multiplication and division problems. • I can recall multiplication and division facts for 2, 3, 4, 5, 6 and 9. • I can use different strategies to solve multiplication problems. • I can use different strategies to solve division problems.	

Conceptual context

At this stage, it is important to check that pupils have a strong understanding of multiplication as being the joining of equal groups and can confidently use skip counting to solve multiplication problems. They need now to have knowledge of multiplication facts for 2, 3, 4, 5 and 10, and be able to draw on these to derive new facts, as well as beginning to use facts such as doubles and halves to solve problems. Pupils need a secure understanding of division in both sharing and grouping contexts and have experience in solving problems using repeated addition and subtraction.

The next stage of development is to extend recall of multiplication and division facts to 6 and 9, exploring how we can use knowledge of facts for 3 and 10 to help. Building on previous experience of multiplying and dividing by 10 and 100, pupils will now learn to multiply whole numbers by multiples of 10, 100 and 1000. This requires a strong knowledge of place value for numbers up to 5 digits, and it is very important to check that pupils are secure with the meaning and value of each digit at this stage. Place value materials can be used to help pupils make connections to their recall of number facts to solve problems with multiples of 10; for example, in solving 50 × 3 we could use bundles of ten sticks to represent the problem:

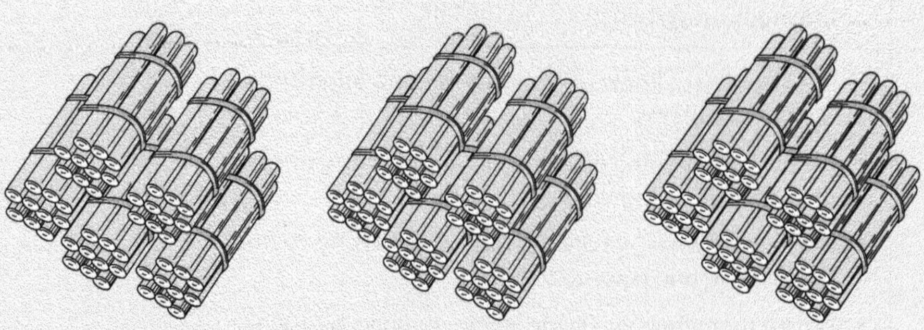

In this way pupils are encouraged to connect with their knowledge that 3 × 5 = 15, seeing the problems as 3 × 5 tens = 15 tens = 150.

Similarly, for dividing multiples of 10, 100 and 100, materials such as money could be used to help promote these connections:

For example, when exploring 600 ÷ 3, we are encouraging pupils to see that they could group the hundreds into 3 groups, using the known fact 6 ÷ 3 = 2 to work out the answer as 2 hundreds, or 200.

Pupils will next extend their understanding of the inverse relationship between multiplication and division and use the strategy of reversing division problems into multiplication problems to solve them. Arrays are a useful way to promote the understanding of inverse properties and pupils should have experience of making, drawing and recording arrays as both multiplication and division sentences.

Pupils require understanding of partitioning as splitting factors into tens and ones, and will now extend this strategy to solve 3-digit by 1-digit problems and introduce grid notation as well as brackets to record thinking. It is important that pupils have wide-ranging experience using both forms of notation to connect the thinking they used, and have opportunities to share and discuss the strategies used.

Partitioning is also now used to solve division problems. Pupils are encouraged to partition by numbers they know will be exactly divisible, rather than by place value, and this therefore requires a strong knowledge of multiplication and division facts, with instant recall. Understanding of factors as whole numbers that exactly divide into another whole number, and multiples as the product of multiplying whole numbers, will also be helpful at this stage. Brackets and grid notation are both introduced, and pupils should have wide experience of using both, and be encouraged to share and discuss which they find more efficient to record their thinking. Arrays or squared paper could be used to model ways to partition using known number facts, e.g. for 48 ÷ 3 encouraging pupils to use known facts such as 18 ÷ 3 and 30 ÷ 3.

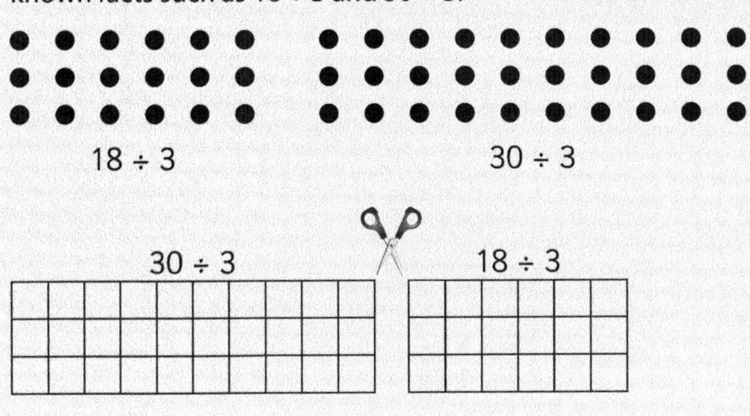

The strategy of rounding and compensating requires pupils to again apply knowledge of multiplication facts, and also to have a secure understanding of rounding to the nearest 10. Materials such as rows of cubes, ten frames or place value materials can be used to promote the idea that we can round to the nearest 10, then adjust accordingly to make the problem easier:

e.g. 18 × 3 = (20 × 3) − (2 × 3)

It is important that pupils have experience using concrete materials. Using concrete materials can promote a secure understanding of the process involved and help to avoid misconceptions. Physical materials allow pupils to clearly see that they have adjusted by adding 3 sets of 2, and need to then compensate by this amount to get the correct answer. Pupils should be encouraged to use brackets or empty number lines to record their thinking.

Pupils are also encouraged at this stage to be able to interpret problems in context and decide on the correct operation or operations they need to use to solve the problem. Unit 4.9 includes problems using a combination of addition, subtraction, multiplication and division that will require pupils to analyse the calculations they need to work out and record each process used to solve the problem.

When multiplying decimal fractions and dividing whole numbers resulting in decimal fractions, it is important to review understanding of place value, fractions and multiplying and dividing by 10 and 100 in previous units. Pictorial or concrete representations such as decimats could be used to assess understanding of decimal fractions to 1 place. Pupils need to have a secure understanding of what happens to each digit as they multiply or divide by 10, 100 and 1000, and that the decimal place does not move. Place value grids are a useful way of exploring this with pupils, with opportunities to apply a range of examples.

When solving division problems involving remainders, pupils will need to use the context of the problem when making a decision on the remainders. They need to know that the context of a problem can affect how they write their answer. For example, when dividing a sum of money, it is efficient to express the remainder in decimal form; however, when dividing a period of time, it may be more useful to express the remainder as a fraction. Pupils need opportunities to explore a range of contexts relevant to daily life and to understand that remainders can be expressed as a whole number, fraction or decimal fraction depending on the particular problem being solved.

Finally, pupils are encouraged to review all the different strategies they have learned for solving multiplication and division problems, and to explore which strategy might be more efficient for a particular problem. It is important that they have opportunities to work with partners and in groups to share different strategies and discuss which work best for different problems. For example, when working out 19 × 4, they could use partitioning to solve it by splitting the factors by place value, or use rounding and compensating to round the 19 to 20.

$$\begin{aligned} 19 \times 4 &= (10 \times 4) + (9 \times 4) \\ &= 40 + 36 \\ &= 76 \end{aligned}$$

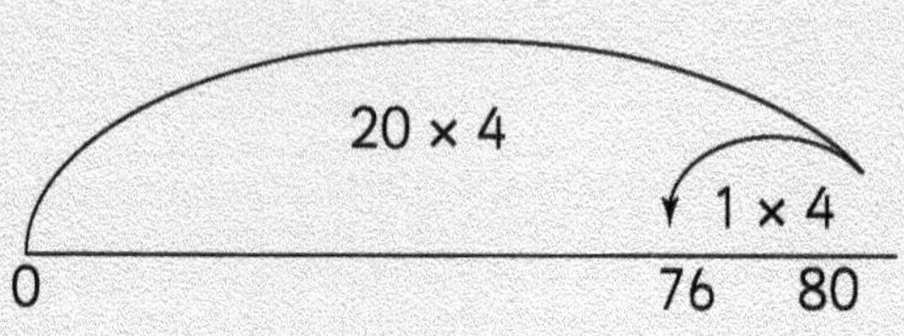

$$\begin{aligned} 19 \times 4 &= (20 \times 4) - (1 \times 4) \\ &= 80 - 4 \\ &= 76 \end{aligned}$$

Think Boards are a very useful way to record and share a range of strategies used. Pupils should be encouraged to develop a wide range of strategies to solve problems, and to think about what makes a strategy efficient.

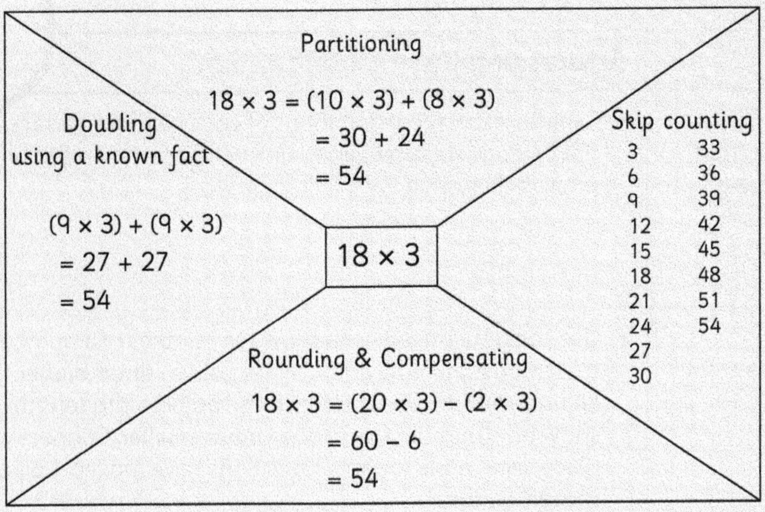

Resources

- Arrays
- Squared paper
- Money – mock £10 notes and £100 notes for multiplying and dividing by 10 and 100
- Empty number lines
- Place value materials, such as sticks, counters or cubes
- Place value grids
- Ten frames
- Cubes
- Decimats

Key vocabulary

- **Array**
- **Associative Principle**
- **Brackets**
- **Commutative property**
- **Distributive principle**
- **Factor**
- **Halving**
- **Inverse property**
- **Multiple**
- **Operation**
- **Partitioning**
- **Product**
- **Remainder**
- **Repeated addition**
- **Repeated subtraction**
- **Reversing**
- **Skip counting**

Misconceptions

Potential misconception	Addressing the misconception
Units 4.1 and 4.2 Rote understanding of 'times tables' always recited in order, without understanding of each fact as a number of equal groups. This can result in pupils being unable to apply their knowledge to solve problems in context.	• Explore visual representations of each fact using arrays, dot cards or squared paper. Link with repeated addition so pupil is clear what each fact represents.
Units 4.3 and 4.4 Using rules such as, 'add a zero' or 'take away a zero' rather than understanding of number getting 10 times bigger or 10 times smaller.	• Use concrete materials to show ten bundles of a number to demonstrate how number has got ten times bigger. Similarly, for division, bundle loose objects into tens to show how number has got ten times smaller. Connect with place value grids to show how digits have moved 1 place to the left or right. • Explain that these rules will not apply when we move on to decimal fractions; it is important we can show understanding by explaining what has happened to the numbers.
Unit 4.5 Lack of understanding of inverse relationship – for example thinking that the inverse of $12 \div 3$ is 12×3 or 3×12.	• Use arrays or squared paper to represent problems visually and allow pupils to explore a range of problems to see connections. Give opportunities for pupils to use concrete materials to physically form groups to multiply and divide.
Unit 4.6 Lack of understanding of place value when partitioning, e.g. incorrectly partitioning 123×3 as $(10 \times 3) + (2 \times 3) + (3 \times 3)$.	• Check understanding of what each digit represents up to 5-digit numbers. • Use place value materials to model multiplying each digit visually.
Unit 4.7 Insecure knowledge of multiplication and division facts for partitioning.	• Review knowledge of multiplication and division facts and ensure pupils have a wide range of experiences to practise and make connections with facts, building towards instant recall.
Unit 4.8 Misunderstanding of compensation – adjusting by the amount rounded, rather than the multiple, e.g. for 18×2, working out $(20 \times 2) - 2$ instead of $(20 \times 2) - (2 \times 2)$.	• Use concrete materials, such as cubes, to model how we round to multiply, and then compensate.
Unit 4.9 Misinterpreting correct operation to solve problem.	• Ensure pupils have a wide range of opportunities to explore, share and discuss word problems and interpret the correct operation to solve the problem.
Units 4.10 and 4.11 Moving decimal point when multiplying or dividing with decimal fractions. Mis-generalising that numbers always get bigger when they are multiplied.	• Use materials, such as decimats, to model multiplying decimal fractions and link with place value grids to show how digits move, not decimal point.
Unit 4.12 Not understanding what remainder actually means for problems in context.	• Ensure pupils have a wide range of opportunities to explore problems with remainders in real-life contexts.

Teaching ideas

Unit 4.1 Recalling multiplication and division facts for 6

- Link pupils' knowledge of multiplication facts of 3 using arrays and doubling. For example, show an array for 3 × 5, then add 3 more rows to show 6 × 5. **What do you notice?** Explain that we can double facts for 3 to work out the answer to multiplication facts for 6.
- Discuss ways to help recall of facts that pupils are finding tricky, e.g. using knowledge of facts of 5 or 10 to help with 6 × 9 and 6 × 6.

Unit 4.2 Recalling multiplication and division facts for 9

- Use arrays to show how we can use our knowledge of multiplication facts of 10 to help us recall facts for 9 quickly. For example, make an array showing 4 × 10, then subtract one group of 4 to make 4 × 9.
- Use hundred squares to explore patterns when multiplying by 9, for instance by adding 10 and subtracting 1. Write down all the multiples of 9: **What do you notice?** Encourage pupils to see that if they add both digits together they add up to 9 each time.

Unit 4.3 Multiplying multiples of 10, 100 or 1000

- Use place value materials such as bundles of 10 sticks, or dot strips to show 3 sets of 20.

  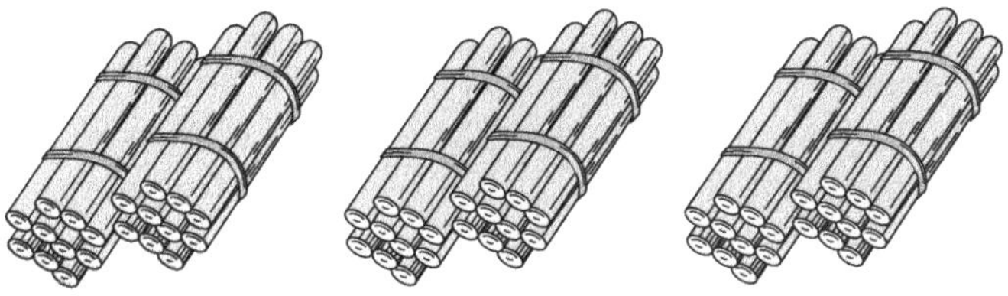

  We are going to work out 3 × 20. **What fact do we know that might help us?**

  Explain that if we know 3 × 2 = 6 we can use this fact to work out the answer.

  3 × 20 = 3 × 2 tens
  = 6 tens
  = 60
- Use hundred squares to repeat example to show 3 × 200.

  3 × 200 = 3 × 2 hundreds
  = 6 hundreds
  = 600
- Repeat for a wide range of examples showing multiples of 10, 100 and 1000, linking to known facts.

Unit 4.4 Dividing multiples of 10, 100 or 1000

- Lay out 8 bundles of 10 sticks. **How many? How do you know?** We are going to divide this 80 by 4. **What fact do you know that might help us?** If we know that $8 \div 4 = 2$ we can use this to help us.

  $80 \div 4 = 8$ tens $\div 4$

  $= 2$ tens

  $= 20$
- Repeat with 8 hundred squares:

  $800 \div 4 = 8$ hundreds $\div 4$

  $= 2$ hundreds

  $= 200$
- Repeat for a wide range of examples showing multiples of 10, 100 and 1000, linking to known facts.

Unit 4.5 Solving division problems by multiplying

- Make an array to show 4 rows of 5. **What does this array show?** Record $4 \times 5 = 20$. **How many groups of 5 in 20?** Point to each row in the array to show that there are 4 groups of 5 in 20. **How would we write this as a division sentence?** Check understanding that we write this as $20 \div 5 = 4$. Repeat for groups of 4, this time pointing at columns to model that $20 \div 4 = 5$. Explain that multiplication and division are opposite operations and introduce the term inverse properties. It is important that pupils understand what this means, although you wouldn't necessarily expect them to use this vocabulary at this stage.
- Explain that we can use this understanding to turn division problems around and use multiplication knowledge to solve them. If we want to work out $30 \div 5$ we can think of it as: **How many groups of 5 are in 30?** Record the problem like this:

  ______ $\times 5 = 30$. We know that $6 \times 5 = 30$, so we also know $30 \div 5 = 6$.

  Repeat for a wide range of examples, using arrays or squared paper if needed.

Unit 4.6 Solving multiplication problems by partitioning

- Use place value materials to show 14 × 3:

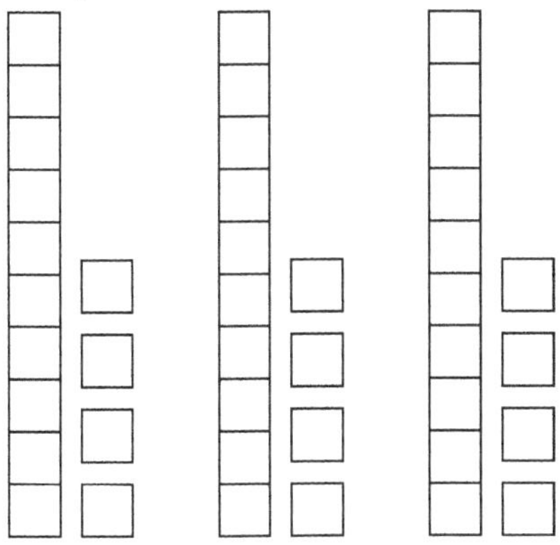

Explain that we can partition the numbers by place value to solve the problem. First physically model splitting the materials into tens and ones.

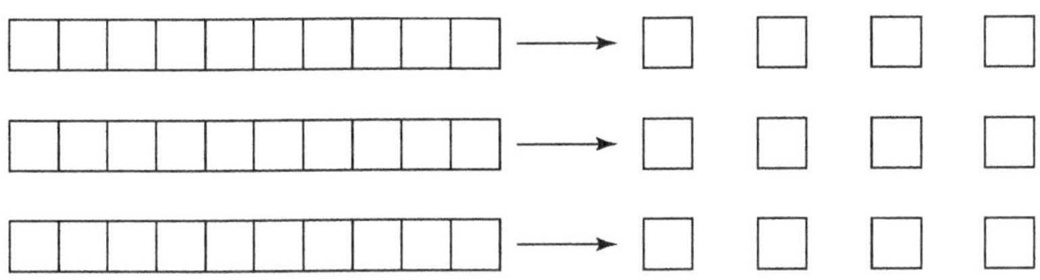

10 × 3 4 × 3

Record 10 × 3 = 30 and 4 × 3 = 12. To get our answer we add these together: 30 + 12 = 42.

Model recording this using brackets and also using a grid.

Give a wide range of examples of 2-digit numbers before moving on to 3-digit numbers. Encourage pupils to record their strategies using bracket and grid notation and discuss which they find more efficient.

Unit 4.7 Using partitioning to solve division problems

- Show a problem such as 52 ÷ 4. Discuss ways we could partition 52 to make it easier. Explain that partitioning by place value doesn't help us here because neither 50 nor 2 are exactly divisible by 4. We could split it into numbers that we know can be divided exactly by 4. **What facts of 4 do you know that could help us here?**

Look for facts such as 40 ÷ 4 = 10. That would leave us 12 ÷ 4, which is 3. We add the 10 + 3 to get our answer, which is 13.

Model recording this using brackets and using a grid. Give a wide range of similar examples and encourage pupils to record their strategies using bracket and grid notation and discuss which they find more efficient.

If pupils are struggling to think of numbers that are divisible, review knowledge of multiplication and division facts for the numbers you are working with.

Unit 4.8 Using rounding and compensation to solve multiplication problems

- Place out 4 rows of 18 cubes. Discuss strategies for working out 18 × 4. Pupils may give suggestions such as splitting by place value. **If I added 2 more cubes to each row, would that make it easier?** Model putting 2 more cubes on each row (a different colour if possible) to give 4 rows of 20. **How many altogether? How do you know?** Link knowledge of 4 × 2 to 4 × 20 to give answer 80. Explain that rounding the answer to the nearest 10 makes it easier to solve.
- Now we need to compensate because the original problem was not 20 × 4, but 18 × 4. Remove the 2 different coloured cubes from the 4 rows. We need to subtract these to work out our answer. Use an empty number line to model 20 × 4 = 80, then compensating by subtracting 2 × 4 to give our answer of 72.
- Repeat with a wide range of similar examples using materials such as ten frames or place value sticks to ensure pupils have a secure understanding of compensating. Encourage the pupils to record their thinking using empty number lines or with bracket notation.

Unit 4.9 Solving problems involving addition, subtraction, multiplication and division

- Give a problem in context, such as: **Isla buys 4 books costing £6 and Amman buys 3 books costing £9. How much do they spend altogether?** Discuss strategies to solve the problem. What will we need to work out first? Explain that we are going to have to use more than one operation to get our answer. Model working out how much Isla spent using multiplication, then how much Amman spent. What will we need to work out next? Check for understanding that we now need to add both amounts together to work out the total. Record each process and the strategies used.
- Give a wide range of examples for pupils to work out using at least two different operations, checking their understanding of the calculations needed at each stage. Ensure pupils have the chance to share their thinking of different problems and to discuss and work through mistakes with the choice of operation used.

Unit 4.10 Multiplying decimal fractions

- Check pupils' understanding of tenths using decimats (or ten frames with ten as the whole). Ask questions such as: **What would 7 tenths look like? What would be one tenth more? How many more tenths to make a whole?**
- Review understanding of multiplying whole numbers by 10, 100 and 1000.
- Give an example such as 0·3 × 10. Use a place value grid to show that the digits move one place to the left and we use zero as a placeholder, just as when we multiply whole numbers by 10. Reinforce that the decimal point does not move, only the digits. Use the place value grid to explore what happens to 0·3 when we multiply it by 100 and 1000.
- Repeat with a wide range of examples and ensure pupils can clearly explain that the decimal fraction gets 10 times bigger when we multiply by 10, rather than rules such as 'move the decimal point'.

Unit 4.11 Dividing whole numbers by 10, 100 and 1000

- Review understanding of dividing whole numbers by 10, 100 and 1000.
- Give an example such as 18 ÷ 10. Use a place value grid to show that the digits move one place to the right, just as when we divide whole numbers by 10. Reinforce that the decimal point does not move, only the digits. Use the place value grid to explore what happens to 180 when we divide it by 100 and 1800 divided by 1000.
- Repeat with a wide range of examples and ensure pupils can clearly explain that numbers get 10 times smaller when we multiply by 10, rather than using rules such as 'taking away a zero'.

Unit 4.12 Solving division problems with remainders

- Review understanding of remainders in division problems. Give a problem in context, such as, **Nuria's teacher has 23 metres of fabric, and she needs 2 metres each to make a costume for the school play. How many costumes can she make?** Discuss how much fabric will be left over. Can she use the remaining fabric? Explain that sometimes we can't use the remainder, so we express it as a whole number. We would record this as 11 r 1.
- Give another problem using a different context, such as: **Finlay shares £34 between himself and his 3 friends. How much do they each get?** In this context, because we are using money it is more useful to record the remainder as £8·50.
- Ensure pupils have opportunities to solve a range of problems in differing contexts relevant to daily life, so that they can explore different ways of expressing remainders, such as using time for a context involving expressing the remainder as a fraction, as well as decimal fractions or whole numbers.

Unit 4.13 Solving multiplication and division problems

- Review understanding of all the strategies we have explored so far for multiplication – partitioning, using doubles or facts we know or rounding and compensating. Use a Think Board to give a problem such as 3 × 17. Ask the pupils to work in pairs to think of 4 different strategies and record their thinking to solve this problem. Look for strategies such as splitting by place value, rounding and compensating, doubling 17 then adding another 17 or using facts such as 3 × 15, then adding 2 × 17. They may also include less efficient strategies such as skip counting. Share and discuss all the strategies identified and ask which they feel is the most efficient and why. You could ask them to draw a thumbs up or smiley face on the most efficient strategy.
- Repeat for strategies for division problems – partitioning, reversing, halving, repeated addition or subtraction.
- Allow lots of opportunities to share and assess different strategies in this way.

Multiplication and division 2C

Section overview

Textbook 2C units	Review units	Prior learning	Benchmarks/ Es&Os
Unit 4.1 Multiplication and division facts for 7	**2A** Units 4.1, 4.2 and 4.8 **2B** Units 4.1, 4.2 and 4.5	• I can recall multiplication and division facts for 2, 3, 4, 5, 6, 9 and 10. • I can use facts I know to solve multiplication and division problems. • I understand the inverse relationship between multiplication and division.	**MTH 2-03c** • Uses multiplication and division facts to the 10th multiplication table. • Multiplies and divides whole numbers by multiples of 10, 100 and 1000. • Multiplies and divides decimal fractions to two decimal places by 10, 100 and 1000. • Multiplies whole numbers by 2-digit numbers. • Multiplies decimal fractions to two decimal places by a single digit.
Unit 4.2 Recalling multiplication and division facts of 8	**2A** Units 4.1, 4.2 and 4.8 **2B** Units 4.1, 4.2 and 4.5	• I can recall multiplication and division facts for 2, 3, 4, 5, 6, 9 and 10. • I can use facts I know to solve multiplication and division problems. • I understand the inverse relationship between multiplication and division.	
Unit 4.3 Multiplication with decimal fractions	**2B** Units 4.3 and 4.10	• I can use facts I know to work out multiplication and division problems. • I know that when we multiply whole numbers by 10, the digits move one place to the left. • I know that when we multiply whole numbers by 100, the digits move two places to the left. • I know that when we multiply whole numbers by 1000, the digits move three places to the left.	
Unit 4.4 Division with decimal fractions	**2B** Units 4.4, 4.7 and 4.11	• I can use facts I know to work out multiplication and division problems. • I know that when we divide whole numbers by 10, the digits move one place to the right. • I know that when we divide whole numbers by 100, the digits move two places to the right. • I know that when we divide whole numbers by 1000, the digits move three places to the right.	
Unit 4.5 Solving multiplication problems	**2A** Units 4.1, 4.2 and 4.8 **2B** Units 4.1 and 4.2 **2C** Units 4.1 and 4.2	• I can recall multiplication and division facts up to 10. • I can use facts I know to work out multiplication and division problems. • I know the value of each digit in numbers up to 6 digits.	

Unit 4.5 Solving multiplication problems	**2A** Units 4.1. 4.2 and 4.8 **2B** Units 4.1, 4.2 **Unit 2.3** **2C** Units 4.1 and 4.2	• I can recall multiplication and division facts up to 10. • I can use facts I know to work out multiplication and division problems. • I know the value of each digit in numbers up to 6 digits.	
Unit 4.6 Solving multiplication problems involving decimal fractions	**2A** Units 4.1, 4.2 and 4.8 **2B** Units 2.7, 2.8, 4.1 and 4.2 **2C** Units 4.1 and 4.2	• I can recall multiplication and division facts up to 10. • I can use facts I know to work out multiplication and division problems. • I know the value of each digit in decimal fractions up to 2 places.	
Unit 4.7 Solving division problems using place value	**2A** Units 4.1, 4.2 and 4.8 **2B** Units 2.3, 4.1, 4.2 and 4.5 **2C** Units 4.1 and 4.2	• I can recall multiplication and division facts up to 10. • I can use facts I know to work out multiplication and division problems. • I know the value of each digit in numbers up to 6 digits. • I understand the inverse relationship between multiplication and division.	
Unit 4.8 Solving multiplication problems using proportional adjustment	**2A** Units 4.1, 4.2 and 4.8 **2B** Units 4.1, 4.2 and 4.5 **2C** Units 4.1 and 4.2	• I can recall multiplication and division facts up to 10. • I can use facts I know to work out multiplication and division problems. • I understand the inverse relationship between multiplication and division.	
Unit 4.9 Solving division problems	**2A** Units 4.1, 4.2 and 4.8 **2B** Units 4.1, 4.2 and 4.5 **2C** Units 4.1 and 4.2	• I can recall multiplication and division facts up to 10. • I can use facts I know to work out multiplication and division problems. • I understand the inverse relationship between multiplication and division.	
Unit 4.10 Solving multiplication problems	**2A** Units 4.1, 4.2 and 4.8 **2B** Units 2.3, 4.1, 4.2 and 4.3 **2C** Units 4.1 and 4.2	• I can recall multiplication and division facts up to 10. • I can use facts I know to work out multiplication and division problems. • I know the value of each digit in numbers up to 6 digits. • I can multiply whole numbers by multiples of 10.	

Unit 4.11 Solving division problems using an algorithm	**2A** Units 4.1, 4.2 and 4.8 **2B** Units 2.3, 4.1, 4.2 and 4.5 **2C** Units 4.1 and 4.2	• I can recall multiplication and division facts up to 10. • I can use facts I know to work out multiplication and division problems. • I know the value of each digit in numbers up to 6 digits. • I understand the inverse relationship between multiplication and division.	
Unit 4.12 Solving division problems involving decimal fractions	**2A** Units 4.1, 4.2 and 4.8 **2B** Units 2.7, 4.1, 4.2 and 4.5 **2C** Units 4.1 and 4.2	• I can recall multiplication and division facts up to 10. • I can use facts I know to work out multiplication and division problems. • I know the value of each digit in decimal fractions up to 2 places. • I understand the inverse relationship between multiplication and division.	
Unit 4.13 Solving multi-step problems using the order of operations	**2B** Unit 3.10 Units 4.9 and 4.13	• I can solve addition and subtraction problems using a range of strategies. • I can solve multiplication and division problems using a range of strategies. • I know that we use brackets to record the calculations we do first. • I can solve two-step problems using different operations.	

Conceptual context

By this point in their learning, pupils should have a strong understanding of multiplication as being the joining of equal groups, and division as sharing equally or forming equal groups, and be able to use a range of strategies to solve problems involving multiplication or division.

As pupils move on to develop proportional reasoning and written algorithms, it is crucial that they have a secure knowledge of multiplication and division facts, with instant recall to draw on and apply. Pupils need many opportunities to practise this knowledge through games and practical experiences, and to make connections with the recall of facts and how they use these in the strategies they are learning. Units 4.1 and 4.2 review recall of multiplication and division facts for 7 and 8, exploring ways to link knowledge of other facts, such as 4, 5 and 10, to help develop recall.

In Units 4.3, 4.4, 4.6 and 4.12, to multiply and divide decimal fractions, pupils need to have a secure understanding of place value. They should be able to read, write and represent numbers up to 6 digits, as well as decimal fractions to 2 places, and understand how to partition by place value to solve problems. Place value grids are a useful resource to explore how each digit is affected by multiplication and division of 10, 100 and 1000, and it is vital that pupils understand that the decimal point does not move, only the digits. Pupils should have understanding of bracket notation, as well as grid methods, with a wide range of experience in using both these methods of recording thinking and strategies.

Using partitioning to solve division problems by looking for numbers we know are divisible is extended, in this book, to partition by place value. Pupils should be encouraged to use a grid method, as well as bracket notation to record how they have worked out problems to make their thinking clear and allow them to share strategies.

In Unit 4.8, proportional adjustment is explored as a strategy to solve multiplication problems. Pupils need strong knowledge of doubling and halving, as well as multiplication and division facts for 3 to use thirding and tripling. Here they are developing the understanding that when we double or triple one factor, and halve or third the other factor, that the answer does not change. Squared paper or an abacus are useful resources to represent how proportional adjustment works, and pupils will need a wide range of experiences using materials to explore this strategy and develop proportional thinking.

Rounding and compensating is a strategy that pupils should have used to solve multiplication problems, and we now extend this by using it as a strategy to solve division problems. Pupils will again need secure recall of multiplication and division facts up to 10, and a solid understanding of rounding to the nearest 10 or 100 to use this strategy effectively. Place value materials, or cubes, can be used to model how we can round the number divisible to the nearest 10 or 100 to make it an easier number to work with, then compensate by the amount adjusted. Pupils will need to use their knowledge of facts to quickly assess if this is an effective strategy for the numbers they are working with, and to understand that they need to adjust by an amount that is also divisible by the number that is the divisor.

In Units 4.10 and 4.11, algorithms are introduced as an alternative method for multiplication and division. At this stage, pupils need to have had wide experience using a range of different mental strategies and have explored a range of written recording methods. It is also essential that they have instant recall or multiplication and division facts up to 10 to use algorithms, as well as a secure understanding of place value. They need to understand the value of each digit and be able to partition numbers confidently by place value. Pupils can then apply the written methods they have learned to problems involving decimal fractions up to two places in Unit 4.12. Again, they must have a secure understanding of reading and writing decimal fractions, and be able to partition whole numbers, tenths and hundredths confidently.

Finally, pupils will extend their understanding of solving multi-step problems in Unit 4.13. They are introduced to the order of operations, where they work from left to right and follow the convention of solving multiplication and division calculations first, then addition and subtraction. Pupils need to understand and use brackets to show which calculations they do first, and to understand that the order they solve the problem in will affect the answer.

Resources

- Squared paper
- Abacus
- Money – mock £10 notes and £100 notes for multiplying and dividing by 10 and 100
- Empty number lines
- Place value materials, such as sticks, counters or cubes
- Place value grids
- Ten frames
- Cubes
- Decimats
- Arrow cards

Key vocabulary

- **Array**
- **Associative principle**
- **Brackets**
- **Commutative property**
- **Distributive principle**
- **Factor**
- **Halving**
- **Inverse property**
- **Multiple**
- **Operation**
- **Order of operations**
- **Partitioning**
- **Product**
- **Remainder**
- **Repeated addition**
- **Repeated subtraction**
- **Reversing**
- **Skip counting**

Misconceptions

Potential misconception	Addressing the misconception
Units 4.3 and 4.4 Pupils may use rules such as, 'add a zero' or 'take away a zero' rather than understanding of number getting 10 times bigger or 10 times smaller.	• Use place value grids to show that digits have moved to the left or right, and reinforce that decimal point does not move. • Use decimats to model numbers getting 10 times bigger when multiplied by 10.
Unit 4.6 Pupils may incorrectly partition decimal fractions or misunderstand the decimal point as separating whole numbers, e.g. $3{\cdot}24 \times 2$ as $(3 \times 2) + (0{\cdot}2 \times 2) + (0{\cdot}4 \times 2)$.	• Review understanding of tenths and hundredths using materials such as decimats. • Review Units 2.7 and 2.8 in Textbook 2B.

Potential misconception	Addressing the misconception
Unit 4.8 Pupils may incorrectly double both factors or halve both factors when using proportional adjustment.	• Use squared paper to make arrays and explore how doubling one factor and halving the other does not affect the answer, e.g. make an array showing 12 × 5 then cut in half, or fold, and reposition to show that answer is equal to 6 × 10.
Unit 4.9 Pupils may incorrectly adjust by simply subtracting the factor, rather than the divided amount, e.g. 72 ÷ 4 as (80 ÷ 4) – 8, rather than (80 ÷ 4) – (8 ÷ 4).	• Use materials such as cubes to show the amount adjusted, and model dividing the amount adjusted by the divisor to correctly compensate.
Units 4.10, 4.11 and 4.12 Pupils may exhibit a lack of understanding or introduce errors with place value when using written methods.	• Review pupils' understanding of the value of each digit and how to partition correctly using place value materials. • Model multiplying each digit using materials to show value. • Link with pupils' previous experience of using grid method so they are clear in how they are partitioning and multiplying each digit.

Teaching ideas

Unit 4.1 Multiplication and division facts for 7

- As multiples of 7 don't have an obvious pattern, encourage pupils to connect with existing knowledge of multiplication and division facts. Model reversing the order of the factors to show that pupils may already know multiplication facts such as 3 × 7, 5 × 7 and 10 × 7.
- Use squared paper or arrays to show how we can split the factors; for instance, in an array of 8 × 8 we could split the 8 into 5 and 3: 8 × 7 = (5 × 7) + (3 × 7).
- Discuss facts that are particularly tricky, and share strategies that might help, such as using 10 × 7 and subtracting 7 to quickly work out 9 × 7.
- Ensure pupils have opportunities to practise recall by playing a range of games to develop knowledge.

Unit 4.2 Multiplication and division facts for 8

- Encourage pupils to use knowledge of multiplication and division facts for 2 and 4 when working on recall on facts for 8. Use squared paper or arrays to model doubling, doubling and doubling again, e.g. if we know the fact 7 × 2 = 14 we can double this answer to get 7 × 4 = 28, then double again to work out 7 × 8 = 56.

- Ask pupils to complete multiplication triangles to practise multiplication and division facts, and cover one of the corners to test a partner, e.g. a triangle showing 6, 8 and 48 could be used to ask questions such as: 48 ÷ 8 = ______ or 8 × ______ = 64.
- Ensure pupils have opportunities to practise recall by playing a range of games to develop knowledge. Packs of cards can be used, such as by shuffling a deck and turning the pile face down then turning over each card at a time and multiplying the value shown by 8, or spreading out a deck face up on a table and giving each other instructions such as: **Find me the card that I would multiply by 8 to make 56**, or **Find me the card that is the answer to 32 divided by 8.**

Unit 4.3 Multiplication with decimal fractions

- Review learning of multiplying by 10, 100 and 1000 and check for understanding that numbers get 100 times bigger when we multiply by 100. Check understanding of decimal fractions using decimats.
- Use a place value grid to model multiplying decimal fractions to 2 places by 10. Show that the decimal point does not move, but each digit moves 1 place to the left and we use zero as a placeholder. Ask questions such as: **What happens to the digit in the tenths column? What about the digit in the hundredth column?**
- Repeat for multiplying by 100 and by 1000, checking understanding that the number gets 100 times bigger when multiplying by 100, and 1000 times bigger when multiplying by 1000.
- Give a wide range of examples for pupils to solve and record.

Unit 4.4 Division with decimal fractions

- Review learning of dividing by 10, 100 and 1000 and check for understanding that numbers get 1000 times smaller when we divide by 1000.
- Use a place value grid to model dividing a number by 10 with the answer giving a decimal fraction, e.g. 32 ÷ 10. **What has happened to the tens? What about the ones?** Reinforce that the decimal point is always fixed between the ones and the tenths. Try different examples showing dividing by 100 and 1000 to establish how the digits move each time we divide by 10, 100 or 1000.
- Give a wide range of examples for pupils to solve and record.
- Try giving problems such as: **I divided a number by 100 and now I have 0·24. What number did I start with?** Model the thinking involved out loud to get the answer of 24. **How have the digits moved?** Encourage pupils to work with a partner and make up problems like this for each other to solve.

Unit 4.5 Solving multiplication problems

- Review understanding of partitioning 3-digit numbers when multiplying. Assess understanding of place value in 4-digit numbers.
- Give an example, such as: **It costs £1837 each for a holiday. How much would it cost for a family of 4 people?** Use arrow cards to model partitioning each digit and multiplying by 4. Use the grid method to record each step. You could also model using bracket notation and discuss which way is easier to use for the numbers involved.
- Give a wide range of examples in context for pupils to solve and record.

Unit 4.6 Solving multiplication problems involving decimal fractions

- Give an example of a decimal fraction to 2 places, such as 4·58. Ask questions such as: **What is this number? How much is the 8 worth? What would one tenth more be? One hundredth less?**
- Review partitioning from previous unit and ask: **How can this help us multiply 4·58 by 3?** Check for understanding that we can use place value to partition decimal fractions when multiplying as we have already learned for whole numbers. Use arrow cards to model partitioning the number into ones, tenths and hundredths. Model using the grid method to multiply each digit by 3 to get our answer. You could also use brackets to record how we partitioned the number, and encourage the pupils to use both forms of notation to record partitioning for different examples.

Unit 4.7 Solving division problems using place value

- Review learning from partitioning numbers to divide them by splitting them into known multiples. Explain that now we are going to use place value to partition numbers to solve division problems. Model an example, such as 165 ÷ 5, and use arrow cards to partition the number into 100, 60 and 5. Show how we can use a grid to divide each value by 5, then add the numbers together to get our answer. Give a range of examples like this where each value can be divided exactly by another number.
- Give a different example, such as 168 ÷ 6, and ask the pupils to partition by place value. **Why is this more difficult? What might be an easier strategy?** Look for understanding that for this particular number, it is easier to partition by known multiples of 6, in this case 120 and 48, rather than partitioning by place value. Record using either a grid or brackets.
- Repeat with a wide range of examples and discuss what strategy will work best for partitioning for particular problems.

Unit 4.8 Solving multiplication problems using proportional adjustment

- Use an abacus and move 14 groups of 5 to the left-hand side to show 14 × 5. **How much is 14 × 5? How many rows of 10 is this?** Record 14 × 5 = 7 × 10. **What do you notice?** Pupils may realise that 14 is double 7 and 5 is half of 10. Repeat with examples such as 8 × 5 and 18 × 5. Explain that when we double one factor, we can half the other factor and the answer doesn't change. Test with more examples using the abacus. Explain that this can help us solve multiplication problems. Give an example such as 22 × 4. Ask pupils to double one factor and halve the other to make an easier problem for us to solve. Record 22 × 4 = 11 × 8 and show both multiplication sentences using the abacus. **Which is easier to solve?**
- Use squared paper or an array to show 3 × 18.

  Now show how we can third each row of 18 (divide rows into 3 groups with 6 in each row). Now leave the first group where it is and move each of the 2 groups down underneath the first group to make a new array showing 9 rows of 6.

  Explain that we can third one of the factors and triple the other in the same way as doubling and halving and the answer will stay the same. Ask: **Which is easier, 3 × 18 or 9 × 6?**
- Repeat with a range of examples using squared paper or arrays to develop the idea of proportional adjustment.

Unit 4.9 Solving division problems

- Give the problem 54 ÷ 3. Represent the 54 using 5 rows of 10 cubes and one row of 4 cubes (or use place value materials).

  Ask if it would make the problem easier if we rounded 54 up to 60 by adding 6 more cubes (use a different colour).

  Now we are working out 60 ÷ 3. Pupils can now work out that 60 ÷ 3 = 20. Model compensating by taking the amount we added and also dividing by 3. Explain that we need to subtract by this amount to get our answer, because the question wasn't 60 ÷ 3, it was 54 ÷ 3.

  Record using brackets or an empty number line:
  54 ÷ 3 = (60 ÷ 3) – (6 ÷ 3) = 20 – 2 = 18
- Repeat with a wide range of examples, allowing pupils to solve, record and show using materials or pictures.

Unit 4.10 Solving multiplication problems

- Give an example such as 22 × 18 and review written methods we have already learned to record solving the problem, i.e. brackets or grid method. Explain that we are going to use an algorithm to solve this problem, and that these can be very useful when problems are very challenging to solve mentally.
- Model setting out the problem vertically, reinforcing the need to make sure we are setting out each digit in the correct place value correctly. Now model multiplying the top number, each digit in turn, by the ones (in this case the 8 in 18). Then model multiplying the top number by the tens (in this case 10). Emphasise the need to set each digit of the answer clearly in the correct place value column. Finally, model adding both numbers together vertically to get our total.

```
                2 2
            ×   1 8
               -----
(22 × 8)      1 7 6
(22 × 10)     2 2 0
               -----
              3 9 6
               -----
```

- Work through a range of different examples to ensure pupils are clear about setting out problems vertically, and use the algorithm at each step of the process.

Unit 4.11 Solving division problems using an algorithm

- Give an example such as 92 ÷ 4 and review strategies that might be used to solve this problem. Explain that we are going to use an algorithm to solve this problem, and that these can be very useful when problems are very challenging to solve mentally.
- Model setting out the problem, like this:

```
   (90 ÷ 4)   (12 ÷ 4)
         2     3
       ---------
    4 |  9  ¹2
```

Add the 10 remaining to the ones

Explain that first we divide the digit in the tens column by 4 and model recording the answer above the 9. Then model adding the remainder to the ones column; in this case we add the 10 remaining to make 12. We then divide the remaining ones and record the answer above the 2. Our answer in this case is 23.

- Work through a range of different examples to ensure pupils are clear about setting out problems, and use the algorithm at each step of the process.

Unit 4.12 Solving division problems involving decimal fractions

- Review learning from previous unit and explain that we are going to use the same algorithm to divide decimal fractions. Give an example, such as 4·32 ÷ 3 and model setting out the problem like this:

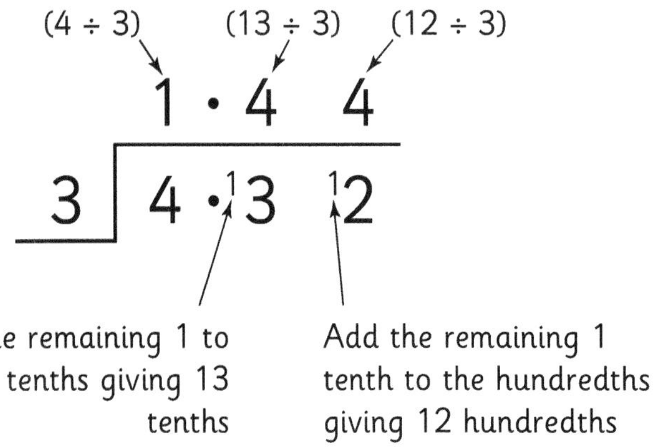

  Model the process of dividing each digit by 3, reinforcing the need for the decimal point to be a fixed point.
- Work through a range of different examples to ensure pupils are clear about setting out problems, and use the algorithm at each step of the process.

Unit 4.13 Solving multi-step problems using the order of operations

- Give an example of a problem involving a range of operations, such as 28 + 8 – 5 × 3. Ask pupils to work out the problem and discuss the answers given. There should be a range of different answers depending on the order pupils have used to work out their answers!
- Order is important! Explain that we use a convention called *the order of operations* to ensure consistency. Work through the problem from the left, and explain we look for multiplication and division calculations first, with each having equal importance. Then we work out addition and subtraction calculations next, with each having equal importance. We use brackets to show which calculations we do first.
- Give a wide range of examples using different operations to ensure that pupils are clear about the order of operations and can use brackets to clearly explain how they worked out each problem.

5 Multiples, factors and primes

Multiples, factors and primes 2A

Section overview

Textbook 2A units	Review units	Prior learning	Benchmarks/Es&Os
Unit 1 Recognising multiples and factors	**2A** Units 4.1, 4.2 and 4.8	• I can read and write number sentences involving multiplication and division. • I can recall multiplication and division facts for 2, 3, 4, 5 and 10 times tables.	**MTH 2-05a** • Identifies multiples and factors of whole numbers and applies knowledge and understanding of these when solving relevant problems in number, money and measurement.

Conceptual context

A multiple is the product of any quantity and a whole number, or a collection of items that is reproduced repeatedly. For example, some multiples of 5 are 10, 20, 15 and 55 (formed by multiplying by 2, 4, 3 and 11).

It is a common misconception that multiples are restricted to multiplication facts. As you can see from the example above, multiples can be generated by multiplying by any whole number. Pupils begin to explore multiples as they learn about patterns and skip count.

A factor is a whole number that exactly divides into another whole number; for example, 3 is a factor of 15 because $3 \times 5 = 15$. In fact another name for a factor is a divisor.

Odd and even numbers are an introduction to factors for pupils. An even collection of items can be shared equally between two people (therefore two is a factor) and with odd collections there will always be one item left over. This then leads to pupils identifying that numbers ending in 2, 4, 6, 8, 0 are even and 1, 3, 5, 7, 9 are odd.

This idea can then be extended to factors of 5 always ending in 0 or 5.

Resources

- Counters or cubes
- Hundred squares
- Numeral rolls or number lines
- Arrays to represent multiplication problems
- Squared paper or grids to represent multiplication problems

Key vocabulary

- **Commutative property**
- **Factors**
- **Multiples**

Misconceptions

Potential misconception	Addressing the misconception
Unit 5.1 The pupil may confuse multiples with factors.	• Use counters to make arrays to show all the factors for different numbers. Encourage pupils to explore all the different possible factors using the counters to make different arrays. Reinforce that factors are the numbers that divide exactly into another number. • Link multiples to skip counting. Use hundred squares or numeral rolls to identify multiples to reinforce understanding that multiples can be made by multiplying a number by any other whole number. • Use vocabulary of multiples and factors when reading and writing multiplication facts.
Unit 5.1 The pupil may think factors or multiples are only numbers in times tables.	• Link multiples to skip counting. Use hundred squares or numeral rolls to identify multiples to reinforce understanding that multiples can be made by multiplying a number by any other whole number. • Make arrays for a range of different numbers and discuss all the possible factors to reinforce understanding that factors are any numbers that divide exactly into another, not just in the range of times tables that may have been learned.

Teaching ideas

Unit 5.1 Recognising multiples and factors

- Show an array of 6 rows of 3. Use concrete materials such as counters or cubes to make arrays for given numbers. Pupils can explore factors by making different arrays.

  Ask: **What multiplication fact does this array represent?** $3 \times 6 = 18$ is represented. Ask: **What factors can you see?** We can see that 3 and 6 are factors of 18. Ask: **What other arrays could you make using these 18 counters?**
- Show an array of 3 rows of 6.

  $1 \times 18 = 18$

  $18 \times 1 = 18$

  $2 \times 9 = 18$

  $9 \times 2 = 18$

  $3 \times 6 = 18$

  $6 \times 3 = 18$

  What other factors of 18 have you found? Record that 1, 2, 3, 6, 9 and 18 are factors of 18. Repeat with a wide range of different examples using concrete materials.
- Use a number line or numeral roll to explore multiples when skip counting. For example, if skip counting in fives, place counters or pegs on each multiple of five as you skip count. Discuss the multiples you have found: **What do you notice? What patterns can you see?** Extend to think about bigger numbers. Ask questions such as: **Is 104 a multiple of 5? How do you know?**

Multiples, factors and primes 2B

Section overview

Textbook 2B units	Review units	Prior learning	Benchmarks/Es&Os
Unit 1 Identifying all factors of a number	**2A** Units 4.1, 4.2, 4.8 and 5.1 **2B** Units 4.1 and 4.2	• I can read and write number sentences involving multiplication and division. • I can recall multiplication and division facts for 2, 3, 4, 5, 6, 9 and 10. • I know that factors are numbers that divide exactly into another number. • I know that multiples are numbers that can be divided by another number with no remainder.	**MTH 2-05a** • Identifies multiples and factors of whole numbers and applies knowledge and understanding of these when solving relevant problems in number, money and measurement.
Unit 2 Identifying multiples of numbers	**2A** Units 4.1, 4.2, 4.8 and 5.1 **2B** Units 4.1 and 4.2	• I can read and write number sentences involving multiplication and division. • I can recall multiplication and division facts for 2, 3, 4, 5, 6, 9 and 10. • I know that factors are numbers that divide exactly into another number. • I know that multiples are numbers that can be divided by another number with no remainder.	

Conceptual context

A factor is a whole number that exactly divides into another whole number; for example, 3 is a factor of 15 because $3 \times 5 = 15$. To identify all the factors of a number we need to investigate which whole numbers divide exactly into that number in a systematic way.

Pupils need to have strong knowledge of multiplication and division facts for the numbers that they are working with at this stage and understand that they need to work systematically to identify all the factors of a given number. They can investigate by using concrete materials such as arrays or grids to explore all the factor pairs of a given number.

To identify multiples of numbers, we can multiply the given number by any whole number; for example, to find multiples of 3 we can multiply by 1, 3, 7 or 100 to find the multiples 3, 9, 21, 300. Exploring patterns of different multiples can help us to solve multiplication and division problems and to estimate if answers are reasonable. We can also begin to explore common multiples by finding multiples that are common to two or more numbers; for example, 10 is a common multiple of 2, 5 and 10.

Resources

- Counters or cubes
- Hundred squares
- Numeral rolls or number lines
- Arrays to represent multiplication problems
- Squared paper or grids to represent multiplication problems

Key vocabulary

- **Commutative property**
- **Factors**
- **Multiples**

Misconceptions

Potential misconception	Addressing the misconception
Units 5.1 and 5.2 The pupil may confuse multiples with factors.	• Use counters to make arrays to show all the factors for different numbers. Encourage pupils to explore all the different possible factors using the counters to make different arrays. Reinforce that factors are the numbers that divide exactly into another number. • Link multiples to skip counting. Use hundred squares or numeral rolls to identify multiples to reinforce understanding that multiples can be made by multiplying a number by any other whole number. • Use vocabulary of multiples and factors when reading and writing multiplication facts.

Potential misconception	Addressing the misconception
Units 5.1 and 5.2 The pupil may think factors or multiples are only numbers in times tables.	• Link multiples to skip counting. Use hundred squares or numeral rolls to identify multiples to reinforce understanding that multiples can be made by multiplying a number by any other whole number. • Make arrays for a range of different numbers and discuss all the possible factors to reinforce understanding that factors are any numbers that divide exactly into another, not just in the range of times tables that may have been learned.

Teaching ideas

Unit 5.1 Identifying all factors of a number

- Ask: **What are factors of a number?** Check understanding of the term 'factor' and review previous learning. Ask: **How do we find all the factors of a number?** Discuss different ideas and methods for finding all the factors.
- Explain that we are going to find all the factors of 24. Give the pupils 24 cubes and ask them to make arrays or rectangles to show all the factors of 24. Model the example 24×1 by placing cubes into one long row. **What factors does this show?** Make sure the pupils understand that this shows 24 and 1. Explore all the factors of 24 by making different rectangles or arrays and model recording them as factor pairs, for example, **1 and 24, 2 and 12,** etc. Ask: **How will we know we have found all the factors?** Discuss ways we could work systematically to record all the factors; for example, starting from 1 and working through each whole number to identify or eliminate factors.
- Repeat with a wide range of examples using concrete materials or squared paper to make all the possible rectangles of given numbers.

Unit 5.2 Identifying multiples of numbers

- Ask pupils to give you a multiple of 5. Check understanding of multiples from answers: **How do you know this is a multiple of 5? Can you write a multiplication fact to show the multiple you have identified?**
- Use a hundred square or number line to explore different multiples of 4 and place counters or pegs to show the multiples you find. **What do you notice about all the multiples of 4? Will 28 be a multiple of 2? What about 39? How do you know?** Discuss multiples of 10. **What patterns do you notice?**
- Link knowledge of multiplication and division facts to understanding of multiples, e.g. 36 is a multiple of 9 because I know $4 \times 9 = 36$. **Is 70 a multiple of 10? How do you know? What else is it a multiple of? How do you know?**
- Try giving YES or NO cards to pupils and give them statements such as, **29 is a multiple of 5 – yes or no?** Pupils hold up YES or NO cards. Ask them to justify their answer.

Multiples, factors and primes 2C

Section overview

Textbook 2C units	Review units	Prior learning	Benchmarks/Es&Os
Unit1 Using knowledge of multiples and factors to work out divisibility rules	**2A** Units 4.1, 4.2, 4.8 and 5.1 **2B** Units 4.1, 4.2, 5.1 and 5.2 **2C** Units 4.1 and 4.2	• I can recall multiplication and division facts up to 10. • I know that factors are numbers that divide exactly into another number. • I know that multiples are numbers that can be divided by another number with no remainder.	**MTH 2-05a** • Identifies multiples and factors of whole numbers and applies knowledge and understanding of these when solving relevant problems in number, money and measurement.
Unit 2 Using knowledge of multiples and factors to solve problems	**2A** Units 4.1, 4.2, 4.8 and 5.1 **2B** Units 4.1, 4.2, 5.1 and 5.2 **2C** Units 4.1 and 4.2	• I can recall multiplication and division facts up to 10. • I know that factors are numbers that divide exactly into another number. • I know that multiples are numbers that can be divided by another number with no remainder.	

Conceptual context

Knowledge of multiples and factors can help us work out divisibility rules to quickly determine if a number can be divided by another number. For example, if we know that all even numbers are multiples of 2, then all even numbers are divisible by 2.

Pupils need a strong knowledge of multiplication and division facts up to 10, which they can apply to test a wide range of examples to determine simple rules of divisibility. They will also need wide experience of exploring multiples and factors using concrete materials such as arrays and grids, which they can apply to develop divisibility rules. Divisibility rules can also be used to find factors of a number.

Knowledge of multiples and factors can also be used to help solve multiplication and division problems, and to estimate if an answer is reasonable or not. For example, if we know that all multiples of 5 end in either 5 or 0, then we know that $92 \div 5 = 18$ is not a reasonable answer because 92 is not a multiple of 5 so cannot be exactly divisible by 18 and therefore must have a remainder.

Prime numbers are not introduced until Third Level; however, pupils may begin to explore these through their learning of multiples and factors and should understand that a prime number is any number greater than 1 that has exactly two factors.

Resources

- Counters or cubes
- Hundred squares
- Numeral rolls or number lines
- Arrays to represent multiplication problems
- Squared paper or grids to represent multiplication problems

Key vocabulary

- **Commutative property**
- **Factors**
- **Multiples**

Misconceptions

Potential misconception	Addressing the misconception
Units 5.1 and 5.2 The pupil may confuse multiples with factors.	• Use counters to make arrays to show all the factors for different numbers. Encourage pupils to explore all the different possible factors using the counters to make different arrays. Reinforce that factors are the numbers that divide exactly into another number. • Link multiples to skip counting. Use hundred squares or numeral rolls to identify multiples to reinforce understanding that multiples can be made by multiplying a number by any other whole number. • Use vocabulary of multiples and factors when reading and writing multiplication facts.

Potential misconception	Addressing the misconception
Units 5.1 and 5.2 The pupil may think factors or multiples are only numbers in times tables.	• Link multiples to skip counting. Use hundred squares or numeral rolls to identify multiples to reinforce understanding that multiples can be made by multiplying a number by any other whole number. • Make arrays for a range of different numbers and discuss all the possible factors to reinforce understanding that factors are any numbers that divide exactly into another, not just in the range of times tables that may have been learned.
The pupil may look only at the last digit for all divisibility rules; for example, thinking if a number is even it must be divisible by 4.	• Take time to explore and model different divisibility rules for 2, 3, 4, 5, 9 and 10. • Encourage pupils to test with a wide range of examples before determining divisibility rules.

Teaching ideas

Unit 5.1 Using knowledge of multiples and factors to work out divisibility rules

- Ask questions such as: **What can you tell me about multiples of 2? What have you learned about multiples of 10? Is 42 a multiple of 5? How do you know?** Check understanding of multiples.
- Explain we can use the knowledge of multiples we have learned to quickly check if a number is exactly divisible by another. Explain that we call these 'divisibility rules'. For example, if we know that all multiples of 2 are even numbers, we therefore know that a number would have to be even to be divisible by 2. Test with a range of examples: **Is 15 divisible by 2? What about 32? 104?**
- Ask pupils: **What would be a divisibility rule for 10?** Discuss answers and link to knowledge of multiples. **Is 55 divisible by 5? What about 72? How do you know?** Discuss a divisibility rule for 5.
- Explore multiples of 3. Model finding the reduced number by adding all the digits together. Explain that if the reduced number is a multiple of 3 we know that the number is divisible by 3. Test with a wide range of examples.
- Explore and model divisibility rules for 4 and test with a wide range of examples.

Unit 5.2 Using knowledge of multiples and factors to solve problems

- Ask: **What are all the factors of 20?** Check understanding of factors and how to identify them. Use arrays to represent all the factors found and investigate if we have identified them all.
- Explain that we can use our knowledge of factors and multiples to help us solve problems and estimate if answers are reasonable. Give problems such as: **If we have 24 pupils on a class trip, what are all the different ways we could group them?** Explore answers and link to knowledge of factors.
- Give a multiplication sentence such as $17 \times 5 = 81$. **Is this a reasonable answer? How do you know?** Link to their knowledge of multiples. Repeat with a wide range of examples and ask the pupils to come up with their own examples of reasonable and unreasonable answers.

6 Fractions, decimal fractions and percentages

Fractions, decimal fractions and percentages 2A

Section overview

Textbook 2A units	Review units	Prior learning	Benchmarks/Es&Os
Unit 6.1 Comparing fractions	**1B** Units 5.1, 5.2, 5.4 and 5.5 **1C** Units 5.3 and 5.6	• I can make and identify equal parts. • I can count fractions on a number line. • I can compare fractions (of the same whole). • I can order fractions by size.	**MNU 2-07a** **MNU 2-07b** **MTH 2-07c** • Uses knowledge of equivalent forms of common fractions, decimal fractions and percentages (for example, $\frac{3}{4} = 0{\cdot}75 = 75\%$) to solve problems. • Calculates simple fractions of quantity and uses this knowledge to solve problems; for example, find $\frac{3}{5}$ of 60. • Creates equivalent fractions and uses this knowledge to put a set of the most commonly used fractions in order. • Expresses fractions in their simplest form.
Unit 6.2 Identifying equivalent fractions pictorially	**1B** Unit 5.6 **1C** Unit 5.5 **2A** Unit 6.1	• I can find simple equivalent fractions. • I can create equivalent fractions by splitting into equal parts. • I can compare and order fractions.	
Unit 6.3 Identifying and creating equivalent fractions	**1B** Unit 5.6 **1C** Unit 5.5 **2A** Units 6.1 and 6.2	• I can find simple equivalent fractions. • I can create equivalent fractions by splitting into equal parts. • I can compare and order fractions. • I can identify equivalent fractions pictorially.	
Unit 6.4 Simplifying fractions	**1B** Unit 5.6 **1C** Unit 5.5 **2A** Unit 6.3	• I can find simple equivalent fractions. • I can create equivalent fractions by splitting into equal parts. • I can compare and order fractions. • I can identify and create equivalent fractions.	
Unit 6.5 Writing decimal equivalents to tenths	**1C** Unit 5.9 **2A** Units 2.5 and 6.3	• I can share one whole equally. • I can read and write decimal fractions. • I can identify and create equivalent fractions. • I understand that the decimal point separates whole objects from parts of objects.	
Unit 6.5 Writing decimal equivalents to tenths	**1C** Unit 5.9 **2A** Units 2.5 and 6.3	• I can share one whole equally. • I can read and write decimal fractions. • I can identify and create equivalent fractions. • I understand that the decimal point separates whole objects from parts of objects.	

Unit 6.5 Writing decimal equivalents to tenths	**1C** Unit 5.9 **2A** Units 2.5 and 6.3	• I can share one whole equally. • I can read and write decimal fractions. • I can identify and create equivalent fractions. • I understand that the decimal point separates whole objects from parts of objects.	
Unit 6.6 Comparing numbers with one decimal place	**1C** Unit 5.5 **2A** Units 6.1, 6.3 and 6.5	• I can compare and order fractions. • I can write decimal equivalents to tenths. • I can identify and create equivalent fractions. • I understand that the decimal point separates whole objects from parts of objects.	
Unit 6.7 Calculating a simple fraction of a value	**1B** Units 5.2 and 5.8 **1C** Units 5.7 and 5.9	• I can identify equal parts. • I can find part of a set. • I can find a fraction of an amount. • I can share one whole equally. • I can use bar models to solve problems.	

Conceptual context

Fractions are very often taught using whole number terminology, leading to pupils thinking about them as two whole numbers with one located over the other rather than a single, relational entity. In order to develop conceptual understanding of fractions, it is important to develop two key images of fractions through *iteration* and *partitioning*.

Iteration

To understand fractions through iteration, start with a unit fraction; for example, one quarter. How can we tell if something is one quarter? A part (or amount) is one quarter if four lots of it makes one whole:

- *This is one quarter of a cake because four lots of it makes a whole cake.*
- *These seven pupils are one quarter of the class because four lots of seven makes the whole class of twenty-eight pupils.*

This ensures that pupils attribute the correct size to fractional pieces because the part is being duplicated and underpins the understanding that a fraction is always a fraction of something.

Partitioning

To understand fractions through partitioning, start with a whole. How can we tell if something is one quarter? It is one quarter if it is the size of a part when you take the whole and split it into four equal parts:

- *This is one quarter of a cake because I have cut the whole cake into four equal slices.*
- *These seven pupils are one quarter of the class because the whole class of twenty-eight pupils have been split into four equal groups.*

These concepts are mutually compatible and should be developed in conjunction with each other. We can create quarters by partitioning a whole into four equal parts and we can check that each part is one quarter by iterating it four times to make the whole.

Many of the misconceptions pupils have about fractions are due to the fact that they are pushed on to the abstract representation too soon. They often read $\frac{1}{3}$ as 'one over three' without fully understanding what this looks like visually. This chapter predominately looks at the concept of equivalence – finding fractions that are equal in size. It is vital that pupils are initially allowed to investigate this concept using concrete materials and visual representations, rather than being told 'whatever you do to the top, you do to the bottom'. Pupils must understand that when we find equivalent fractions, the size of the 'portion' remains exactly the same. The object or amount is being split into smaller (or larger) equal parts so we get more (or less) of those parts. This conceptual understanding can often be lost or confused when we simply rely on multiplication (or division) of the numerator and denominator in order to find equivalents. The same is the case when finding a fraction of a value; for example, 'What is three fifths of 20 apples?'. It is important for pupils to understand that the whole set is being split into five equal groups of four apples and that we are only interested in three of these groups rather than being told to 'multiply by the numerator and divide by the denominator'.

Resources

- A4 or A5 paper
- Printable fraction wall
- Modelling clay (to represent cookies, chocolate bars, etc.)
- Fraction cards
- Base 10 materials
- Counting stick
- Card

Key vocabulary

- **compare**
- **decimal fraction**
- **decimal place**
- **decimal point**
- **equivalent**
- **simplify**

Misconceptions

Potential misconception	Addressing the misconception
Units 6.5 and 6.6 Pupils think fractions and decimals are negative numbers.	• Represent fractions using concrete materials (e.g. fraction cards, modelling clay, etc.) and visuals (e.g. bar models).
Unit 6.1 Pupils think, for example, that $\frac{1}{3}$ is bigger than $\frac{1}{2}$ of an object or amount.	• Avoid use of the word 'over' when naming fractions. • Use fraction cards and bar models to represent fractions visually.
Units 6.1, 6.2, 6.3, 6.4 and 6.7 Pupils consider fraction names to be absolute labels rather than relational labels.	• Give answers that relate to the whole amount, i.e. *one quarter of the bar or two thirds of the pizza* rather than just *one quarter* or *two thirds*.
Unit 6.4 Pupils regard equivalence and simplification as two unrelated concepts.	• Visually represent each concept using bar models to ensure pupils understand inherent links.
Units 6.5 and 6.6 Pupils do not understand the value of each digit or what the zero represents in a decimal number with zero units, e.g. 0·4.	• Use materials to demonstrate the scale of each digit, e.g. base 10 materials and 'pupil made' tenths (see activity outlined in Teaching Ideas, Unit 6.5). • Make tenths with the pupils by cutting up one whole (use card, chocolate bars, etc.) • Make decimal numbers using materials/visual representations (see activities outlined in Teaching Ideas, Units 6.5 and 6.6).
Units 6.5 and 6.6 Pupils do not understand that the column to the right of the decimal place is ten times smaller (and just 'add/subtract zeros' or 'move digits to the left or right').	• Explain the purpose of the decimal point – to separate digits representing whole things from digits representing parts of things. Use base 10 materials to represent 'wholes' (to the left of the decimal point) and pupil made tenths to represent 'fractions' (to the right of the decimal point).

Teaching ideas

Unit 6.1 Comparing fractions

- Set out tables with different numbers of chairs at each (two chairs at one table, four at another, etc.).
- Put a cookie in the middle of each table and explain that the cookie is to be shared equally between the pupils sitting at that table.
- Ask the pupils questions like: **Which table would you prefer to sit at? At which table will you get the biggest/smallest piece of cookie? What fraction of a cookie would you get at each table?** etc.

- Repeat this activity with different numbers of cookies at each table, e.g. three cookies between four pupils, two between five, five between six, etc. **How might we cut the cookies so that each pupil at the table gets an equal share?**
- Repeat this activity with different sized cookies, chocolate bars, etc. Discuss how difficult it is to compare the shares when the objects being shared are different sizes.
- Provide pupils with equally sized pieces of paper.
- Challenge them to make one half, one quarter, one eighth, one third and one sixth by folding and cutting the paper.
- Can pupils make equal parts by folding and then cut out one of those equal parts each time?
- Ask pupils questions like: **Which fractions are easiest to make? Can some fractions help us make others?**
- Demonstrate that we can make eighths by halving, halving the halves, then halving the quarters. We can make sixths by halving the thirds, etc.

Unit 6.2 Identifying equivalent fractions pictorially

- Give pupils different sets of fraction cards (pupil 1: one half; pupil 2: three quarters; pupil 3: three sixths; pupil 4: six eighths; etc.).
- Challenge pupils to find their matching partners. Ask the pupils questions such as: **What fraction do you have? Who has the same size as you?** etc.
- When attempting the 'Let's practise' questions (page 86), pupils may find it difficult to name the fractions. For example, they may say that the name of the fraction in 1 a) is one half. While technically this is correct, we want them to understand that the name of each slice is determined by the number of equal parts that make up the whole. Six equal parts means they are each called sixths. Amman is getting three of these slices so we can say that he is getting three sixths of the pizza.
- Provide pupils with modelling clay or paper to represent a chocolate bar and instruct them to cut them into eight equal parts.
- **What fraction is each part?** (*one eighth*)
- **Remove three parts. What fraction of the chocolate bar do you have left?** (*five eighths*)
- Repeat using different numbers of equal parts, e.g. fifths, twelfths, etc.
- When attempting the 'Challenge' activity, print off the fraction wall and encourage pupils to cut it into strips to begin with.
- Model making an 'equal to' sign by cutting out equivalent fractions and positioning them one above the other, e.g. one half = two quarters:

one half

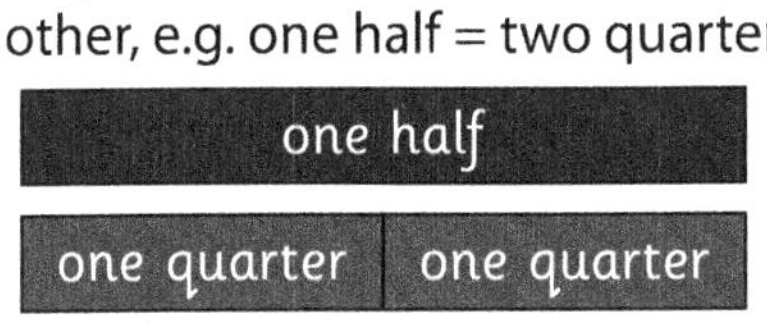

two quarters

- Challenge the pupils to make a display of 'equal to' signs by cutting out different sets of equivalent fractions, for example:

Equivalent fractions

Unit 6.3 Identifying and creating equivalent fractions

- Provide pupils with A4 or A5 paper and instruct them to make four sets of halves by folding each piece into two equal parts. Colour one of each of the halves and write 'one half' on one them.
- Ask the pupils to fold one piece of paper so that the halves are split into two equal parts (thus making quarters):

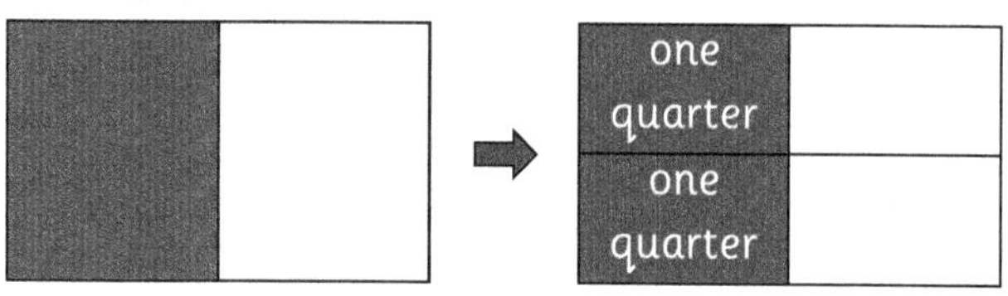

- Ask the pupils: **What have we made now? How many quarters are the same as one half?**
- Repeat with the other halves but folding them into three and four equal parts:

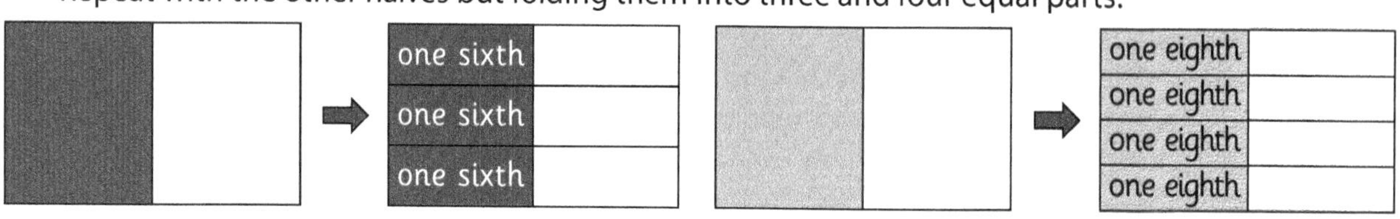

- Encourage pupils to name each part by writing, for example, 'one quarter' on each part.

Unit 6.4 Simplifying fractions

- Organise pupils into pairs.
- Give pupils blocks of modelling clay and tell them they have to get eight sixteenths each. **How can you ensure that each part is equal?** (Half, half, half and half again.)
- Ask questions like: **Is there a simpler way to do that so that we end up with exactly the same amount of clay?** (Rather than cutting the whole amount into 16 parts and getting 8 each, they could have cut it into 2 parts and get 1 part each.)
- Repeat activity with one pupil getting three twelfths and the other nine twelfths.

- It is important that pupils understand that the amount of clay they get is unchanged. They are having to cut fewer parts therefore they are making the task simpler.
- When attempting the 'Challenge' activity, print off the fraction wall and encourage pupils to cut it into strips to begin with. Model making an 'equal to' sign by cutting out equivalent fractions and positioning them one above the other, e.g. one half = two quarters. Challenge the pupils to make a display of 'equal to' signs by cutting out different sets of equivalent fractions.

Unit 6.5 Writing decimal equivalents to tenths

- Use base 10 materials to demonstrate visually the scale of 1, 10, 100, 1000.
- Ask pupils questions such as: **What is the relationship between 1 and 10; 10 and 100; 100 and 1000?** (and vice versa). It is important that pupils understand that each number is getting 10 times bigger (or 10 times smaller).
- Demonstrate how each of the numbers are constructed using the concrete resources; for example, to make 100 we need one of the hundred 'flats', no tens and no ones.
- Discuss what numbers the pupils know of that are smaller than one.
- Take examples of different fractions and write them on the board. Explore the idea that we can come up with unlimited examples of fractions.
- At this point it is important that pupils understand the function of the decimal point. It is there to separate whole numbers from equal parts or fractions.
- The decimal system is restrictive when it comes to representing fractions because each place gets 10 times smaller (or larger). The first place to the right of the decimal point must be ten times smaller than one so has to represent tenths.
- Make tenths with the pupils using thick card and cutting out square centimetres. Demonstrate that ten of these tenths is equivalent to one.
- Give pupils the opportunity to use base 10 materials and card tenths to make examples of numbers to one decimal place, e.g. 0·7, 1·3, 24·5, etc.

Unit 6.6 Comparing numbers with one decimal place

- Use base 10 materials (and tenths made from card) to make the following numbers:

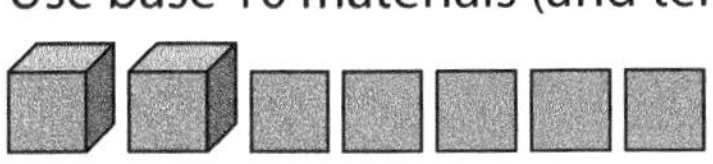

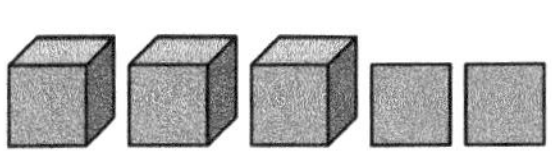

- Challenge pupils to write each number (using decimals) and decide which is bigger. Can they justify their choice?

- Write or display the following numbers: 2·7, 1·5, 4·2, 3·6. Challenge pupils to make the numbers using the materials and order them from smallest to largest.
- Use materials to make numbers that contain a zero, e.g. 20·3, 105·8, 14·0, etc. Challenge pupils to write each number.
- Repeat the activity but this time write numbers and challenge pupils to make them using the materials:
  - 10·4
  - 207·9
  - 26·0
  - 30·0

Unit 6.7 Calculating a simple fraction of a value

- Draw or display the following bar model:

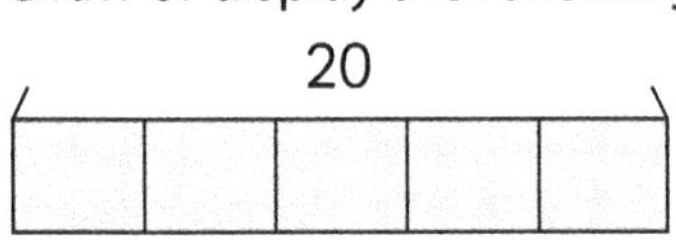

- Tell the pupils that you have 20 sweets split equally between five packets. Ask questions like: **What fraction of the sweets are there in each packet? How many sweets do I have in each packet? How many sweets do I have altogether in three packets?** and **What fraction of the total sweets is this?**
- Repeat using different quantities and bar models split into a different number of parts.
- Display bar models like the following:

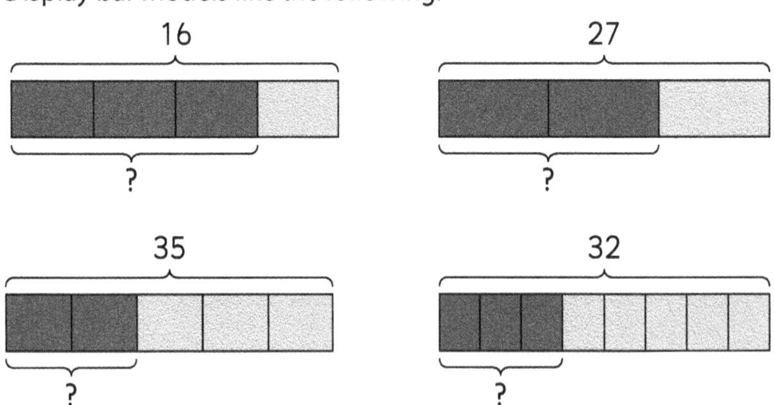

- Challenge pupils to work out the missing value for each bar. (Encourage the use of manipulatives to share equally between the total number of parts, then screen the parts that we are not interested in.)

Fractions, decimal fractions and percentages 2B

Section overview

Textbook 2B units	Review units	Prior learning	Benchmarks/ Es&Os
Unit 6.1 Identifying equivalent fractions	**1C** Unit 5.5 **2A** Units 6.1, 6.2 and 6.3	• I can create equivalent fractions by splitting into equal parts. • I can compare and order fractions. • I can identify equivalent fractions pictorially. • I can identify and create equivalent fractions.	**MNU 2-07a** **MNU 2-07b** **MTH 2-07c** • Uses knowledge of equivalent forms of common fractions, decimal fractions and percentages (for example, $\frac{3}{4} = 0{\cdot}75 = 75\%$) to solve problems. • Calculates simple fractions of quantity and uses this knowledge to solve problems; for example, find $\frac{3}{5}$ of 60. • Creates equivalent fractions and uses this knowledge to put a set of the most commonly used fractions in order. • Expresses fractions in their simplest form.
Unit 6.2 Calculate equivalent fractions	**2A** Units 4.3, 6.2 and 6.3 **2B** Unit 6.1	• I can identify equivalent fractions pictorially. • I can identify and create equivalent fractions. • I can solve problems using multiplication.	
Unit 6.3 Compare and order fractions	**2A** Units 6. and 6.3 **2B** Units 6.1 and 6.2	• I can compare and order fractions. • I can order fractions on a number line. • I can identify and create equivalent fractions. • I can calculate equivalent fractions.	
Unit 6.4 Decimal equivalents to hundredths	**1C** Unit 5.9 **2A** Units 2.5 and 6.5 **2B** Unit 6.1	• I can share one whole equally. • I can read and write decimal fractions. • I can write decimal equivalents to tenths. • I can identify and create equivalent fractions.	
Unit 6.5 Decimal equivalents to simple fractions	**2A** Units 2.5, 6.4 and 6.5 **2B** Unit 6.1 and 6.4	• I can read and write decimal fractions. • I can write decimal equivalents to tenths. • I can identify and create equivalent fractions. • I can simplify fractions. • I can find decimal equivalents to hundredths.	
Unit 6.6 Add and subtract fractions	**1B** Unit 5.3 **1C** Unit 5.4 **2B** Units 6.1 and 6.2	• I understand fractional notation. • I can add and subtract like fractions. • I can identify and create equivalent fractions. • I can calculate equivalent fractions.	
Unit 6.7 Calculating a fraction of a value	**1B** Unit 5.2 **1C** Units 5.7 and 5.9 **2A** Unit 6.7	• I can identify equal parts. • I can find part of a set. • I can find a fraction of an amount. • I can use bar models to solve problems. • I can calculate a simple fraction of a value.	

Unit 6.8 Comparing numbers with two decimal places	**2A** Units 6.1, 6.3 and 6.6 **2B** Unit 6.4	• I can compare and order fractions. • I can write decimal equivalents to tenths and hundredths. • I can identify and create equivalent fractions. • I can compare numbers with one decimal place.	
Unit 6.9 Percentages	**1B** Unit 5.3 **2B** Units 6.1 and 6.2	• I can name fractions and understand fractional notation. • I can calculate equivalent fractions. • I can identify and create equivalent fractions.	
Unit 6.10 Converting fractions to percentages	**1B** Unit 5.3 **2B** Units 6.1, 6.2 and 6.9	• I can name fractions and understand fractional notation. • I can calculate equivalent fractions. • I can identify and create equivalent fractions. • I understand and can use the term percentage (%).	
Unit 6.11 Percentage calculation	**2A** Unit 3.7 **2B** Units 6.7, 6.9 and 6.10	• I can solve part-part-whole problems using a bar model. • I can calculate a fraction of a value. • I understand and can use the term percentage (%). • I can convert fractions to percentages.	

Conceptual context

Understanding equivalence is the key to developing a deep understanding of many of the concepts across fractions, decimals and percentages. In order to understand decimals and percentages it is essential that pupils can find and calculate equivalents to tenths and hundredths. A deep understanding can be achieved if pupils are able to visualise fractions (using bar models) and use these visual representations to find equivalents and simplify. Often pupils' understanding of equivalence is confused when they are moved too quickly to the abstract representation and have to rely on processes that they don't fully understand. The 'goal' of progression in understanding fractions is not to move pupils away from visual representations and onto the abstract. These visual representations are essential to develop conceptual understanding of fractions at third and fourth levels and beyond. Decimals and percentages are simply alternative representations of fractions. It is vital that pupils' understanding of this is developed in order that they can make links between them. Bar modelling allows pupils to visualise these intrinsic links rather than, for example, relying on memorising that *three quarters = 0·75 = 75%*.

Previously, equivalent fractions have been found by splitting fractions into equal parts so, for example, one half split into two equal parts becomes two quarters, five equal parts becomes five tenths, etc. Not all equivalents can be found in this way. Often, we need to simplify a fraction first. For example, if we want to find the number of tenths equivalent to three sixths, we cannot simply split them into equal parts. Pupils may wrongly conclude that this can't be done if we rely on this methodology.

Many misconceptions related to equivalence derive from moving to the abstract process without the necessary conceptual understanding in place, i.e. 'whatever you do to the top, you do to the bottom'. This should be avoided as it can lead pupils to incorrectly conclude that the size of the fraction is increasing when we multiply both the numerator and denominator by the same number and decreasing when we divide. The key understanding we are aiming to develop in Unit 6.2 is that equivalent fractions are created by splitting the whole into a greater number of smaller parts; therefore, we need to get more parts in order that we get exactly the same share. We can do this by *multiplying* the total number of parts and taking the equivalent number of these parts.

In textbook 2A (Unit 6.1), fractions were compared visually using bar models; for example, we can see clearly that three fifths is larger than one quarter of a bar:

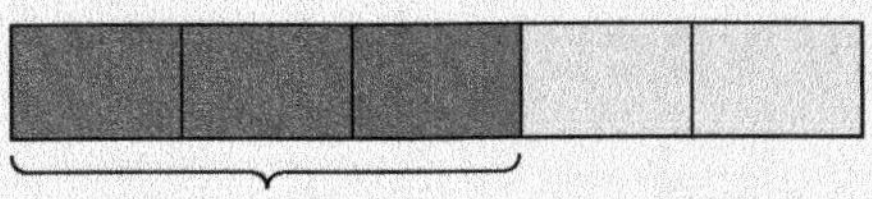

three fifths

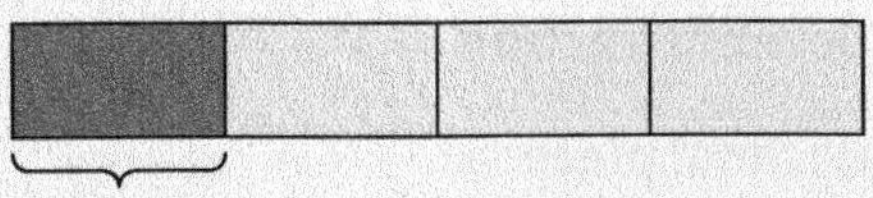

one quarter

This isn't always conclusive when comparing fractions that appear similar in size, e.g. comparing two thirds with three quarters. As such, it is important to develop strategies for comparing fractions conclusively. This can be achieved through equivalence. We can't say definitively that three quarters of a bar is larger than two thirds, but we can when we compare their equivalents, i.e. comparing nine twelfths to eight twelfths.

The purpose of Unit 6.4 is to develop pupils' understanding of decimal numbers to two decimal places. Pupils should now understand how to write tenths and hundredths in their decimal form and appreciate the scale of the first two decimal places.

The aim of Unit 6.5 is to develop this understanding so that we can convert any fraction into a decimal fraction, i.e. not limited to just tenths and hundredths. This will be done by applying their understanding of equivalence.

Adding and subtracting like fractions was introduced in Book 1C (Unit 5.4) using real objects and fraction cards. The aim of Unit 6.6 is to develop this understanding through the use of visual representations (bar models) and begin to think about adding and subtracting fractions using equivalence. The most common misconception when adding fractions is that pupils simply add both numerators and both denominators. For example, $\frac{1}{4}+\frac{2}{4}=\frac{3}{8}$. In order to avoid this, it is important to add and subtract fractions using visual representations. In book 1C, the abstract fractional notation was not used but rather the fractions were written in words. Most pupils will find it easier to add and subtract fractions when presented with a question like, *Finlay has three sixths of a cookie and Amman has two sixths of a cookie. How much do they have altogether?* rather than $\frac{3}{6}+\frac{2}{6}=?$.

Unit 6.9 introduces the term *percent* for the first time. It is important that pupils understand that any fraction, for example one half, can be thought of as one *out of* two equal parts. A percentage is an alternative way of writing a fraction *out of* one hundred so relates specifically to hundredths:

Forty-five hundredths (or $\frac{45}{100}$) = 45 out of 100 = 45 percent (or 45%) N.B. The percentage sign simply represents a fraction out of 100.

The aim of Unit 6.10 is to extend on the learning in Unit 6.9 by developing the understanding to convert fractions to percentages and percentages to fractions in their simplest form. It is a common misconception to think of fractions and percentages as two unrelated concepts. It is essential that pupils understand that a percentage is simply an alternative representation of a fraction out of one hundred.

Resources

- A4 or A5 paper
- Printable fraction wall
- Modelling clay (to represent cookies, chocolate bars, etc.)
- Fraction cards
- Base 10 materials
- Counting stick
- Card

Key vocabulary

- **compare**
- **decimal fraction**
- **decimal place**
- **decimal point**
- **denominator**
- **equivalent**
- **numerator**
- **percent**
- **simplify**

Misconceptions

Potential misconception	Addressing the misconception
Units 6.4, 6.5 and 6.8 Pupils regard fractions and decimals as two separate concepts and are unable to make links between the two.	• Represent decimals using concrete materials and visuals in order to see for example that one tenth is the same as 0·1. • Show fractions and decimals on the same number line.
Units 6.1, 6.2, 6.3, 6.5, 6.9 and 6.10 Pupils regard equivalence and simplification as processes and fail to understand that the size of the portion remains constant.	• Use paper, modelling clay, etc. to create equivalent fractions and represent these visually using bar models.
Units 6.1, 6.2, 6.3, 6.5, 6.9 and 6.10 Fractions multiply (and therefore get bigger) when calculating equivalent fractions.	• As above. • Avoid moving to the abstract process too soon, i.e. multiplying numerator and denominator by the same number. • Continue to represent equivalent fractions visually after calculating equivalent fractions (Unit 6.2) has been covered.

Potential misconception	Addressing the misconception
Units 6.4, 6.5 and 6.10 Pupils consider hundredths to be greater than tenths (because one hundred is greater than ten).	• Give pupils the opportunity to make tenths and hundredths and 'build' numbers using the materials. • Use bar models to demonstrate scale visually.
Units 6.4 and 6.5 Pupils think that numbers with more decimal places are bigger than those with less, e.g. 1·6 is less than 1·34.	• Give pupils the opportunity to make tenths and hundredths and 'build' numbers using the materials. • Use bar models to demonstrate scale visually.
Unit 6.6 Pupils add (or subtract fractions) by adding the numerators together and the denominators together, e.g. they think $\frac{1}{2}$ plus $\frac{1}{3}$ equals $\frac{2}{5}$.	• Use concrete materials, fraction cards, etc. when adding or subtracting fractions. • Develop understanding that fractions can only be added when they belong to the same set; therefore, finding equivalents is necessary when adding unlike fractions. • Use bar models to represent addition / subtraction of fractions visually.
Units 6.9 and 6.10 Pupils regard fractions and percentages as two separate concepts and are unable to make links between the two.	• Percentages can only be introduced once pupils understand equivalence to hundredths. • Teach percentage as a fraction out of 100, i.e. it is simply an alternative representation of a fraction.

Teaching ideas

Unit 6.1 Identifying equivalent fractions

- Arrange pupils into two groups of five. Cut a block of clay into sixths and tell them that each group is going to get three sixths to share equally between them. **What fraction of the block of clay will you each get?** Allow pupils to investigate and conclude that they can solve this in different ways.
- Solution 1 – they may cut each sixth into five equal parts in order to produce fifteen thirtieths and take three parts each. Each of their three parts can then be combined to produce one tenth each. Each group will therefore have five tenths.
- Solution 2 – a simpler solution to this problem may be for each group to combine their three sixths to make one half of the original block of clay. Each half can then be split equally to make five parts so each group has five tenths of the original block.
- Model this methodology using bar models (or alternatively challenge the pupils to demonstrate their solution visually):

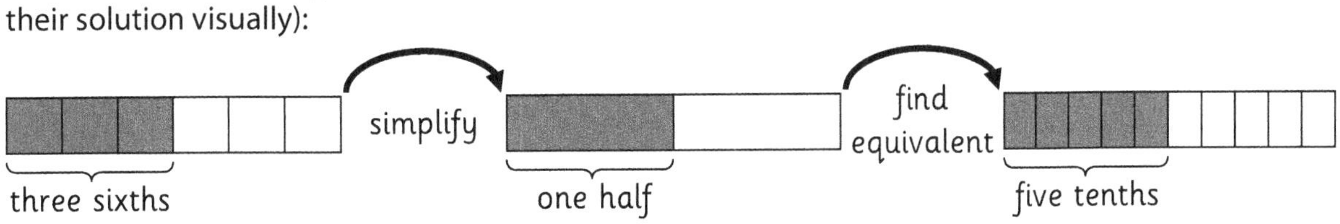

Unit 6.2 Calculating equivalent fractions

- Provide pupils with paper and ask them to fold it in half. Unfold and colour one of the halves.
- Instruct them to fold their halves in half (then unfold to see how many equal parts they have created).
- Ask questions like: **What has happened to the total number of parts?** (*they've doubled*), **What has happened to the number of parts that you have coloured?** (*they've also doubled*).
- Repeat activity by folding the halves into three, four, five, etc. equal parts.
- Demonstrate that we can find these equivalents by multiplying the total number of parts (denominator) and the number of coloured parts (numerator).

Unit 6.3 Comparing and ordering fractions

- Offer pupils the choice of getting two thirds or seven twelfths of a bar of chocolate and ask them to justify their choice (some pupils may opt for the seven twelfths because they are getting more parts).
- Draw or display bar models representing each fraction side by side (in order to make it difficult to compare visually):

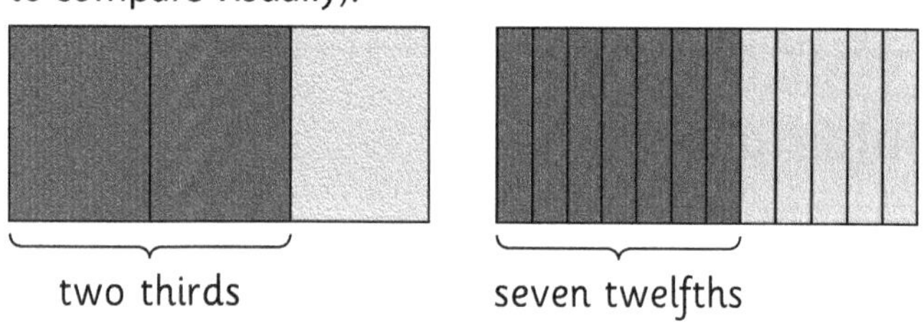

- Challenge pupils to prove that they have chosen correctly.
- Ask questions like: **Are we able to able to compare these directly? What can be done so that we say for definite which is larger?**
- Demonstrate that we can prove definitively that two thirds is larger by splitting each third into four equal parts, thus creating the equivalent eight twelfths:

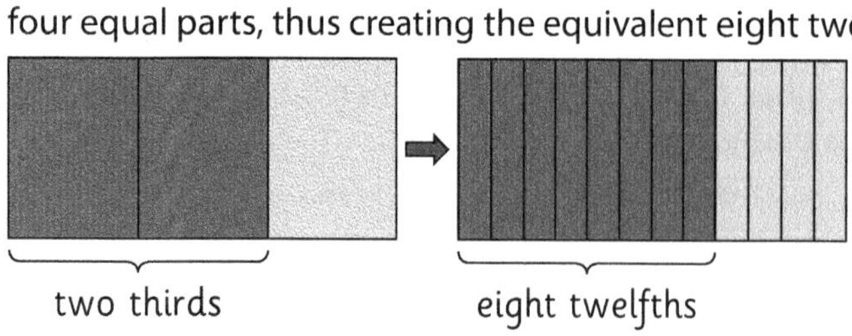

- Repeat the above activity using fractions where one denominator **is not** a factor of the other, e.g. compare two thirds with three fifths:

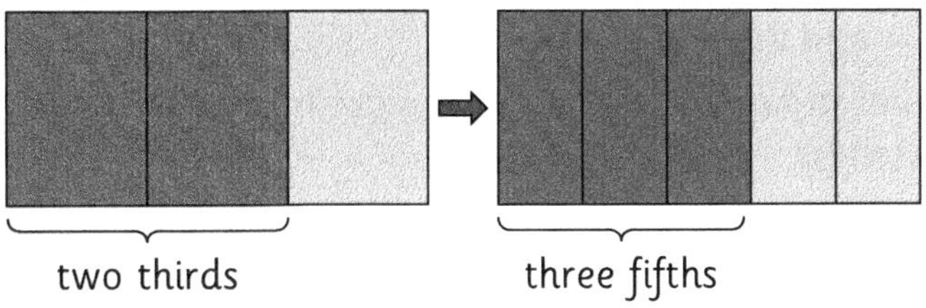

- Ask questions like: **Are we able to able to compare these directly? What can be done so that we say for definite which is larger? Can we convert thirds into fifths or fifths into thirds?**
- Demonstrate that we can only prove definitively which is larger if we find a common equivalent for thirds and fifths. Ask questions like: **What can thirds and fifths both be converted into?**

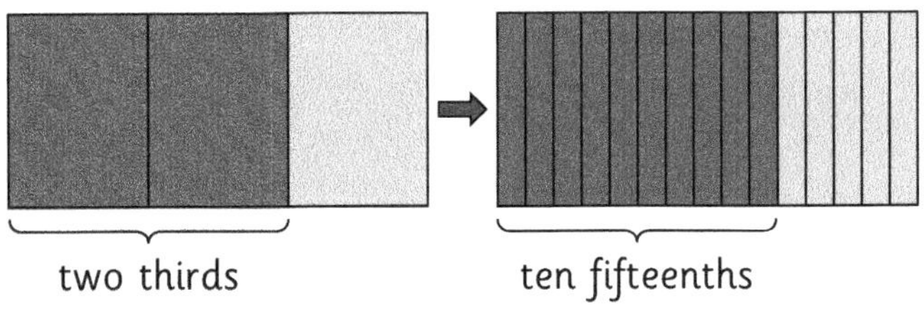

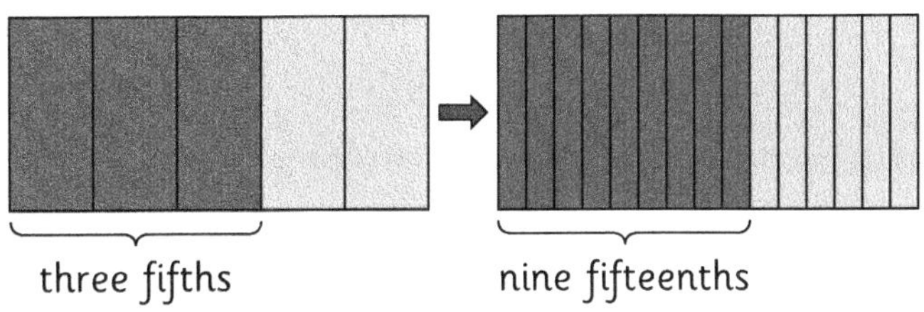

- Thirds and fifths can both be converted into fifteenths. We can now say definitively that two thirds of a bar is larger than three fifths because ten fifteenths is larger than nine fifteenths.

Unit 6.4 Decimal equivalents to tenths and hundredths

- In textbook 2A (Unit 6.5), pupils' conceptual understanding of decimals was developed by making tenths in order to give them a sense of scale.
- The activity outlined in the teaching guide should now be continued for hundredths.
- Make hundredths with the pupils by making tenths (draw square centimetres on thick card) and cutting them into ten equal parts (it will be easier if they make a strip of tenths and cut the last one into ten equal parts):

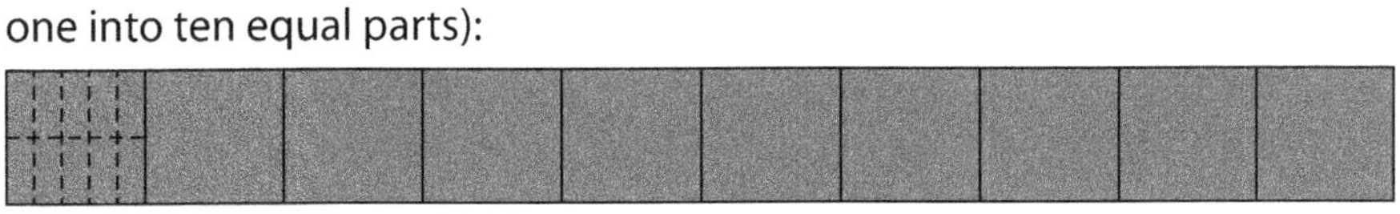

- Ask questions like: **What are ten hundredths equivalent to?** (one tenth), **What are one hundred hundredths equivalent to?** (one whole).
- Give pupils the opportunity to use resources to make examples of numbers to two decimal places, e.g. 0·53, 1·26, 24·71, etc.

Unit 6.5 Decimal equivalents to simple fractions

- Ask pupils: **True or false: one half = 0·1 and three fifths = 0·3?**
- Discuss what these decimal fractions mean and use resources to make both numbers, i.e. 0·1 means one tenth and 0·3 means three tenths.
- Discuss the fact that there is no *place* for halves or fifths in the decimal system, i.e. each place increase (or decreases by a scale of ten):

Th	H	T	U		tth	hth	
1	0	0	0	•	0		one thousand
	1	0	0	•	0		one hundred
		1	0	•	0		ten
			1	•	0		one
			0	•	1		one tenth
			0	•	0	1	one hundredth

- Challenge pupils to find the equivalent to both fractions using tenths (or hundredths).
- Explore the concept that one half is equivalent to five tenths (and two fifths is equivalent to four tenths). Now that we have found their equivalents in tenths, we can write both as decimal fractions: i.e.
  - o one half $\left(\frac{1}{2}\right)$ = five tenths $\left(\frac{5}{10}\right)$ = 0·5
  - o two fifths $\left(\frac{2}{5}\right)$ = four tenths $\left(\frac{4}{10}\right)$ = 0·4
- **How could we write one quarter as a decimal fraction? Are we able to convert quarters into tenths?** *We cannot convert quarters into tenths so we will need to try converting into hundredths. One quarter is equivalent to twenty-five hundredths (which is the same as two tenths and five hundredths). So, we can write 0·25 as the decimal equivalent to one quarter.*
- **How could we write four eighths as a decimal fraction? Are we able to convert eighths into tenths or hundredths?** *(No)* **Is there anything else we can do?** *We cannot convert eighths into tenths or hundredths but we can simplify this fraction. Four eighths can be simplified to make one half. One half is equivalent to five tenths. So, we can write 0·5 as the decimal equivalent to four eighths.*
- **Can we write one third as a decimal fraction? Are we able to convert thirds into tenths or hundredths?** *(No)* **Is there anything else we can do?** *We cannot simplify this fraction. If we shared out ten tenths between thirds we can put three tenths in each and have one left over. We could split this into ten hundredths. If we shared out the hundredths between each third we can put three tenths in each and have one left over. We would need to split this into ten thousandths, share out three to each and have one left over. This process would go on and on forever because there would always be one part left over.*

Unit 6.6 Adding and subtracting fractions

- This unit aims to extend the learning developed in Book 1C (Unit 5.4) by adding and subtracting fractions using bar models to visualise.
- Provide pupils with fraction cards (scaled to represent the size of each fraction).

N.B. Multiple copies of each will need to be provided.

- Contextualise the activity by telling the pupils that these represent, for example, bars of chocolate, half bars, etc.
- Organise pupils into pairs and instruct one pupil to take three half bars of chocolate and the other to take four half bars of chocolate.
- **How much chocolate do you have altogether?** *(seven half bars of chocolate)*
- Demonstrate how we can represent this using a bar model.
- Repeat using different examples, for example:
  - Three twelfths of a bar and five twelfths of a bar.
  - One third of a bar and two thirds of a bar.
  - Five eighths of a bar and six eighths of a bar.
- Ask questions like: **Amman has four sixths of a pizza. He eats three sixths. How much pizza does he have left?** *(one sixth of a pizza)*
- Demonstrate how we can represent this using a bar model.
- Repeat using different examples, for example:
  - Starts with six eighths and eats two eighths.
  - Starts with seven halves and eats three halves.
  - Starts with one whole and eats three quarters.
- When attempting question 2 in this unit it may be helpful for some pupils to replace the abstract fractional notation with words, e.g. *two fifths + one fifth = ?*
- The Challenge question requires that pupils apply their understanding of equivalence in order to add and subtract fractions with different denominators. Again, it may be advantageous to support pupils by replacing the fractional notation with words.

Unit 6.7 Calculating a fraction of a value

- This unit aims to extend the understanding of calculating a fraction of a value developed in Book 2A (Unit 6.7) using larger numbers and more complex word problems whereby the answer is not simply found by calculating the fraction of the amount found in the question. For example, the answer to question 3 a) will not be found by simply working out one quarter of 1120 km.
- Ask the pupils the following: **Two fifths of the pupils in a school are boys. How many girls are in the school if there are 440 pupils altogether?**

- In order to solve this correctly, pupils will have to apply their mathematical reasoning skills and analyse the problem using a bar model.
- When analysing a problem using a bar model it is helpful to focus on the context rather than the numbers first.
- Start by drawing the whole bar and ask: **What does the whole bar represent?** i.e. *the total number of pupils in the school:*

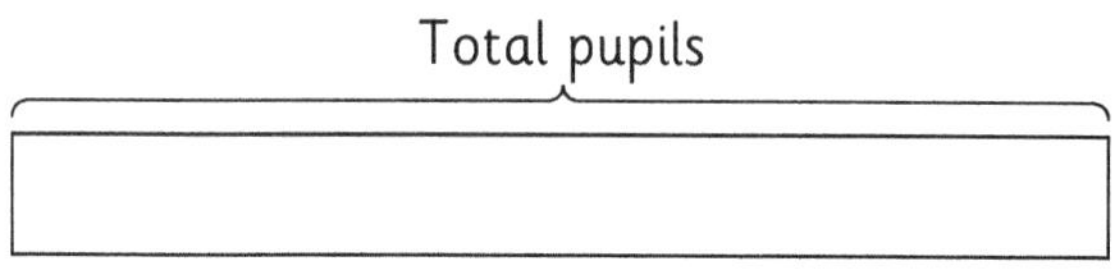

- Next, ask the question: **How many parts should we split the bar into?** Encourage pupils to justify, i.e. *five parts because we are dealing with fifths:*

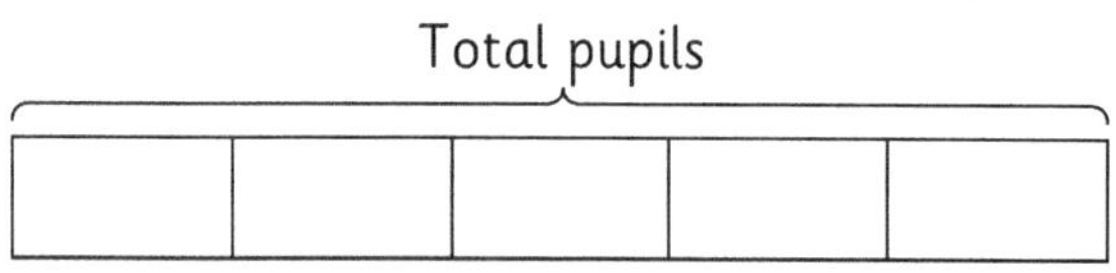

- Then, ask the question: **What do the parts represent?** i.e. *the number of boys and girls:*

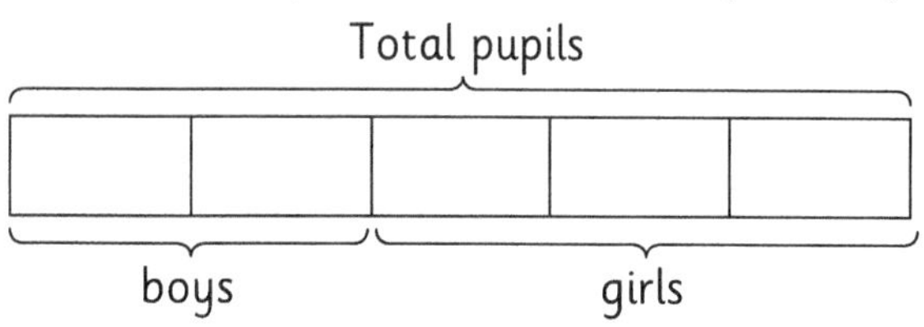

- Now, ask the question: **What values do we know and what are we trying to find out?** i.e. *We know the total number of pupils is 440 and we are trying to find out the number of girls:*

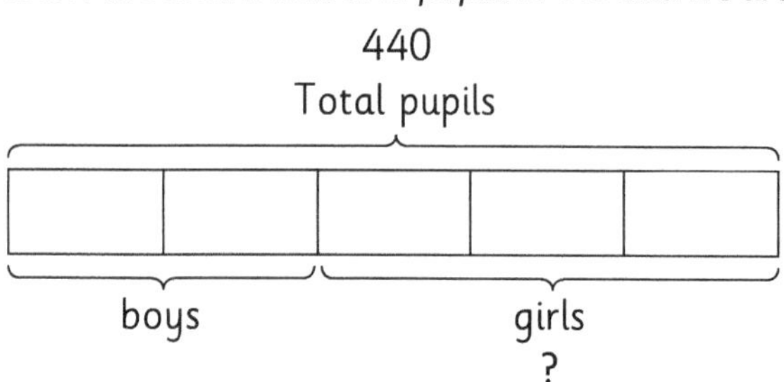

- Once the question has been analysed effectively, pupils can decide on a suitable strategy to solve the problem.
- Ask pupils further examples:
  - **Nuria has saved £320. She spends three eighths of her savings on a new bike. How much money does she have left?**
  - **A theatre has 780 seats. Five sixths of the tickets for a show have been sold. How many seats will be empty at the show?**
  - **Isla buys a new jacket in a sale, saving one quarter off the full price. What was the original cost of the jacket if she paid £45 in the sale?**

Unit 6.8 Comparing numbers with two decimal places

- Write or display 2·7 and 2·34. Ask pupils to discuss which number is bigger and justify their decision. Challenge them to make each number using base 10 materials (and tenths and hundredths made from card).
- Write or display the following numbers. 3·6, 3·16, 3·46, 3·06. Challenge pupils to make the numbers using the materials and order them from smallest to largest.
- Use materials to make numbers that contain a zero, e.g. 20·34, 25·08, 17·80, etc. Challenge pupils to write each number.
- Repeat the activity but this time write the numbers and challenge pupils to make them using the materials:
  - 5·27
  - 30·70
  - 23·08
  - 65·56

Unit 6.9 Percentage

- Write or display the word 'percent' and ask pupils to discuss: **Have you heard this word before?** (phone / tablet battery, sale discounts, etc.) **What does percent mean?** (per means *out of* and cent means *one hundred*).
- Write the sign for percentage: %
- Ask pupils: **Does this look like anything else we know?**
- Discuss the fact that it is written in the same way that we write a fraction (e.g. $\frac{1}{2}$) and is made out of the digits that make 100 (it is a fraction out of 100).
- **Sweets come in bags of 100. Nuria has eaten 5 sweets; Isla has eaten 14 sweets; Amman has eaten 23 sweets; Finlay has eaten 36 sweets. What fraction of their sweets have each of the pupils eaten?**
  - Nuria: five hundredths or $\frac{5}{100}$
  - Isla: fourteen hundredths or $\frac{14}{100}$
  - Amman: twenty-three hundredths or $\frac{23}{100}$
  - Finlay: thirty-six hundredths or $\frac{36}{100}$
- Discuss, for example, that **five hundredths means 5 *out of* 100 sweets. *Percent* means *out of* so we can write: $\frac{5}{100}$ = 5%.**
- **Write each of the fractions of sweets eaten by Isla, Amman and Finlay as a percentage.**
- Display various examples of fractions out of one hundred and challenge pupils to identify:

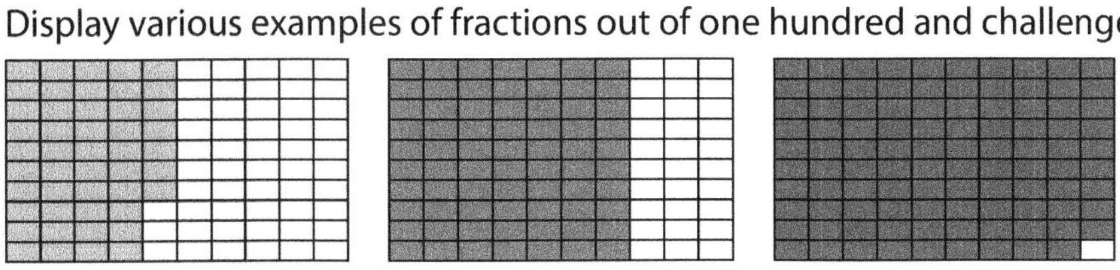

- **How would each fraction be written as a percentage?**
- The challenge question focuses on converting any fraction to a percentage (explored in greater detail in Unit 6.10) by applying their understanding of equivalence.
- Ask pupils: **How many fractions can you find that are equivalent to one half?**
- Pupils may find, for example, $\frac{2}{4}, \frac{3}{6}, \frac{4}{8}, \frac{5}{10}, \frac{10}{20}, \frac{25}{50}, \frac{50}{100}$, etc.
- Discuss the meaning of these, e.g. $\frac{2}{4}$ means 2 *out of* 4, $\frac{3}{6}$ means 3 *out of* 6, etc.
- $\frac{50}{100}$ means 50 *out of* 100. *Percent* means *out of* one hundred so we can say that one half *is equivalent to* 50%.

Unit 6.10 Converting fractions to percentages

- Choose a fraction, e.g. one quarter, and challenge pupils to find as many equivalent fractions as they can. Ask pupils: **Can you find an equivalent out of one hundred?**
- Now that we have found that one quarter is equivalent to twenty-five hundredths (or 25 *out of* 100) we can simply write it as 25%.
- Repeat using different examples of fractions, for example:
  - One fifth
  - One tenth
  - Three tenths
  - Eight twentieths
- Question 2 requires pupils to convert a percentage into a fraction in its simplest form.
- Write or display a variety of percentages. Challenge pupils to convert them to fractions and express them in their simplest form, e.g. 20%, 60%, 15%, 64%, etc.

Unit 6.11 Percentage calculation

- Calculating a percentage of a value is no different to calculating a fraction of a value (see Unit 6.7). When calculating a percentage, the fraction in question is expressed *out of* one hundred.
- Draw or display the following bar model:

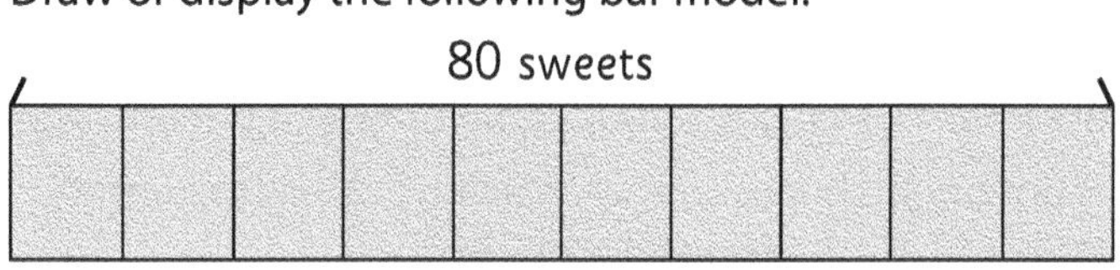

- Tell the pupils that you have 80 sweets split equally between ten packets.

- Ask questions like: **What fraction of the sweets are there in each packet? What is this expressed as a percentage? How many sweets do I have in each packet? How many sweets do I have altogether in four packets? What fraction of the total sweets is this?** and **What percentage of the total sweets is this?'**

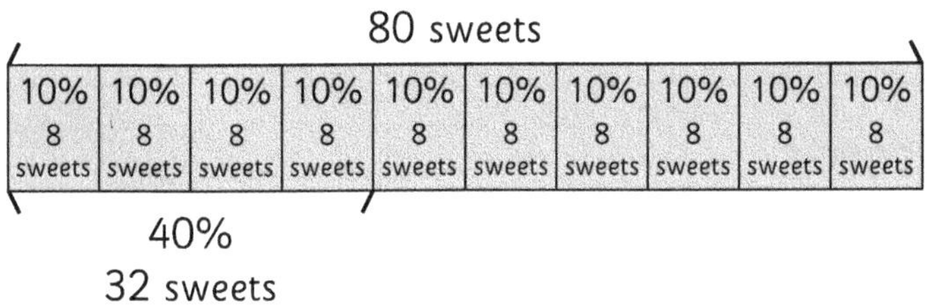

- Repeat using different quantities and finding different percentages (stick to multiples of 10% for now)
- Display bar models like the following:

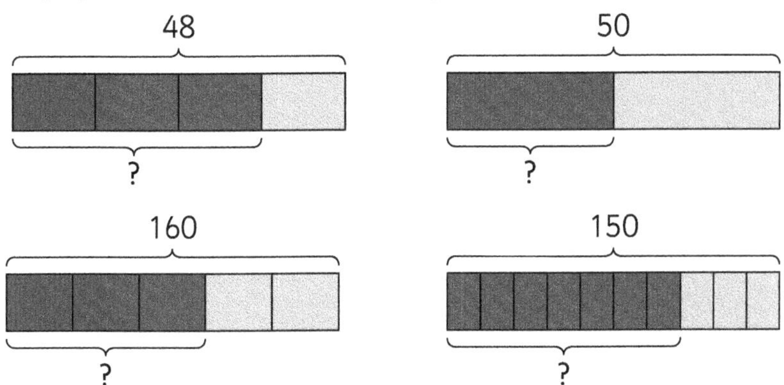

- Challenge pupils to work out the missing value for each bar, asking questions like: **What percentage of the whole amount is this?**

Fractions, decimal fractions and percentages 2C

Section overview

Textbook 2C units	Review units	Prior learning	Benchmarks/ Es&Os
Unit 6.1 Converting fractions	**1B** Units 5.3 and 5.7 **1C** Unit 5.8 **2A** Unit 6.4 **2B** Unit 6.1	• I understand and can use fractional notation. • I can count in fractions. • I can share one whole. • I can identify and create equivalent fractions. • I can simplify fractions. • I can use bar models to solve problems.	**MNU 2-07a** **MNU 2-07b** **MTH 2-07c** • Uses knowledge of equivalent forms of common fractions, decimal fractions and percentages; for example, $\frac{3}{4} = 0{\cdot}75 = 75\%$, to solve problems. • Calculates simple percentages of a quantity, and uses this knowledge to solve problems in everyday contexts; for example, calculates the sale price of an item with a discount of 15%. • Calculates simple fractions of a quantity and uses this knowledge to solve problems; for example, find $\frac{3}{5}$ of 60. • Creates equivalent fractions and uses this knowledge to put a set of most commonly used fractions in order.
Unit 6.2 Comparing and ordering fractions	**2A** Unit 6.4 **2B** Units 6.1, 6.2 and 6.3	• I can simplify fractions. • I can identify and create equivalent fractions. • I can calculate equivalent fractions. • I can compare and order fractions.	
Unit 6.3 Simplifying fractions (using common factors)	**2A** Units 5.1 and 6.4 **2B** Units 5.1 and 6.2	• I can simplify fractions. • I can calculate equivalent fractions. • I understand and can find factors.	
Unit 6.4 Adding and subtracting fractions	**2B** Units 6.1, 6.2 and 6.6 **2C** Unit 6.1	• I can identify and create equivalent fractions. • I can calculate equivalent fractions. • I can add and subtract like fractions. • I can convert fractions.	
Unit 6.5 Converting decimal fractions to fractions	**2B** Units 6.2, 6.4 and 6.5 **2C** Unit 6.3	• I can calculate equivalent fractions. • I can write decimal equivalents to tenths and hundredths. • I can find decimal equivalents to simple fractions. • I can simplify a fraction.	
Unit 6.6 Calculating a fraction of a fraction	**2B** Units 6.1 and 6.7	• I can identify and create equivalent fractions. • I can calculate a fraction of a value. • I can analyse and solve problems using a bar model.	

Unit 6.7 Dividing a fraction by a whole number	**2B** Units 6.1 and 6.7 **2C** Unit 6.6	• I can identify and create equivalent fractions. • I can calculate a fraction of a value. • I can calculate a fraction of a fraction. • I can analyse and solve problems using a bar model.	• Expresses fractions in their simplest form.
Unit 6.8 Dividing a whole number by a fraction	**2B** Unit 4.5 **2C** Unit 6.7	• I can solve division problems by grouping. • I can solve division problems using a bar model. • I can divide a fraction by a whole number.	
Unit 6.9 Equivalents: fractions, decimals and percentages	**2A** Unit 6.4 **2B** Units 6.5 and 6.10 **2C** Unit 6.5	• I can simplify a fraction. • I can find decimal equivalents to simple fractions. • I can convert fractions to percentages. • I can convert decimal fractions to fractions.	
Unit 6.10 Solving fraction, decimal and percentage problems	**2B** Units 6.7 and 6.11 **2C** Unit 6.9	• I can calculate a fraction of a value. • I can calculate a percentage of a value. • I can convert fractions, decimals and percentages. • I can use bar models to solve problems.	

Conceptual context

Following on from Book 2B, equivalence continues to be the underlying theme within this chapter. A sound understanding of equivalence will support pupils with all of the key concepts covered in Book 2C. As stated previously, most of the misconceptions regarding fractions are as a result of moving to the abstract too quickly. While this chapter may introduce some of these abstract processes, e.g. *simplifying fractions by dividing the numerator and denominator by the same value*, it is important that we do not leave behind concrete and pictorial representations. Concrete and pictorial representations of fractions are necessary to support conceptual understanding throughout the third and fourth level and beyond.

When converting fractions (Unit 6.1), avoid using an abstract process such as converting $2\frac{1}{2}$ into an improper fraction by 'multiply the 2 (whole number) by the 2 (denominator), then add the 1 (numerator) to get 5 and put this over the 2'. Pupils may be able to follow this process relatively easily and produce the correct answer of $\frac{5}{2}$, but this will not support conceptual understanding. When converting from a mixed number to an improper fraction, start by drawing a bar model to represent $2\frac{1}{2}$:

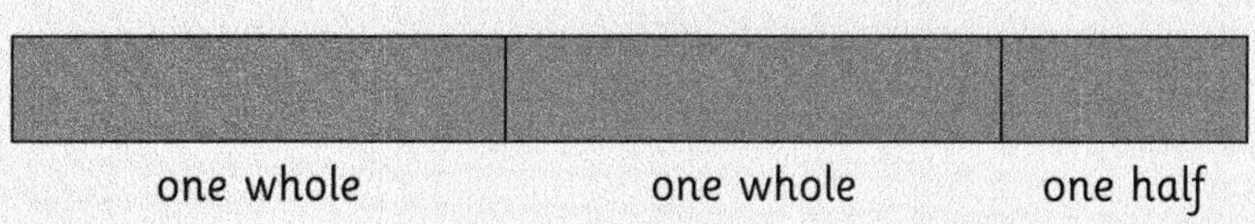

Then split the wholes into two equal parts to make halves (five halves altogether or $\frac{5}{2}$):

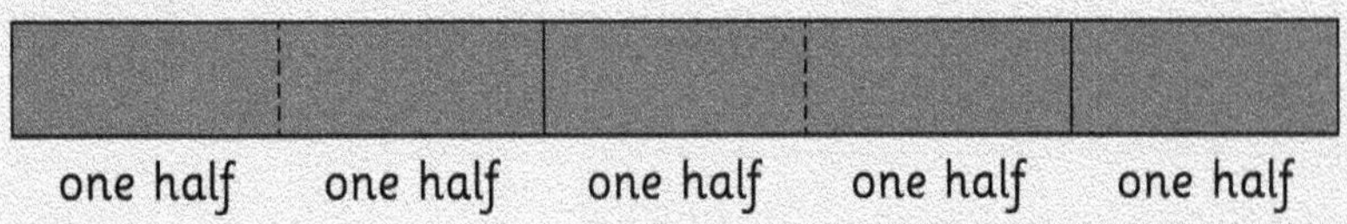

When converting from an improper fraction to a mixed number, start by drawing a bar model to represent five halves:

Then combine pairs of halves to make two wholes and one half (two and a half or $2\frac{1}{2}$):

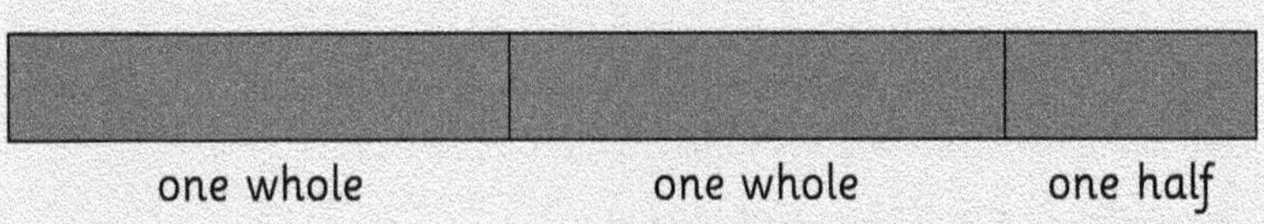

When simplifying a fraction (Unit 6.3) we are finding a simpler way to make a portion of exactly the same size. For example, rather than cutting a pizza into twelve equal parts and taking six of those parts, it is far simpler to cut the pizza into two equal parts and take one of the parts. The key understanding we are aiming to develop here is that simplified fractions are created by splitting the whole into a smaller number of larger parts; therefore, we need to get fewer parts in order that we get exactly the same share. We can do this by *dividing* the total number of parts and taking the equivalent number of these parts.

Dividing by a fraction (Unit 6.8) can often be confusing when introduced in an abstract way that pupils find difficult to understand. When solving, for example, $2 \div \frac{2}{3}$, the abstract explanation is to 'invert the fraction and multiply', thus becoming $2 \times \frac{3}{2}$, without any clear explanation as to why we use this process. In order to help understanding of division by a fraction, it is helpful to think of division by grouping: 12 divided by 3 can be explained using the question, *'How many teams of 3 can we make with 12 players?'* The answer to this is **4 teams of 3** (not simply 4) which we find by making one team of three, then another, and so on. So, $2 \div \frac{2}{3}$ can be thought of as, *'How many portions of two thirds of a pizza can we get from two pizzas?'* If there are two pizzas and each person is to get two thirds of a pizza, then the first thing we need to do is cut the pizza into thirds (making six thirds altogether). If we share out two thirds of the pizza at a time, then we can feed three people.

It is important that pupils understand the link between fractions, decimal fractions and percentages (Units 6.9 and 6.10), i.e. that these are simply different formats for expressing the same thing; they are equivalent. A **percentage** is a fraction out of one hundred and a **decimal fraction** is a fraction expressed in tenths, hundredths, etc. in order to fit the decimal number system. These should be represented visually in order that pupils develop conceptual understanding rather than relying on memorisation of equivalents for common fractions, i.e. $\frac{1}{2} = 0{\cdot}5 = 50\%$.

Resources

- Resource sheet a
- A4 or A5 paper
- Printable fraction wall
- Modelling clay (to represent cookies, chocolate bars, etc.)
- Fraction cards
- Base 10 materials
- Counting stick
- Card

Key vocabulary

- **common equivalent**
- **compare**
- **decimal fraction**
- **decimal place**
- **decimal point**
- **denominator**
- **equivalent**
- **Mixed number**
- **numerator**
- **percent**
- **Proper fraction**
- **simplify**

Misconceptions

Potential misconception	Addressing the misconception
Units 6.1 and 6.2 Pupils may think fractions cannot be greater than one.	• Count fractions with pupils where the fractions are represented using concrete materials and / or visuals making sure you go beyond one whole, e.g. Count up to twelve quarters and beyond (see Unit 6.1).
Unit 6.2 When comparing fractions, pupils may think only the value of the numerator is considered, e.g. the assumption is made that $\frac{7}{12}$ is larger than $\frac{4}{5}$ because 7 is larger than 4.	• Draw visual representations (bar models) when comparing fractions in order that pupils appreciate that the denominator affects the size of each equal part (therefore understanding that twelfths are significantly smaller than fifths).

Potential misconception	Addressing the misconception
Units 6.3 and 6.5 Pupils may think fractions get smaller when calculating simplified fractions (due to the use of division).	• Use paper, modelling clay, etc. to create simplified fractions and represent these visually using bar models. • Avoid moving to the abstract process too soon, i.e. dividing the numerator and denominator by the same value. • Continue to represent simplified fractions visually after *Simplifying fractions* (Unit 6.2) has been covered.
Unit 6.4 Pupils add (or subtract fractions) by adding the numerators together and the denominators together, e.g. they think $\frac{1}{2}$ plus $\frac{1}{3}$ equals $\frac{2}{5}$	• Use concrete materials, fraction cards, etc. when adding or subtracting fractions. • Develop understanding that fractions can only be added when they belong to the same set; therefore, finding equivalents is necessary when adding, unlike fractions. • Use bar models to represent addition or subtraction of fractions visually.
Unit 6.6 Pupils may think calculating a fraction of a fraction is regarded as an abstract multiplication process.	• Set problems within a context, e.g. *I have half a bag of sweets. I eat one third of them. What fraction of a bag of sweets do I eat?* • Represent these problems visually (see Let's Learn Unit 6.6). • Avoid setting abstract questions, e.g. $\frac{1}{2} \times \frac{1}{3} = ?$
Unit 6.7 Pupils may think the denominator is divided when dividing a fraction by a whole number, e.g. $\frac{1}{4} \div 2 = \frac{1}{2}$ (the 4 is divided by 2 to get 2).	• Set problems within a context, e.g. *I have a quarter of a chocolate bar. I split it into two parts. What fraction of the chocolate bar is each part?* • Represent these problems visually (see 'Let's Learn' Unit 6.7). • Avoid setting abstract questions, e.g. $\frac{1}{2} \div 2 = ?$
Unit 6.8 Pupils may have an inability to understand the concept of division by a fraction.	• Avoid 'invert and multiply'. • Ensure understanding of division by grouping and set problems within a context, e.g. *How many portions of half a cake can I cut from two cakes?* • Ensure answers relate to the fraction, e.g. *4 halves or 4 half portions of cake (not just an answer of 4).*
Units 6.9 and 6.10 Pupils regard fractions, decimal fractions and percentages as separate concepts and are unable to make links between them.	• Represent fractions, decimal fractions and percentages using bar models in order to visually appreciate the links.

Teaching ideas

Unit 6.1 Converting fractions

- Show pupils an image of, for example, ten half pizzas:

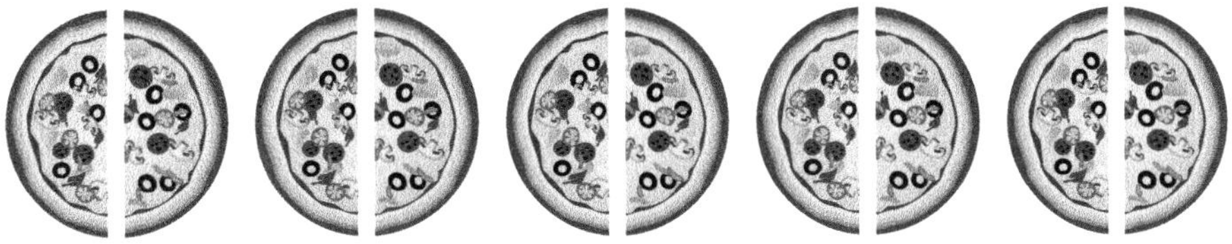

- **Count the half pizzas with me** (*One half, two halves, three halves, etc.*)
- Ask questions like: **What is six half pizzas the same as?** *(three whole pizzas)* **What is five half pizzas the same as?** *(two and a half whole pizzas)* **How many pizzas would we need to order if eight people wanted half a pizza each?** *(four whole pizzas).*
- Repeat the counting activity; however, this time count the 'wholes' *(one half, one, one and a half, two, two and a half, etc).*
- Introduce the terms **mixed numbers** (*when we use a mix of whole numbers and fractions, e.g. two and a half pizzas*) and **improper fractions** (*when we use only the fractions, e.g. five half pizzas*).
- Repeat using different fraction portions, e.g. quarters, thirds, etc.

Unit 6.2 Comparing and ordering fractions

- In textbook 2B (Unit 6.3), fractions were compared using equivalence (where the denominator for one fraction was a factor of the denominator for the other, e.g. comparing $\frac{3}{4}$ to $\frac{7}{12}$).
- This unit aims to extend this understanding by being able to compare any fractions using equivalence including improper fractions. E.g. compare $\frac{3}{4}$ and $\frac{5}{7}$ or $\frac{7}{3}$ and $\frac{9}{4}$.
- Ask the pupils: **Would you rather have half or three fifths of a bag of sweets? How could we say for definite which is more?** (*It is difficult to compare one fraction type with another*) **How could our understanding of equivalent fractions help us?**
- In order to compare any fraction, it will be useful to be able to make any fraction accurately.
- Provide pupils with paper and challenge them to make fifths (they will discover that making fifths, sevenths, thirds, etc. are very difficult to make accurately).
- Allow the pupils to review and compare their results. Ask questions like: **How many parts have you made?** (some pupils may have made six parts by making five folds) **Are all of your attempts at fifths the same size? Why did you find it difficult?** (requires judgement because we can't rely on making halves)
- *Let's make fifths accurately (See Resource sheet a for illustrated instructions).*
- Starting from one end of a piece of A4 paper, use a ruler to mark five lines an equal distance apart.

- Cut a strip from the opposite end of the piece of paper.
- Line up one corner of the strip with the corner of the A4 paper and the opposite corner positioned exactly on the fifth line.
- Mark the strip at each interval where the lines meet (these will be an equal distance apart so you have accurately marked fifths).
- Make halves in the same way or simply by folding a strip of paper accurately.
- Ask the pupils: **Can we convert halves into fifths or fifths into halves?** (*No*)
- Ask the pupils: **Can we convert fifths and halves into a common equivalent?** (*A common equivalent can be found by splitting the fifths into two equal parts and the halves into five equal parts.) See 'Let's learn' page 99.*
- We can now say definitively that six tenths is larger than five tenths of the bag so we should have chosen three fifths over the half.

Unit 6.3 Simplifying fractions

- In Book 2B (Unit 6.3), pupils learned to calculate equivalent fractions by increasing the number of parts that makes up one whole (denominator) and increasing the number of parts we are interested in (numerator) in proportion to one another.
- For example:

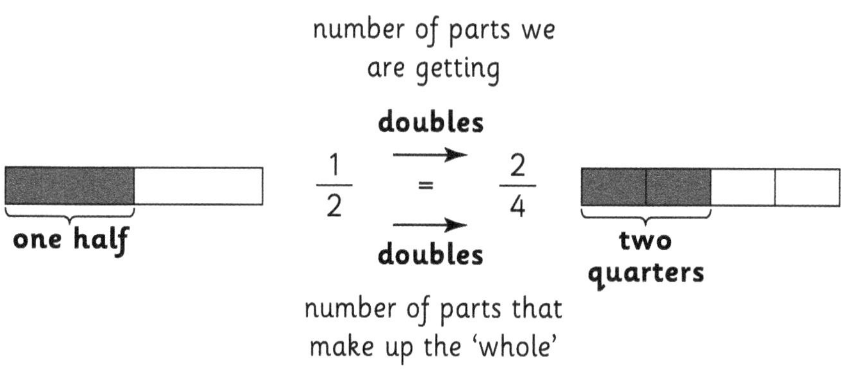

- The concept of simplifying a fraction follows the exact same logic but in reverse.
- Organise pupils into fours.
- Give groups blocks of modelling clay and instruct them to: **Cut the clay to get three twelfths each.**
- Ask questions like: **Is there a simpler way to do that so that we end up with exactly the same amount of clay?** (*Rather than cutting the whole amount into 12 parts and getting 3 each, they could have cut it into 4 parts and get 1 part each.*)

- Ask questions like: **What has happened to the total number of parts?** (*there are 3 times less*) **What has happened to the number of parts that you each get?** (*it's also 3 times less*) **What's happened to the size of each part?** (*they are three times bigger*) **Have you ended up with less modelling clay?** (*it's exactly the same amount*).
- Demonstrate that we can find a simplified fraction by dividing the total number of parts (denominator) and the number of parts we are getting (numerator) by the same amount, for example:

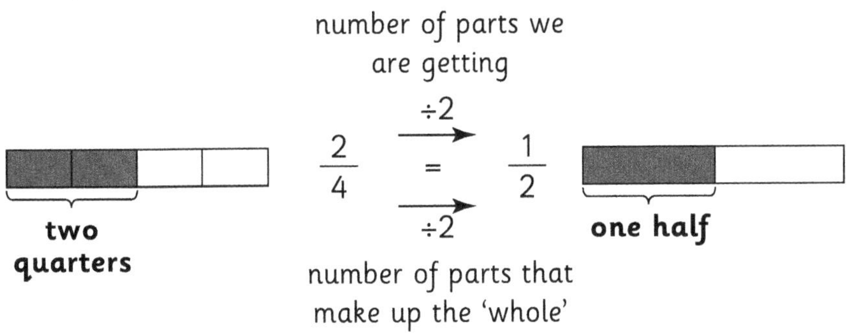

Unit 6.4 Adding and subtracting fractions

- This unit aims to extend the learning developed in Book 2B (Unit 6.6) by adding and subtracting fractions using equivalence (including mixed numbers and improper fractions).
- Share the following problem with the pupils: *Amman has three quarters of a bar of chocolate and Nuria has three eighths of a bar. How much chocolate do they have altogether?*
- Organise pupils in pairs and provide them with fraction cards, paper and pencils.
- Challenge pairs to:
  - **Solve it**: using the resources.
  - **Prove it**: drawing pictures / diagrams (to represent the resources).
  - **Explain it**: in their own words (*e.g. we couldn't add three quarters and three eighths to begin with because they aren't alike. We had to change three quarters into six eighths before we could add them together*).
  - **Calculate it**: write the number story, i.e. $\frac{3}{4} + \frac{3}{8} = \frac{6}{8} + \frac{3}{8} = \frac{9}{8}$ (written using digits or words).

- **See the diagram below**:

Solve it

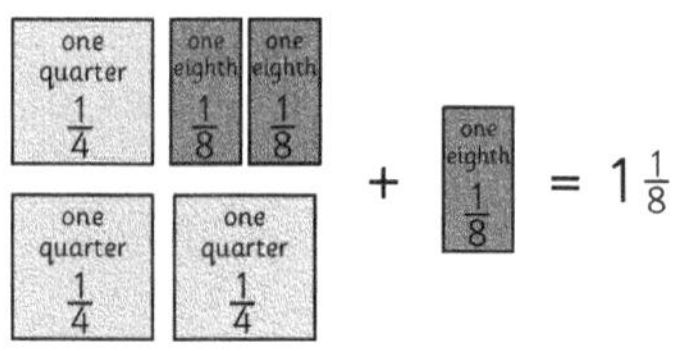

Prove it

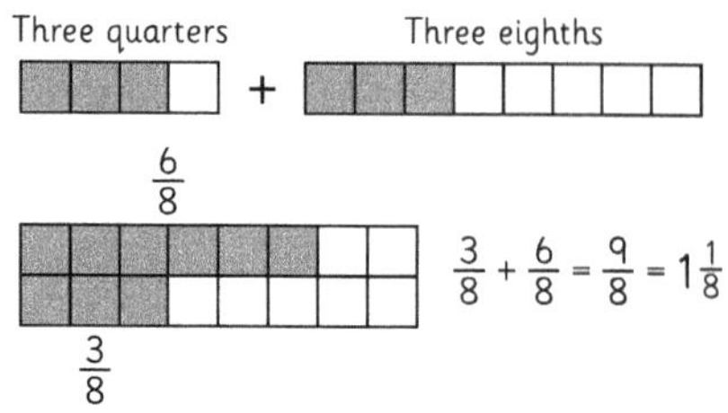

Explain it

Three quarters are the same as $\frac{6}{8}$

Then you can add $\frac{6}{8}$ to the $\frac{3}{8}$ to get $\frac{9}{8}$

$\frac{9}{8}$ is the same as $1\frac{1}{8}$

Calculate it

$$\frac{3}{8} + \frac{6}{8} = \frac{9}{8} = 1\frac{1}{8}$$

- Demonstrate how the three quarters needs to be converted into eighths before they can be added together.
- Repeat activity using one half and one third (both will have to be converted into sixths).

Unit 6.5 Converting decimal fractions to fractions

- In Book 2B (Unit 6.5), pupils learned to convert simple fractions to a decimal equivalent. In this unit, pupils are learning to do the reverse, i.e. convert a decimal fraction to a fraction in its simplest form.
- Write or display the numbers 0·9 and 0·81.
- Ask questions like: **How would we write these as fractions?** ($\frac{9}{10}$ *and* $\frac{81}{100}$) **Which number is larger? Can you prove it?** ($\frac{9}{10}$ *can be converted into* $\frac{90}{100}$)
- Write or display the numbers 0·6 and 0·15.
- Ask questions like: **How would we write these as fractions?** ($\frac{6}{10}$ *and* $\frac{15}{100}$), **Can they be simplified?** ($\frac{3}{5}$ *and* $\frac{3}{20}$) **Explain how they can be simplified.** (*Pairs of tenths can be combined to make fifths; sets of five hundredths can be combined to make twentieths.*)
- Write or display the number 0·64.
- Ask questions like: **How would we write this as a fraction?** ($\frac{64}{100}$) **How many different ways can this be simplified?** ($\frac{32}{50}$; $\frac{16}{25}$)
- Repeat using different decimal numbers that can be simplified, e.g. 0·24, 0·45, 0·8, etc.

Unit 6.6 Calculating a fraction of a fraction and Unit 6.7 Dividing a fraction by a whole number

- Units 6.6 and 6.7 essentially deal with the same concept, i.e. splitting a fraction of something into smaller parts.
- We can think of this in two ways:
  - Finding a fraction of a fraction: *I have half a pizza. I eat half of it. How much do I eat?*
  - Dividing a fraction by a whole number: *I have half a pizza. I cut it into two equal slices. What size is each slice?*
- Both of these questions are asking the same thing and it is important that pupils are able to make this link, i.e. *finding half of something is exactly the same as dividing it into two equal sized parts.*
- Confusion can arise when finding a fraction of a fraction if it is taught in an abstract way, i.e. set out as a plain multiplication: $\frac{1}{2} \times \frac{1}{2} = ?$
- In this example, pupils are told to multiply the numerators together ($1 \times 1 = 1$) and multiply the denominators together ($2 \times 2 = 4$) to get an answer of $\frac{1}{4}$ which they can get without understanding the process.
- In order to develop understanding, our aim is to enable pupils to visualise the problem using a bar model:

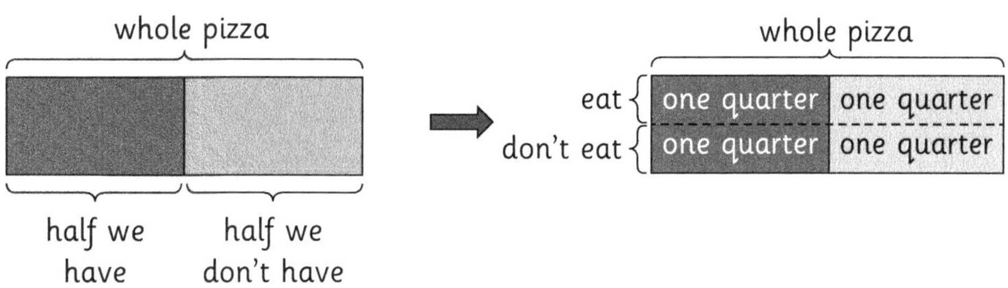

- When we cut the half pizza in half we create quarters. One of the quarters is eaten.
- Provide pupils with equal sized pieces of paper (A4 or A5) and ask them to make halves (by folding and colouring one part).
- Split the class into two groups. Ask one group to quarter their halves (by folding them into four equal parts) and the other group to divide their halves by four. Ask the groups to compare their results.
- Ask questions like: **What do you notice about the parts you have created?** *(they are the same size)* **What fraction is each part?** *(one eighth)*, etc.
- Question 1.c) (Unit 6.6) asks the pupils to find two thirds of half a bar. This will require them to split the half into three equal parts (making sixths) and count two of these parts (two sixths or one third).

Unit 6.8 Dividing a whole number by a fraction

- Ask pupils: **How many portions of one sixth can we get from two whole cakes?** (Provide paper or modelling clay to represent the cakes.)
- The answer to this must relate to the context, i.e. *twelve portions* or *twelve sixths* (not simply *twelve*).
- The number story $4 \div \frac{1}{2} = 8$ is abstract and has little meaning when not set in a context.
- Instead, ask questions like: **I want to divide 4 bars of chocolate into halves. How many half bars can I make?** *(8 half bars of chocolate)* **A glass holds one third of a litre. How many glasses can be filled with 5 litres of water?** *(15 glasses)*
- Question b) in the Challenge box asks pupils to divide five bottles of orange juice into glasses that hold two fifths of a bottle.
- Each bottle is split into fifths (so that we can take two fifths at a time).
- Each glass is split into two parts equal in size to two fifths of a bottle because this is the capacity of each glass:

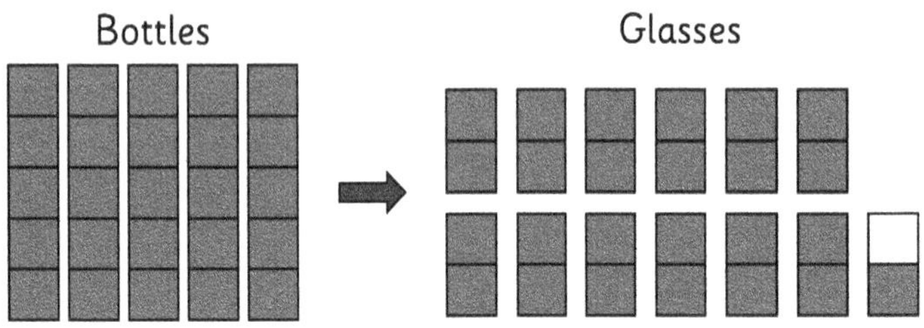

- As two fifths of a bottle are poured out at a time, we can fill twelve whole glasses with one fifth of a bottle remaining. When we pour this final one fifth into a glass it will fill half the glass. Therefore, the answer is *twelve and a half glasses*.

Unit 6.9 Equivalents: fractions, decimals and percentages

- Write or display $\frac{3}{5} = 0{\cdot}3 = 30\%$ and ask pupils to discuss: **Is this true or false? Draw a bar model to represent each in order to prove your thoughts.**

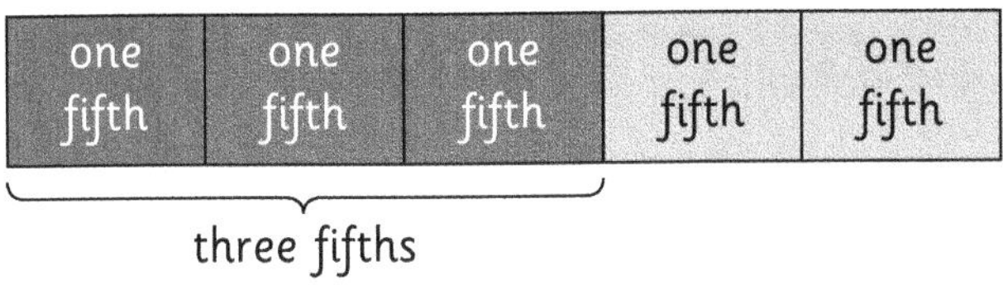

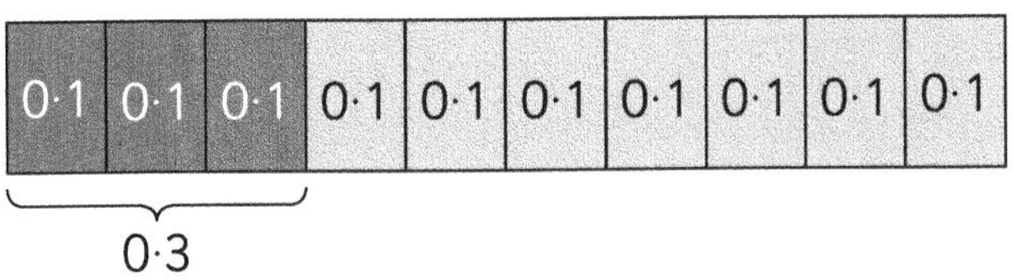

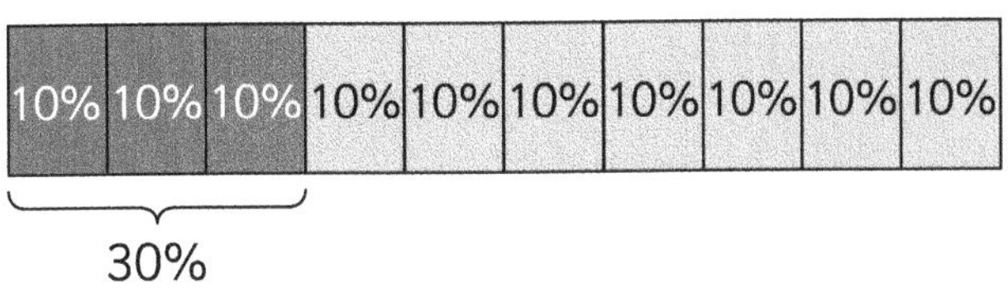

- The bar models for each clearly show that this is false.
- Challenge pupils to: **Convert three fifths into both a decimal fraction and a percentage using your bar models to demonstrate.**

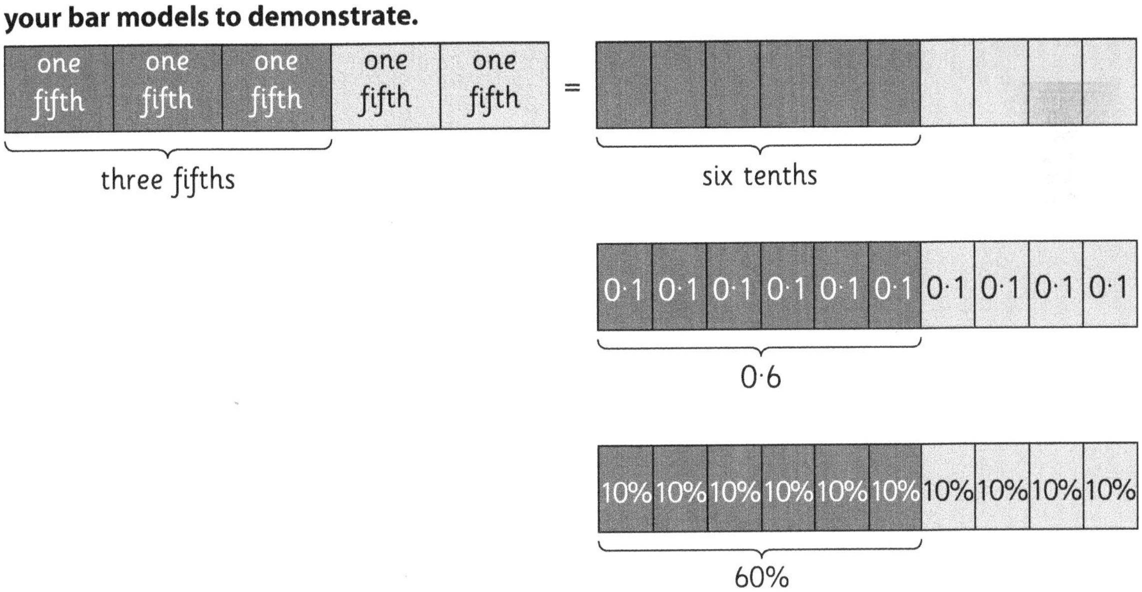

- Demonstrate that, when converting fifths to both a decimal and a percentage, we can convert to tenths (rather than having to convert to hundredths for a percentage) by halving each fifth. Each part represents 0·1 and 10%.

Unit 6.10 Solving fraction, decimal and percentage problems

- The aim of this unit is to develop the understanding that fraction, decimal fraction and percentage problems are interchangeable, i.e. a problem may be presented as a percentage: **What is 25% of £50?** However, in order to solve this some pupils may find it easier to think of this as: **What is a quarter of £50?**
- Display images of fractions representing hundredths.
- Ask questions like: **Can you write these as a fraction, a decimal fraction and a percentage?**
- Repeat, using images of fractions representing tenths.
- Repeat, using images of fractions representing halves, quarters, fifths, twentieths, twenty-fifths and fiftieths.
- Display an image of a bar model, for example:

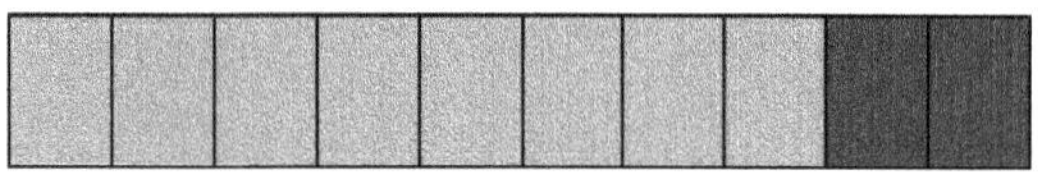

- Demonstrate how you can make up a problem to fit, for example:

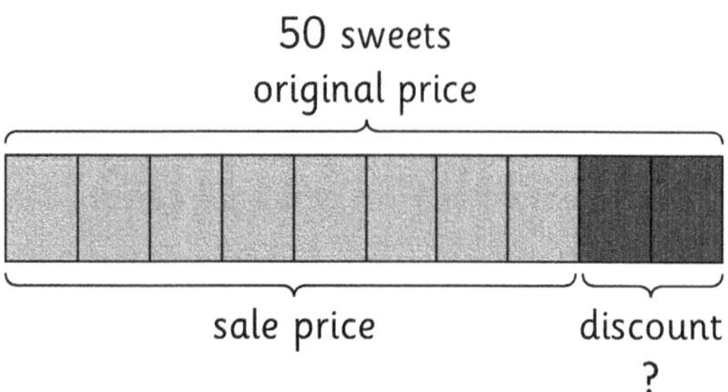

- ***A pair of trainers usually costs £50. How much money will you save if you buy them in a sale with 20% off?***

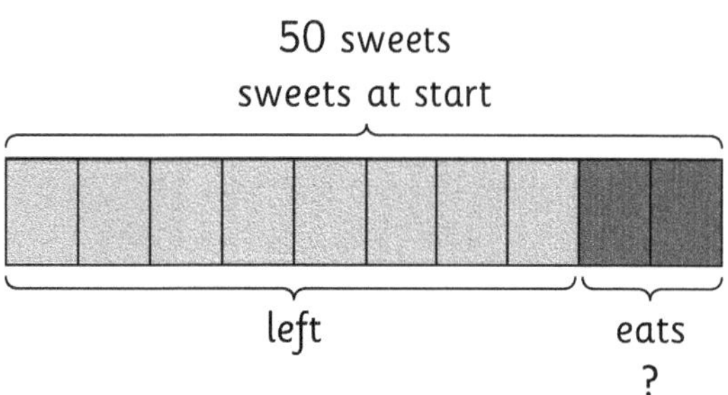

- ***Mark had 50 sweets. He ate two tenths of them. How many sweets did he eat?***
- Challenge pupils to make up their own problems to fit the following bar models (encourage them to produce a fraction, decimal and percentage problem):

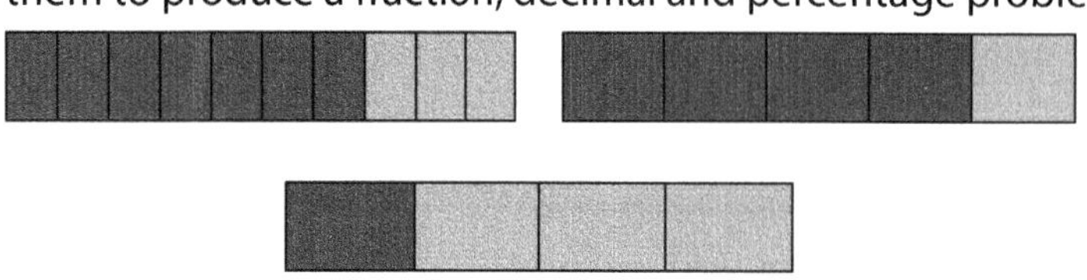

7 Money

Money 2A

Section overview

Textbook 2A units	Review units	Prior learning	Benchmarks/Es&Os
Unit 7.1 Writing amounts using decimal notation	**1C** Unit 6.1	• I can write amounts accurately in different ways, e.g. *two pounds and thirty-six pence = 236p = £2·36.*	**MNU 2-02a** • Adds and subtracts whole numbers and decimal fractions to two decimal places. **MNU 2-09a** • Carries out money calculations involving the four operations. • Compares costs and determines affordability within a given budget.
Unit 7.2 Budgeting	**1C** Units 6.2 and 6.3	• I can calculate how much money has been spent in total using appropriate calculations. • I can calculate how much money is left over. • I can record my answers using decimal notation.	**MNU 2-09a** • Carries out money calculations involving the four operations. • Compares costs and determines affordability within a given budget.
Unit 7.3 Saving money	**1C** Units 6.1, 6.2 and 6.3	• I understand the difference between want and need. • I can demonstrate an awareness of the different ways that goods can be paid for. • I recognise that there are different ways that you can save money. • I have an awareness of the products and services that banks provide.	**MNU 2-09b** • Demonstrates understanding of the benefits and risks of using bank cards and digital technologies.
Unit 7.4 Profit and loss	**1C** Units 6.1, 6.2 and 6.3	• I can read and write in decimal notation. • I can calculate amounts using decimal notation.	**MNU 2-09c** • Calculates profit and loss accurately; for example, when working with a budget for an enterprise activity.

Conceptual context

Teaching pupils to become financially astute involves much more than the development of mathematical skills such as coin recognition and calculating the percentage increase and decrease of a price. Developing *financial knowledge and understanding* will help pupils understand the concept of money, the nature of money and its functions and uses. Guiding pupils through day-to-day money management will enhance their *financial skills and competence*. Raising pupils' awareness of the wider impact of money and personal finance decisions for their own future, their families and their community teaches *financial responsibility*.

Teaching pupils to make sound financial decisions is essential to help them be able to identify and make the best use of opportunities in a fast-changing world. Pupils who are financially capable can use credit and debit cards, manage loans, buy goods and services online and make the best decisions regarding all their financial transactions.

Resources

- Plastic coins and notes
- Coin value Resource sheet 7a
- Dice marked £100, £100, £75, £50, £50, £10
- Budget Sheet Resource sheet 7b
- Saving money Resource sheet 7c

Key vocabulary

- **Afford**
- **Budget**
- **Compare**
- **Currency**
- **Decimal notation**
- **Hundredth**
- **Least**
- **Loss**
- **Profit**
- **Receipt**
- **Tenth**

Misconceptions

Potential misconception	Addressing the misconception
Unit 7.1 Pupils recognise the coins and can tell you how much they have without understanding the value of each coin.	Go over each coin with pupils and ask them to write down its decimal value, write it as a fraction and explain what it is equivalent to. See teaching ideas 7.1.
Unit 7.1 Pupils believe that coins with larger numbers on the face, have greater value than a pound coin. For example, a 20p coin is worth more than £1 coin.	Ensure that the pupils understand the relationship of pence and pounds. Use coins to show that 100 pence are equivalent to one pound in value.

Teaching ideas

Unit 7.1 Writing amounts using decimal notation

- Using plastic coins and notes, show pupils various combinations and ask them to tell you how much you have. For example, show pupils three £1 coins and two 20p and ask them to tell you how much you have: £3·20. This should help you to establish their general understanding of the concept of the value of money.
- Next, show pupils the following amounts and ask them to write their answers on a whiteboard using the correct decimal notation.

- You are looking for them to write £2·73 but they may have one of the following: two pounds and seventy-three pence, or 273p.
- Explain to pupils how the coins relate to ones, tenths and hundredths and emphasise that the decimal part refers to a fraction of a whole pound as opposed to a whole number of 1p, 2p, 5p coins.
- Now, ask pupils to name all the coins that they know. Then go through the coins to check their understanding. Either on a whiteboard or using Resource sheet 7a complete the following with pupils:

Fraction form:

$\frac{1}{100}$

Word form:

One pence

Coin:

Money form:

1p

Place value:

0	0	•	0	1
Tens	Ones		$\frac{1}{10}$	$\frac{1}{100}$

- Continue this with other coins until you are satisfied that pupils understand the value of each coin. Pupils need to understand this prior to being able to add, subtract, multiply and divide amounts of money.

Unit 7.2 Budgeting

- Ask pupils what a budget is and give them the opportunity to discuss this with a partner. You could ask them if they have heard this word being used, where they have heard it, and in what context.
- Ask pupils to give examples of when a budget would be useful. Ask them if they have ever saved up for something, and what did they do. Introduce the next activity by putting them into pairs and asking them to think about something that they would want to buy. Tell them that each pair needs to save £50 for food, £100 for the item they want to buy and enough savings to pay the 'emergency charge' at the end of 4 weeks (4 rolls of the dice). Each pair will need a dice marked £100, £100, £75, £50, £50, £10; and a copy of the Budget Sheet (Resource sheet 7b) for each pair. Remind pupils how to add and subtract amounts such as £100, £75, £50 and £25.
- Roll the money dice four times. Each time, pupils will record the amount in their earnings column, then decide how they are going to allocate that money. They can split the sum of money as they choose across the three columns (food, item they want to buy, and savings). Each pair will roll the dice four times and after the fourth time they will add up the totals in each of the columns. You roll the dice one last time: this represents the 'emergency charge'. Each pair of pupils must subtract the emergency charge from their total savings. Explain to pupils that this charge is to cover any unforeseen expenses that may occur – repairs to household items, loss of earnings due to sickness, etc. Ask pupils to discuss how much they have in each column and if they managed to have £50 for food, £100 for the item they were saving for and still pay the 'emergency charge'. Discuss the benefits of having money available to deal with emergencies. Ask pupils what they would do if they didn't have any money saved or available to pay for emergencies. Encourage pupils to discuss the benefits of budgeting and planning how and what they spend their money on.

Unit 7.3 Saving money

Ask pupils if they have ever bought something and then realised that they could have paid less for it in another shop or online. Discuss why this happens.

In pairs, ask pupils to consider the different ways they can think of to save money when buying things – going to different shops, bulk buying, multipacks…

Show pupils Resource sheet 7c and ask them to work out which option is the Best Buy and why. Pupils should be looking not only at the price of the crisps but also the weight in each bag in order to work out the best deal. Encourage them to consider the benefits of buying more to save money versus buying only what you need.

Unit 7.4 Profit and loss

- Introducing the concept of profit and loss can be complicated for pupils to understand. Start by asking pupils to discuss what their understanding of profit and loss are. Ask them if they have heard these terms, where they have heard them and what they think they mean.
- Discuss with pupils why making a profit in business is important. Talk about the fact that businesses make a profit by making the selling price (cost for the public to buy it), more than it has cost to make the product. The money that is left over, or extra, is called the *profit*. A business can make more profit by making the product as cheap as they can and then selling it for as much as people are willing to pay. However, sometimes a company cannot sell their products for more than it costs them to make it and they even have to sell it for less. When this happens it is called a *loss*, which means that they have lost money and it has actually cost the company money. Businesses need to make the price high enough to cover all their expenses and generate a profit, but not so high that people will not buy the product or will buy the item somewhere else from a different company. Often, companies will compare the prices of other similar items made by other companies to make sure they set competitive prices.
- Ask pupils what will happen if a business makes a loss. Discuss the possible implications that might occur. For example: it might have to cut costs, change the selling price, people might lose their jobs, the business may have to change the way it makes the product, or the business might have to close.

Money 2B

Section overview

Textbook 2B units	Review units	Prior learning	Benchmarks/Es&Os
Unit 7.1 Money problems using the four operations	**2A** Unit 7 **2B** Units 4.4, 4.6 and 4.10	• I can identify and use coins and notes up to and including £20·00. • I can work out change using a variety of coin and note combinations. • I can use £ and p notation when using money.	**MNU 2-02a** • Adds and subtracts whole numbers and decimal fractions to two decimal places. **MNU 2-09a** • Carries out money calculations involving the four operations. • Compares costs and determines affordability within a given budget.
Unit 7.2 Budgeting	**2A** Units 7.2 and 7.3	• I understand that by comparing the cost of items I can save money. • I can calculate the total amount spent using the four operations. • I can record calculations and answers using decimal notation.	**MNU 2-09a** • Carries out money calculations involving the four operations. • Compares costs and determines affordability within a given budget.
Unit 7.3 Profit and loss	**2A** Unit 7.4	• I understand the term profit. • I understand the term loss. • I can calculate the profit and loss using the four processes. • I can record my answers using the correct decimal notation.	**MNU 2-09b** • Demonstrates understanding of the benefits and risks of using bank cards and digital technologies.
Unit 7.4 Discounts	**2A** Unit 7.3	• I understand how to compare costs of the products that I buy. • I can demonstrate an awareness of the different ways to save money. • I recognise that there are different ways to purchase items to save money. • I can discuss the products and services that banks provide.	**MNU 2-09c** • Calculates profit and loss accurately, for example, when working with a budget for an enterprise activity.
Unit 7.5 Credit, debit and debt	**2A** Unit 7.3	• I understand that there are different ways to pay for something. • I can recognise the terms credit and debit. • I recognise the term debt. • I can discuss the products and services that banks provide.	**MNU 2-09a** • Carries out money calculations involving the four operations. • Compares costs and determines affordability within a given budget.

Conceptual context

Teaching pupils to become financially astute involves much more than the development of mathematical skills such as coin recognition and calculating the percentage increase and decrease of a price. Developing *financial knowledge and understanding* will help pupils understand the concept of money, the nature of money and its functions and uses. Guiding pupils through day-to-day money management will enhance their *financial skills and competence*. Raising pupils' awareness of the wider impact of money and personal finance decisions for their own future, their families and their community teaches *financial responsibility*.

Teaching pupils to make sound financial decisions is essential to help them be able to identify and make the best use of opportunities in a fast-changing world. Pupils who are financially capable can use credit and debit cards, manage loans, buy goods and services online and make the best decisions regarding all their financial transactions.

Although pupils are exposed to money from a very early age, there are some suggestions that they do not use money as much as perhaps the pupils of 50 years ago did. How often do we see 6 and 7-year-old pupils going to the shop to get the messages today? Money as a physical object is also seen less by children. At the supermarket more and more people use bank cards to pay for their goods. Does this suggest to pupils that buying stuff appears to cost nothing? In this book, using bank cards such as debit and credit cards is explored. Making links with a local credit union can help to bring this subject alive for pupils who may not bank regularly.

Resources

- Plastic coins and notes
- Coin value Resource sheet 7a
- Dice marked £100, £100, £75, £50, £50, £10
- Budget Sheet, Resource sheet 7b

Key vocabulary

- Afford
- Budget
- Compare
- Credit
- Currency
- Debit
- Debt
- Decimal notation
- Discount
- Expenses
- Hundredth
- Least
- Loss
- Profit
- Receipt
- Sale
- Tenth

Misconceptions

Potential misconception	Addressing the misconception
Unit 7.1 Pupils recognise the coins and can tell you how much they have without understanding the value of each coin.	Go over each coin with pupils and ask them to write down its decimal value, write it as a fraction and explain what it is equivalent to.

Teaching ideas

Unit 7.1 Money problems using the four operations

- Write the following question on the board:

  What change would you get from £10 if you bought a comic that cost £3·20 and a packet of sweets that cost £1·85? Ask pupils how they would calculate this. Ideally, you are looking for them to perform an addition calculation to work out the total spend before subtracting this from £10.

$$\begin{array}{r} £3{\cdot}20 \\ +\ £1{\cdot}85 \\ \hline £5{\cdot}05 \\ \hline \end{array}$$

  or £3·20 + £1 + 80p + 5p

- To calculate the change from £10, pupils can subtract or round to the nearest pound and compensate by adding on:

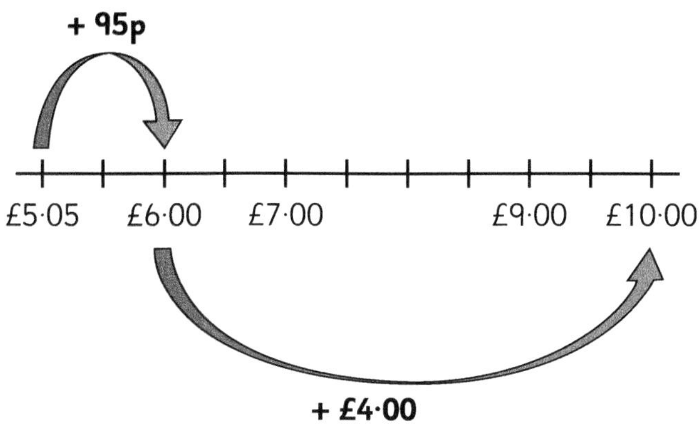

95p + £4·00 = £4·95

- Allow pupils the opportunity to explore different ways to calculate this, showing them both the formal addition/subtraction and the chunking method.
- Now, present pupils with the following problem:

  How much would you spend if you bought the following items – 4 apples that cost 36p each, 5 oranges that cost 25p each, and 6 bananas that cost 20p each?

  Ask pupils what operations they would use and how they will work it out.

$$\begin{array}{r} 36 \\ \times \quad 4 \\ \hline 144\text{ p} \end{array} \qquad \begin{array}{r} 25 \\ \times \quad 5 \\ \hline 125\text{ p} \end{array} \qquad \begin{array}{r} 20 \\ \times \quad 6 \\ \hline 120\text{ p} \end{array} \qquad \begin{array}{r} 144 \\ 125 \\ +120 \\ \hline 389\text{ p} \end{array}$$

- Discuss with pupils the different calculations required and that they then need to convert the total from pence into pounds and pence: **£3·89**.

Unit 7.2 Budgeting

- Ask pupils to share their understanding of what a budget is and give them the opportunity to discuss this with a partner. You could ask them if they have heard this word being used, where they have heard it and in what context.
- Ask pupils to give examples of when a budget would be useful. Ask them if they have ever saved up for something, and what did they do. Ask pupils what they would do if someone gives them £10/£100/£100, what could they do with it?

- Discuss how the pupils might save up for something and the various ways of saving money including pupil's savings accounts from banks and building societies. Ask pupils to consider the different factors they look at when they are buying an expensive item like a laptop or a bike. Do they just buy the first one they see? Do they buy the cheapest one? Do they go around different shops and gather information? Take time to explore the different aspects that may influence their decision:
  - The make/model of the item
  - The features that it has – speed, memory, size, colour, warranty…
  - Price
  - Discount
  - What they need it for
  - How often they will use it
- Allow pupils to discuss what is important to them when they are buying something and the various factors that influence them.

Unit 7.3 Profit and loss

- Review the concept by asking pupils to discuss what their understanding of profit and loss is. Ask them if they have heard these terms, where they have heard them and what they think they mean.
- With pupils, discuss the reasons why making a profit in business is important. Talk about the fact that businesses make a profit by making the selling price (cost for the public to buy it), more than it has cost them to make the product, highlighting that the money that is left over (or extra) is called profit. A business can make more profit by making the product as cheap as they can and then selling it for as much as they can. However, sometimes a company cannot sell its product for more than it costs them to make it and they even have to sell it for less. When this happens it is called a loss, which means that they have lost money and it has actually cost the business money. Businesses need to make the price high enough to cover all their expenses and generate a profit, but not so high that people will buy the item somewhere else. Often, companies will compare the prices of similar items being sold by other companies to make sure they set realistic prices. Ask pupils what expenses are and why these make a difference to the price.
- Ask pupils what will happen if a business makes a loss, and discuss the possible implications. For example: the business might have to cut costs, change the selling price of the product/service, employees might lose their jobs, the business may have to change the way they make the product, or the business might have to close.

Unit 7.4 Discounts

- Ask pupils what a discount is and where you might get one.
- Explain to pupils that a discount is when an item is purchased for less than the normal price. This is also known as a sale price. When a discount is applied, there is often a percentage of the normal price, or the discount percentage, deducted from the normal price. The new, lower price is also known as the discounted price. When you are interested in buying an item, often people wait until the item is on sale, or offered at a discount. This way you can save your money. In some stores they offer discounts every day, but you should also check the quality of the item, as there may be a reason you are getting the item at a discount.
- Explain to pupils that fractions and percentages are often used to advertise discounts:

  $\frac{1}{2}$ price   20% off $\frac{1}{4}$ of the original price
- Ask pupils to calculate the sale price of a pair of trainers that were originally £50 but are now 50% cheaper. Are they familiar with the fact that 50% is the same as $\frac{1}{2}$? Can they identify the discount amount and the new price?
- With pupils, discuss the cost of using a mobile phone. Then ask each pair of pupils to look at two different mobile phone offers and get them to calculate which would be the best choice for them. They will need to be able to explain their choice.
- Consider the following:
  - For what reason might a phone company offer customers a free phone?
  - How can you calculate the monthly rental if the phone company bills every quarter?

Unit 7.5 Credit, debit and debt

- Begin by putting each of the words credit, debit and debt on the board and asking pupils to explain their understanding of each of them.
- Discuss the concept of starting to use money in different ways as we get older, with some pupils having their own bank cards and pre-paid cards, therefore they no longer use coins and notes to pay for things but instead use plastic cards. Explain that there are two main types of card: a debit card that uses your own money, and a credit card that borrows money from the bank.
- A debit card is the card you receive from the bank when you open a bank account. Many banks now offer them to children and a debit card acts like a key to access your money. With a debit card, you are using your own money, so you will not be charged. However, when you have spent all of your money the account will be empty. A credit card is something that you apply for at a bank when you want to be able to borrow money. If you are accepted, you will be given a limit up to which you can borrow, based on how much you can afford to pay back. However, using a credit card can be expensive as there is a charge for borrowing the money, so you not only have to pay back the money you borrow but also the extra money that you are charged. This is one way that banks make money: it is an example of how the banks, as businesses, create profit.
- Ask pupils if they can think of the advantages and disadvantages of using a credit and a debit card.

Money 2C

Section overview

Textbook 2C units	Review units	Prior learning	Benchmarks/Es&Os
Unit 7.1 Money problems using the four operations	2B Unit 3, 7.1	• I can solve problems involving decimal notation using the four operations. • I can demonstrate an understanding of notes and coins and their values. • I can use £ and p notation when using money.	**MNU 2-02a** • Adds and subtracts whole numbers and decimal fractions to two decimal places. **MNU 2-09a** • Carries out money calculations involving the four operations. • Compares costs and determines affordability within a given budget.
Unit 7.2 Budgeting	**2B** Unit 7.2	• I know by comparing costs we can save money. • I can keep track of the total amount spent using the four operations. • I can record calculations and answers using decimal notation.	**MNU 2-09a** • Carries out money calculations involving the four operations. • Compares costs and determines affordability within a given budget.
Unit 7.3 Profit and loss	**2B** Unit 7.3	• I understand the terms profit and loss. • I can calculate profit and loss. • I can record my answers using the correct vocabulary and decimal notation.	**MNU 2-09b** • Demonstrates understanding of the benefits and risks of using bank cards and digital technologies.
Unit 7.4 Discounts	**2B** Unit 7.4	• I understand how to compare costs in order to save money. • I recognise that there are different options when purchasing items. • I can demonstrate an understanding of how different discounts can save money. • I understand that retailers use discounts to encourage customers to purchase products.	**MNU 2-09c** • Calculates profit and loss accurately; for example, when working with a budget for an enterprise activity.
Unit 7.5 Hire purchase	**2B** Unit 7	• I understand that there are different ways to pay for something. • I can recognise the terms credit, debit and debt. • I understand and can explain the products and services that banks provide.	**MNU 2-09c** • Calculates profit and loss accurately; for example, when working with a budget for an enterprise activity.

Conceptual context

Teaching pupils to become financially astute involves much more than the development of mathematical skills such as coin recognition and calculating the percentage increase and decrease of a price. Developing *financial knowledge and understanding* will help pupils to understand the concept of money, the nature of money and its functions and uses. Guiding pupils through day-to-day money management will enhance their *financial skills and competence*. Raising pupils' awareness of the wider impact of money and personal finance decisions for their own future, their families and their community teaches *financial responsibility*.

Teaching pupils to make sound financial decisions is essential to help them be able to identify and make the best use of opportunities in a fast-changing world. Pupils who are financially capable can use credit and debit cards, manage loans, buy goods and services online and make the best decisions regarding all their financial transactions.

In Book 2C, using bank cards (e.g. debit, loyalty and credit cards) is explored. Making links with a local credit union can help to bring this subject alive for pupils who may not bank regularly. Financial capability can also be addressed in other curricular areas, e.g. Social Studies, Literacy, Technologies and Expressive Arts.

Resources

- Plastic coins and notes
- Coin value Resource sheet 7a
- Dice marked £100, £100, £75, £50, £50, £10
- Budget Sheet, Resource sheet 7b

Key vocabulary

- **Afford**
- **Budget**
- **Compare**
- **Credit**
- **Currency**
- **Debit**
- **Debt**
- **Decimal notation**
- **Discount**
- **Expenses**
- **Hire purchase**
- **Hundredth**
- **Least**
- **Loss**
- **Profit**
- **Receipt**
- **Sale**
- **Tenth**

Misconceptions

Potential misconception	Addressing the misconception
Unit 7.1 Pupils recognise the coins and can tell you how much they have without understanding the value of each coin.	Go over each coin with pupils and ask them to write down its decimal value, write it as a fraction and explain what it is equivalent to.

Teaching ideas

Unit 7.1 Money problems using the four operations

- Ask pupils the following question: **If I had £10·00 and I have £3·25 change, how much did I spend?** With pupils, discuss how they can use the information that they have to calculate the answer and what operations they could use.

To calculate the change from £10, pupils can either do a column subtraction or round to the nearest pound:

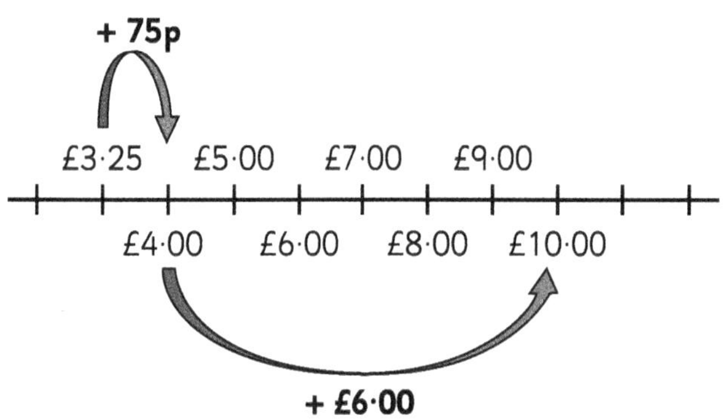

75p + £6·00 = £6·75

- Now ask pupils how they would calculate the following: **Lewis gets £3·50 per week for tidying his room. How long will it take him to save up for a game that costs £32?** Discuss the different ways that pupils could calculate this. Explore multiples of 10 as an option.

Unit 7.2 Budgeting

- Discuss with pupils that before they are ready to create a budget, they must think carefully about the things that they want and the things that they need. Ask them to list their wants and their needs and share these lists with a partner.
- Go over the fact that *needs* are the things we must have to live, such as food, clothing and somewhere to live. However, *wants* are things we would like to have, for example a new game or toy. Basically, *wants* are things we can live without.

- Ask the group if they can think of occasions when they may get a boost to their savings (e.g. birthday money, a bonus at work, etc.). Then ask if they can think of times when a sudden extra bill may arrive (damage to the house/car due to an accident, having to replace something that is broken, etc.). Explore with the group why saving and budgeting is important based on the discussion about unexpected bills.

Classroom currency

Ask pupils to create their own classroom currency.

- Ask pupils to think about all of the things that they use on a daily basis within the classroom (e.g. desks, chairs, jotters, pencils, rubbers, sharpeners) then ask them to give these items a value based on the classroom currency they have created. Now, ask them to think about all the daily jobs they do in the classroom and consider how much they should be paid for these roles. Once you have agreed as a class on a fair price for each item and a reasonable salary for the roles, ask pupils to work in pairs to create a budget so they can purchase all of the items they will need to do their work.

Unit 7.3 Profit and loss

- Ask pupils to write a definition of profit and loss in their maths jotters. Encourage them to share their understanding with a partner before opening the discussion up to the whole class. At this stage, you want to ensure that pupils clearly understand that, in order to be successful, a business needs to make a profit; this means that the selling price is higher than the amount it costs to make an item. Pupils should be able to explain that if you buy something and then sell it for more than you paid for it, you have made a profit.
- Pupils should also have an understanding that if a business sells an item for less that it cost them to make it, they have made a loss. Making a loss means that the business is not generating money; it therefore needs to consider alternative ways to be successful.

Classroom currency

- Extending the activity from 7.2, you could ask pupils how they could create a profit using their classroom currency.
- One idea could be that a group or individual who has carried out a variety of different jobs could buy up all the pencils and then sell them for more that he/she paid for them.
- Profit and loss can also be explored through an enterprise activity, e.g. the school tuck shop or perhaps the parent council could come into class to explain how they raise funds.

Unit 7.4 Discounts

- Ask pupils to write down a definition to explain what a discount is and where you would get one.
- Discuss with pupils that a discount is any reduction that is added to the original price of an item that results in it being cheaper than it was previously. When a discount is applied, there is often a percentage of the normal price, or the discount percentage, deducted from the normal price. The new, lower price is also known as the discounted price. When you are interested in buying an item, often people wait until the item is on sale, or offered at a discount. This way you can save your money. In some stores they offer discounts every day but, you should also check the quality of the item, as there may be a reason you are getting the item at a discount.
- Explain to pupils that fractions and percentages are often used to advertise discounts:

  $\frac{1}{2}$ price   25% off $\frac{1}{3}$ of the original price
- Ask pupils to calculate the sale price of a pair of trainers that were originally £80 but are now 25% cheaper.

  Are they familiar with the fact that 25% is the same as $\frac{1}{4}$?

  Can they identify the discount amount and the new price?

  £80 ÷ 4 = £20 therefore the discount is £20

  £80 – £20 = £60 therefore the sale price is £60

Classroom currency

- Extending the activity from 7.3, you could ask pupils how they could introduce a discount or loyalty scheme using their classroom currency.
- One idea could be that you could trade in equipment and receive a discount when buying something else. Another idea could be a loyalty card, so each time you buy a pencil you receive a stamp and for every 6 stamps you receive a free pencil.

Unit 7.5 Hire purchase

- Introduce the concept of hire-purchase or 'HP' to pupils. A customer can purchase goods by making a small deposit or down-payment. The outstanding balance is then calculated, a percentage of interest is added to this, and the customer pays this sum over a given number of equal monthly instalments.
- Ask pupils why they think someone might choose this payment option. You could also ask them what the advantages and disadvantages might be. Point out that by selecting this payment method, the customer will pay more than the original price unless there is an interest-free deal.
- Share with pupils that the hire-purchase agreement was developed in the United Kingdom in the 19th century to help customers who had a cash-flow problem to make an expensive purchase they otherwise would not be able to make.
- Ask pupils to work out the following:

  Mr Irwin wants to buy a new bike. The shop has one advertised for £475 with an interest-free deal. If he pays a £75 deposit and agrees to pay the balance over 10 months, how much will he pay per month?

  £475 – £75 = £400

  £400 ÷ 10 = £40 therefore Mr Irwin would pay £40 per month

Classroom currency

- Extending the activity from 7.4, you could ask pupils how they could incorporate hire purchase.
- One idea could be that, instead of buying your desk and chair, you could put down a deposit and pay weekly.

8 Time

Time 2A

Section overview

Textbook 2A units	Review units	Prior learning	Benchmarks/Es&Os
Unit 8.1 Telling the time to the minute – 12-hour clock	1C Units 7.4 and 7.5	• I can tell the time in 5-minute intervals or less using analogue and digital 12-hour clocks. • I can record 12h time using a.m. and p.m. notation. • I can convert digital time to analogue and vice-versa. • I can use digital and analogue clocks to show and read minutes to and past the hour.	**MNU 2-10b** **MNU 2-10c** • Reads and records time in both 12-hour and 24-hour notation and converts between the two. • Knows the relationships between commonly used units of time and carries out simple conversion calculations; for example, changes $1\frac{3}{4}$ hours into minutes.
Unit 8.2 Converting between 12h and 24h time	1C Units 7.4 and 7.5	• I can make links between 24h and 12h clock (hours only). • I can use 24h clock to show and read time (hours only). • I can write the time from an analogue clock – 12h and 24h. • I know that 1 day = 24 hours.	**MNU 2-10b** • Reads and records time in both 12-hour and 24-hour notation and converts between the two. • Knows the relationships between commonly used units of time and carries out simple conversion calculations; for example, changes $1\frac{3}{4}$ hours into minutes.
Unit 8.3 Converting minute intervals to fractions of an hour	1C Unit 7	• I can read $\frac{1}{4}$ past, $\frac{1}{2}$ past and $\frac{1}{4}$ to on analogue and digital clocks. • I know that an hour can be divided equally into 2 halves. • I know that an hour can be divided equally into 4 quarters.	**MNU 2-10b** • Reads and records time in both 12-hour and 24-hour notation and converts between the two. • Knows the relationships between commonly used units of time and carries out simple conversion calculations; for example, changes $1\frac{3}{4}$ hours into minutes. • Selects the most appropriate unit of time for a given task and justifies choice.

Unit 8.4 Calculating time intervals or durations	**1C** Unit 7	• I can tell the time in 5-minute intervals using analogue and digital clocks. • I know basic time facts including – 60 minutes in 1 hour, 24 hours in 1 day. • I can use and interpret a variety of timetables. • I can estimate time intervals or durations using the appropriate unit of measure.	**MNU 2-10a** **MNU 2-10b** • Reads and records time in both 12-hour and 24-hour notation and converts between the two. • Knows the relationships between commonly used units of time and carries out simple conversion calculations; for example, changes $1\frac{3}{4}$ hours into minutes. • Uses and interprets a range of electronic and paper-based timetables and calendars to plan events or activities and solve real-life problems. • Calculates durations of activities and events including situations bridging across several hours and parts of hours using both 12-hour clock and 24-hour notation. • Estimates the duration of a journey based on knowledge of the link between speed, distance and time. • Chooses the most appropriate timing device in practical situations and records using relevant units, including hundredths of a second. • Selects the most appropriate unit of time for a given task and justifies choice.
Unit 8.5 Speed, time and distance calculations	**1C** Unit 7	• I recognise key vocabulary when measuring time including – speed, distance, miles, kilometres. • I know basic time facts including – 60 minutes in 1 hour, 24 hours in 1 day. • I can use and interpret a variety of timetables. • I can estimate time intervals or durations using the appropriate unit of measure.	**MNU 2-10a** **MNU 2-10c** • Knows the relationships between commonly used units of time and carries out simple conversion calculations; for example, changes $1\frac{3}{4}$ hours into minutes. • Uses and interprets a range of electronic and paper-based timetables and calendars to plan events or activities and solve real-life problems. • Estimates the duration of a journey based on knowledge of the link between speed, distance and time. • Chooses the most appropriate timing device in practical situations and records using relevant units, including hundredths of a second. • Selects the most appropriate unit of time for a given task and justifies choice.

Conceptual context

As pupils learn about time, they will explore:

1. The duration of an event – the passing of time.
2. Telling the time.
3. Measuring and calculating speed, distance and time.

Before starting this chapter, pupils will have already started to tell the time and make links to their everyday routine. They will also have used a calendar and/or timetable to track events and develop the concept of passing time by measuring the length of time it takes for familiar tasks. To help pupils grasp the concept of time, incorporate vocabulary such as morning, noon, afternoon, evening, and night times into your general discussion. Let the pupils know when activities occur during specific times of the day – 'you eat breakfast in the morning,' or 'you go to bed at night,' to assist in learning these periods of time. *Recorded time*, which is the time at which an event takes place, is referred to as: o'clock, a.m. and p.m., the 24-hour system and so on. *Time-intervals* is the length of time an activity lasts for and is measured in units such as seconds, minutes, hours, days, weeks, years, decades and so forth.

As pupils develop their understanding of time, they will master three concepts: duration, telling the time and speed calculations. They will start to learn about duration and about telling the time informally through their interactions with other people both in and out of school. Pupils need to understand the concept of duration for smaller time periods, using non-standard units of time and then, usually, in minutes. Once the pupils have the necessary number skills, they can begin to tell the time in minutes after the hour. They should be exposed to and use a range of displays to show and record time to strengthen the links between written, 12-hour and 24-hour time.

To develop the concept of duration, pupils can make water timers, use pendulums, or count 'the number of claps' as non-standard measurements of time. Once pupils can measure and talk about time through non-standard units of time, they are ready to learn to use standard units of time. At this point, use vocabulary such as minute, hour and second. The pupils should also start to read the time from both analogue and digital clocks. Again, opportunities to develop these concepts can come from the pupil's own experiences. They could be using different ways of measuring the length of the same events and need to agree on a format to become consistent in their measurements. There could be discussion about the accuracy and importance of measuring time, such as being on time to class.

It is likely that pupils will be able to read time to the nearest hour, half hour or even quarter of an hour before understanding the concept of time units because of their early exposure to adult conversation and encouragement. You can develop this early understanding to reading and telling the time by the number of minutes after the hour. Pupils should start to record their answers on their own clock faces (using concrete materials) by drawing clock faces, as digital time and written in words.

Once pupils can use minutes and seconds with confidence, you can reinforce the concept of duration through timing activities. Pupils will have experience of various timing devices such as analogue clocks, digital clocks, watches, sand timers, simple sundials and stopwatches.

As pupils become confident at reading and telling the time, they can begin to solve problems such as estimating, measuring and calculating how long single and multiple events take. **(Note that they will need at least advanced counting or partitioning strategies to solve problems that involve minutes and hours. If there are repeated actions, they will need to use repeated addition, skip counting, multiplication or division.)** This can be built up to reading information from timetables written in 24-hour time and using this information to plan journeys and other events. Links with schools overseas lead naturally to an exploration of the concept of time zones.

Unlike other areas of measurement time is not metric, thus the units of measure can be confusing – 60 seconds in a minute, 60 minutes in an hour, 24 hours in one day – which can make calculating time intervals using subtraction tricky. Using formal subtraction would not be advised; however, 'chunking' along a number line will make it easier to visualise and calculate.

Practical activities can be used to introduce the concept of speed, with obvious links to Science, Health and Wellbeing and PE. The pupils will realise that speed is made up of two measurements: the distance travelled and the time taken to travel that distance. As they develop their understanding, they should be able to use the link between speed, distance and time to estimate then calculate one unknown given the other two. **(Note that substituting into formulae or expressions is a Third Level Mathematics skill.)**

Resources

- Analogue clock faces
- Sand timers
- Stopwatch
- Dice
- Individual whiteboards and whiteboard pens
- Blank analogue clock faces (see Resource sheet 8a)
- Blank digital clock faces (see Resource sheet 8b)
- 12-h/24-h timeline (see Resource sheet 8c)

Key vocabulary

- **Analogue**
- **Anticlockwise**
- **Clockwise**
- **Convert**
- **Destination**
- **Digital**
- **Durations**
- **Interval**
- **Quarter**

Misconceptions

Potential misconception	Addressing the misconception
Units 8.1, 8.2, 8.3 and 8.4 Pupils may not be clear that the hands go around the clock twice in one day.	Using an analogue clock can cause confusion because the hands go around the face twice in one day. It is therefore advisable to make this clear. In addition, allow time to discuss the different types of watches and clocks that pupils may use.
Units 8.1, 8.2, 8.3 and 8.4 Pupils can often confuse minutes past and minutes to the hour.	Take time at the outset using the analogue clock face to establish understanding. Revise the value of the numbers on the clock face – you could ask pupils to count in 5s from 12 around the clock face, emphasising the change after $\frac{6}{30}$ minutes past to minutes to.
Units 8.1, 8.2, 8.3 and 8.4 Pupils might confuse the minute hand and the hour hand.	Using an analogue clock, ensure that pupils are aware that the short hand points to the hour and the longer hand points to the minute.
Units 8.1, 8.2, 8.3 and 8.4 Pupils might not be confident about the direction that the clock hands rotate a round the clock face.	Using an analogue clock, explain that the hands move in a clockwise direction, which means that they move towards the right-hand side, down and around the face of the clock.

Teaching ideas

Unit 8.1 Telling the time to the minute - 12-hour clock

- Review key terms with pupils before they begin the unit, including: analogue, digital, 12h, 24h, a.m. and p.m.
- Ask the following questions:

  What is the difference between the minute and the hour hand?

  What is the time duration between each number on the clock face?

  Time can be recorded in different ways. Can you name them?
- Using an analogue clock face, show the pupils 8:20 a.m. and ask them to write down the time in digital and in words on their whiteboard.
- Clarify how pupils know the difference between the hours and the minutes. Encourage them to write twenty *minutes past* eight including the key vocabulary to demonstrate their understanding.
- In the same way, show 11:45, 12:05 and 6:35 and ask pupils to write in digital and in words.

Unit 8.2 Converting between 12h and 24h time

- Look at 12-hour/24-hour timeline (see Resource sheet 8c) and ask pupils to discuss their observations.
    - ✓ Do they see that the 24-hour time is adding 12 onto the 12-hour time?
    - ✓ Do they recognise that a.m. in 24-hour time is preceded with a 0?
    - ✓ Do they notice that 24-hour time does not have a . in between the numbers?
- Ask the following questions: **What is the difference between the way the time is written in 12h and 24h and can you see a pattern or a rule for writing 24h time?**
- Using an analogue clock, show pupils a time and read it out stating whether it is a.m. or p.m. and ask them to write it in digital form on their whiteboard. Ask them to explain their answer – ensuring that they have fully understood what they have written (reinforce the questions above).
- In pairs, ask pupils to show an analogue time to their partner and get their partner to write down the digital time on their whiteboard and vice versa. Encourage the correct use of vocabulary to reinforce pupil understanding.

Unit 8.3 Converting minute intervals to fractions of an hour

- Discuss with pupils their understanding of fractions and ask if they can explain the meaning of $\frac{1}{2}$, $\frac{1}{4}$ and $\frac{3}{4}$.
- Using a blank clock face with no hands, ask pupils what the numbers represent.
- Draw hands on the clock as shown and ask what time this is.

- Ask the pupils how they would write this in digital format. They may show 0615 or 6:15 or something similar. Explore any differences with the group and show the pupils pictures of digital clocks showing the time.
- Discuss with pupils the fact that the '15' is a 'quarter of 60 minutes'.
- Clean the clock face and draw hands on the clock as shown and ask what time this is.

- Ask the pupils how they would write this in digital format – 2:30.
- Discuss with pupils the fact that the '30' is 'half of 60 minutes'.
- Clean the clock face and draw hands on the clock as shown and ask what time this is.

- Ask the pupils how they would write this in digital format.
- Discuss with pupils the fact that the '45' is '$\frac{3}{4}$ of 60 minutes'.
- Give pupils a blank clock face and get them to split it into quarters, then ask them to write into each quarter how many minutes each one represents.
- Using the numbers to reinforce that each one is equal to 5 minutes, stop at the 3 and get pupils to write down 15 / quarter / $\frac{1}{4}$.
- Continue to count in 5s and get pupils to stop at the 6 and write 30 / half / $\frac{1}{2}$.
- Continue to count in 5s and get pupils to stop at 9 and write 45 / three quarters / $\frac{3}{4}$.
- Continue to count in 5s and get pupils to stop at 12 and write 60 / 1 whole / 1.

Unit 8.4 Calculating time intervals or durations

- Start with two clocks that show times within the same hour and ask pupils to find how many minutes there are between the times.

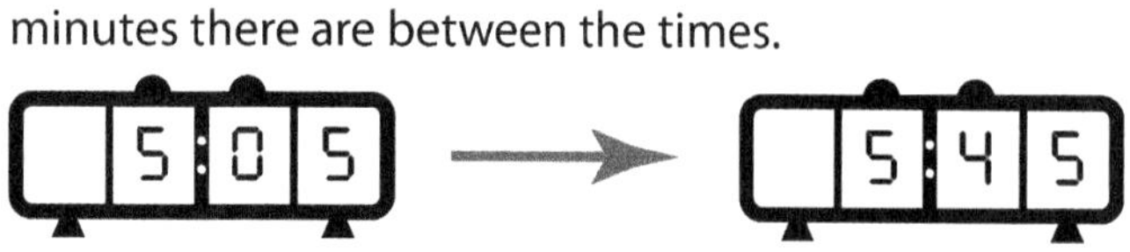

- Discuss with pupils how they can subtract the minutes to calculate the time duration – 45 – 5 = 40 therefore, the time duration is 40 minutes.

- Then show two clocks that show times which bridge the hour and ask pupils to find how many minutes there are between the times.

- Discuss with pupils how they can split the minutes into chunks to calculate the time duration

  10:45 – 11:00 = 15 minutes

  11:00 – 11:25 = 25 minutes
- Now, add the minutes together, 15 + 25 = 40 minutes.
- This can also be shown on a number line.

Unit 8.5 Speed, time and distance calculations

- Show the group pictures of a cheetah, a peregrine falcon, a human and a swordfish. Ask the pupils to vote on which is the fastest, slowest, etc. Ask the pupils to explain their choices. Reveal that the order is:

  Peregrine falcon (200 mph; 320 kph)

  Cheetah (75 mph; 120 kph)

  Swordfish (60 mph; 95 kph)

  Human (28 mph; 45 kph)
- With pupils, discuss their understanding of speed in relation to the rate of travel and the distance travelled in a standard unit of time.
- Ask pupils to generate ideas of where speed would be measured – transport, animals, sport, etc.
- Ask pupils to name different units of distance – metres, kilometres, miles, etc.
- Introduce the concept of how time relates to measuring speed and distance – miles per hour (mph), kilometres per hour (kph).
- With pupils, discuss that there are three factors that are taken into consideration when calculating a journey – distance, speed and time. Now look at 'Let's Learn' section on page 122.

Time 2B

Section overview

Textbook 2B units	Review units	Prior learning	Benchmarks/Es&Os
Unit 8.1 Reading and writing 12h and 24h time	**2A** Unit 8.2	• I can tell the time in 1-minute intervals or less using analogue and digital 12-hour clocks. • I demonstrate an awareness of 12-hour and 24-hour time. • I can convert digital time to analogue and vice-versa. • I can use digital and analogue clocks to show and read minutes to and past the hour. • I can estimate a period of time.	**MNU 2-10b** • Reads and records time in both 12-hour and 24-hour notation and converts between the two. • Knows the relationships between commonly used units of time and carries out simple conversion calculations; for example, changes $1\frac{3}{4}$ hours into minutes.
Unit 8.2 Converting units of time using fractions	**2A** Unit 8.3	• I can read $\frac{1}{2}$ and $\frac{1}{4}$ past on digital and analogue clocks. • I can show $\frac{1}{2}$ and $\frac{1}{4}$ past on digital and analogue clocks. • I can convert digital time to analogue and vice-versa using $\frac{1}{2}$ and $\frac{1}{4}$ past. • I can use digital and analogue clocks to show and read minutes to and past the hour.	**MNU 2-10b** • Knows the relationships between commonly used units of time and carries out simple conversion calculations; for example, changes $1\frac{3}{4}$ hours into minutes. • Selects the most appropriate unit of time for a given task and justifies choice.
Unit 8.3 Calculating time intervals using timetables	**2A** Unit 8.4	• I can place times on a daily timetable. • I can read the times on a timetable. • I can estimate a period of time.	**MNU 2-10a** **MNU 2-10b** • Reads and records time in both 12-hour and 24-hour notation and converts between the two. • Knows the relationships between commonly used units of time and carries out simple conversion calculations; for example, changes $1\frac{3}{4}$ hours into minutes. • Uses and interprets a range of electronic and paper-based timetables and calendars to plan events or activities and solve real-life problems. • Calculates durations of activities and events including situations bridging across several hours and parts of hours using both 12-hour clock and 24-hour notation.

Unit 8.4 Measuring time	**1C** Unit	• I can place times on a daily timetable. • I can understand how long a minute is. • I understand what can be done in a minute. • I can estimate a period of time.	**MNU 2-10a** **MNU 2-10b** • Reads and records time in both 12-hour and 24-hour notation and converts between the two. • Knows the relationships between commonly used units of time and carries out simple conversion calculations; for example, changes $1\frac{3}{4}$ hours into minutes. • Uses and interprets a range of electronic and paper-based timetables and calendars to plan events or activities and solve real-life problems. • Calculates durations of activities and events including situations bridging across several hours and parts of hours using both 12-hour clock and 24-hour notation.
Unit 8.5 Speed, distance and time calculations	**2A** Unit 8.5	• I understand key vocabulary when measuring time – speed, distance. • I know key time facts. • I can interpret and use timetables. • I can estimate time intervals or durations using the appropriate unit of measure – cm, metres, miles, kilometres.	**MNU 2-10a** **MNU 2-10c** • Knows the relationships between commonly used units of time and carries out simple conversion calculations; for example, changes $1\frac{3}{4}$ hours into minutes. • Uses and interprets a range of electronic and paper-based timetables and calendars to plan events or activities and solve real-life problems. • Calculates durations of activities and events including situations bridging across several hours and parts of hours using both 12-hour clock and 24-hour notation. • Estimates the duration of a journey based on knowledge of the link between speed, distance and time. • Chooses the most appropriate timing device in practical situations and records using relevant units, including hundredths of a second. • Selects the most appropriate unit of time for a given task and justifies choice.

Unit 8.6 Time problems		• I can understand basic time facts – *60 secs = 1 min; 60 min = 1 hr…* • I can read and write analogue and digital time. • I understand 12h and 24h time. • I can estimate a period of time.	**MNU 2-10a** **MNU 2-10b** • Reads and records time in both 12-hour and 24-hour notation and converts between the two. • Knows the relationships between commonly used units of time and carries out simple conversion calculations; for example, changes $1\frac{3}{4}$ hours into minutes. • Uses and interprets a range of electronic and paper-based timetables and calendars to plan events or activities and solve real-life problems. • Calculates durations of activities and events including situations bridging across several hours and parts of hours using both 12-hour clock and 24-hour notation.

Conceptual context

This section will continue to develop pupils' knowledge of the concept of time by exploring duration, telling the time and speed/distance/time calculations. To support this, reinforce vocabulary such as morning, noon, afternoon, evening, and night times into your general discussion. Let the pupils know when activities occur during specific times of the day; 'you come to school in the morning,' or 'you go to bed at night,' to assist in learning these periods of time. Discuss the difference between *recorded time*, which is the time at which an event takes place and *time-intervals*, which is the length of time an activity lasts for and is measured in units. When using an analogue clock, make it clear that the hands go around the face twice in one day and allow time to discuss the different types of watches and clocks that pupils may use.

Ensure that pupils know that time is not metric, unlike other areas of measurement, and the units of measure are different – 60 seconds in a minute, 60 minutes in an hour, 24 hours in 1 day – which can make calculating time intervals using subtraction tricky. Using formal subtraction would not be advised; however, 'chunking' along a number line will make it easier to visualise and calculate. Encourage dialogue with pupils about the different instruments that can be used to measure time including some less conventional methods. This will highlight the depth of understanding pupils have about the concept of time.

Consolidating pupils' knowledge of fractions is key to ensuring they will have an understanding of converting units of time using fractions and that each time they are finding $\frac{1}{4}$, $\frac{1}{2}$ and $\frac{3}{4}$ of 60 minutes. When reading and writing times, use a variety of terms to reinforce the relationships for pupils – 15 minutes past 3, quarter past 3, three fifteen – to ensure they recognise that it is the same time. Encourage pupils to name and correctly use different instruments of measure to calculate time so they are exposed to as many different experiences as possible.

Resources

- Analogue clock faces
- Sand timers
- Stopwatch
- Blank analogue and digital clock faces (see Resource sheet 8d)
- Bingo (see Resource sheet 8e)

Key vocabulary

- **Actual**
- **Analogue**
- **Anticlockwise**
- **Clockwise**
- **Convert**
- **Destination**
- **Digital**
- **Durations**
- **Estimate**
- **Hourglass**
- **Instrument**
- **Interval**
- **Quarter**
- **Sundial**

Misconceptions

Potential misconception	Addressing the misconception
Units 8.2, 8.4 and 8.6 Pupils may think that time can only be measured using a clock and in seconds, minutes and hours.	Through questioning and quality discussion, ensure that pupils are familiar with the many different **instruments of measure** that are used – clock, watch, stopwatch, sundial, calendar, diary, etc. Also, ensure that pupils are familiar with the different **units of measure** that are used – milliseconds, seconds, minutes, hours, days, weeks, years, decades…
Unit 8.1 Pupils may not understand that the hands go around an analogue clock once in a day.	As there are only the digits 1–12 on the clock face, it is important to make pupils aware that the hands will rotate around the clock twice in 1 day/24 hours.
Units 8.1, 8.2 and 8.3 Using formal addition or subtraction would not be advised.	Using formal addition or subtraction calculations will not be accurate as you are working within 60, but 'chunking' along a number line will make it easier to visualise and calculate.

Teaching ideas

Unit 8.1 Reading and writing 12h and 24h time

- Ask pupils about how you know whether a time is referring to the morning or the afternoon.
- Discuss the different ways that this is expressed in analogue, digital, 12-hour and 24-hour.
- Using Resource sheet 8d, read out different times using a range of vocabulary and ask pupils to record their answers on the blank clock faces.
- Ask pupils to write down various activities that they do from early morning to later at night. Beside each one, ask them to record the time for each activity in analogue and digital in both 12-hour and 24-hour format.
- Introduce the 'Let's learn' section on page 134.

Unit 8.2 Converting units of time using fractions

- Play Bingo (see Resource sheet 8e) to allow pupils to practice their understanding of units of time.
- Using an analogue clock, show the pupils a time and ask them to tell you what the time will be in $\frac{1}{2}$ an hour or $\frac{1}{4}$ of an hour: **The clock shows 11:15 am. What time will it be in $\frac{3}{4}$ of an hour?** This will recap their understanding of how many minutes are in $\frac{3}{4}$ of an hour and how to add this on to an existing time. Start by adding within the hour and if pupils are confident you can start bridging the hour. When questioning pupils, ensure you use a wide range of vocabulary to clarify understanding: **If the show starts at 1510 and lasts $\frac{1}{2}$ an hour, what time will it finish? The bus leaves at 7 in the morning and the journey lasts $\frac{1}{4}$ of an hour. What time will it arrive?**

Unit 8.3 Calculating time intervals using timetables

- Ask pupils the following questions to generate discussion prior to starting the unit: **What is a time interval? Why do you need to calculate a time interval? When would you use a timetable?**
- Establish pupils' understanding of the different types of timetable: bus, train, daily, TV schedule by showing some images on the board. Ask pupils why these different sorts of timetables help us and how they provide important information to allow us to plan our day.
- Use sequential vocabulary when discussing the times on a timetable: **What time is the *first* bus? What time is the *next* bus? When is the *last* bus?** This will reinforce the pupils' understanding of the order events occur.

Unit 8.4 Measuring time

- Start by getting pupils to generate a work bank of all the ways they know that time can be measured. Explain that this is not a test but an opportunity to share their ideas about prior knowledge. Ask pupils some of the most unusual ways they have measured time: the length of a song they have been listening to, the duration of a programme, the time it takes to play a game, etc.
- Think about an activity that pupils do on a daily basis; for example, tie their laces and ask them to estimate how long this activity will take them. Then, using a stopwatch, time them carrying out this task. Discuss how their estimate compared with the actual time. **Were you surprised by the result? If so, why?**

Unit 8.5 Speed, distance and time calculations

- Discuss the key vocabulary with pupils to ensure they understand that *distance*, *speed* and *time* are units of measure. Take one of the words at a time and explore them: **What is speed? How can distance be measured? What unit of speed would you use to measure the speed of a car/plane/boat/snail/human? Can you name as many units of measure for speed as possible? – miles or kilometres per second, minute, hour, etc.**
- Speed is a way of measuring how quickly an object moves from one place to another. It is equal to the distance travelled divided by the time it has taken. It is possible to find any of these three values by using the other two. Explain the relationship between speed, distance and time and the fact that as long as you have 2 out of 3 values, you can calculate the third.
- Measure the pupils running over a distance of 10 metres. Discuss how you can work out the speed of each pupil. If you have access to equipment such as balloons, toy cars or fans and sail toys the group can set up tracks, measure the distance and time, then calculate the speed of each one. This activity can be varied by changing the size of the balloon/sail/fan and so forth.

Unit 8.6 Time problems

- This unit will consolidate the pupils' knowledge of time facts whilst reinforcing their ability to identify key process words within word problems. Start by asking pupils to answer the following questions: **How many minutes in an hour? How many seconds in a minute? How many hours in a day? How many months in a year? How many days in a week/fortnight?**
- Ask the pupils the following question:

  'Mr Tortolano walks his dog to the park twice every day. It takes him $\frac{1}{2}$ an hour to walk there and $\frac{1}{2}$ an hour to walk back.

  How many minutes does he walk a day?'

Look at the question with the pupils and ask them how they will find the answer. Ask them: **What are the key pieces of information that will help them find the answer**.

'Mr Tortolano walks his dog to the park twice every day. It takes him $\frac{1}{2}$ an hour to walk there and $\frac{1}{2}$ an hour to walk back.

How many minutes does he walk a day?'

Bring their attention to the words that are underlined and ask: If he walks $\frac{1}{2}$ an hour there and $\frac{1}{2}$ an hour back twice a day, how long altogether is he walking? Clarify their understanding that $\frac{1}{2}$ an hour is equal to 30 minutes, therefore it is 30×4 that is the calculation that they need to work out.

Time 2C

Section overview

Textbook 2C units	Review units	Prior learning	Benchmarks/Es&Os
Unit 8.1 Investigating how long a journey will take	**2B** Unit 8.3	• I can use a timetable to identify start and end time. • I can calculate time durations. • I can identify different ways to identify how long a journey will take.	**MNU 2-10a** **MNU 2-10b** • Reads and records time in both 12-hour and 24-hour notation and converts between the two. • Knows the relationships between commonly used units of time and carries out simple conversion calculations; for example, changes $1\frac{3}{4}$ hours into minutes. • Uses and interprets a range of electronic and paper-based timetables and calendars to plan events or activities and solve real-life problems. • Calculates durations of activities and events including situations bridging across several hours and parts of hours using both 12-hour clock and 24-hour notation.
Unit 8.2 Calculating the duration of activities and events	**2B** Units 8.3 and 8.4	• I can calculate time durations. • I can read and write 12h and 24h time. • I can convert digital time to analogue and vice versa. • I can use timetables to calculate time durations.	**MNU 2-10a** **MNU 2-10b** • Reads and records time in both 12-hour and 24-hour notation and converts between the two. • Knows the relationships between commonly used units of time and carries out simple conversion calculations; for example, changes $1\frac{3}{4}$ hours into minutes. • Uses and interprets a range of electronic and paper-based timetables and calendars to plan events or activities and solve real-life problems. Calculates durations of activities and events including situations bridging across several hours and parts of hours using both 12-hour clock and 24-hour notation.

Unit 8.3 Investigating ways speed, time and distance can be measured	**2B** Unit 8.5	• I understand the terms Speed, Time and Distance. • I can use the Distance = Speed × Time calculation. • I can understand the terms mph and kph.	**MNU 2-10a** **MNU 2-10c** • Knows the relationships between commonly used units of time and carries out simple conversion calculations; for example, changes $1\frac{3}{4}$ hours into minutes. • Uses and interprets a range of electronic and paper-based timetables and calendars to plan events or activities and solve real-life problems. • Calculates durations of activities and events including situations bridging across several hours and parts of hours using both 12-hour clock and 24-hour notation. • Estimates the duration of a journey based on knowledge of the link between speed, distance and time. • Chooses the most appropriate timing device in practical situations and records using relevant units, including hundredths of a second. • Selects the most appropriate unit of time for a given task and justifies choice.
Unit 8.4 Calculating time accurately using a stopwatch	**2B** Unit 8.3	• I can estimate a period of time. • I understand different units of time. • I can use the appropriate unit of time for an activity. • I can record the time an activity will take.	**MNU 2-10a** **MNU 2-10b** • Reads and records time in both 12-hour and 24-hour notation and converts between the two. • Knows the relationships between commonly used units of time and carries out simple conversion calculations; for example, changes $1\frac{3}{4}$ hours into minutes. • Calculates durations of activities and events including situations bridging across several hours and parts of hours using both 12-hour clock and 24-hour notation.
Unit 8.5 Converting between units of time	**2B** Unit 8.2	• I understand key time vocabulary. • I understand the different units of time. • I can use different units of time appropriately.	

Conceptual context

These units will deepen the pupils' understanding of time by looking at it as a dimension in which events can be ordered from the past through to the present and into the future. In addition, it is also the measurement of durations of events and the intervals that exist between them.

Explain to pupils that time is not an object or substance they can touch or see. Time has many aspects and represents different things to different people in a variety of different circumstances. People use the word 'time' in a multitude of ways to express a number of different things. There are numerous sayings that relate to time:

'*Time* seemed to stand still.'

'Do you have the *time*?'

'It will take a very long *time*!'

'This is not the first *time* you have done that.'

'It's *time* for dinner.'

'I don't have *time* today.'

'Sing in *time* to the music.'

As pupils become confident at reading and telling the time, they can begin to solve problems such as estimating, measuring and calculating how long both single and multiple events take. This can be extended to reading information from timetables written in 24-hour clock and using this information they can plan journeys and other events. Linking with schools overseas will lead naturally to an exploration of the time zones.

Once pupils are confident at reading, telling and measuring time using seconds, minutes and hours, tenths and hundredths of a second can be introduced. This can be done through discussion of how we measure time at sporting events. (Ensure pupils have place-value knowledge of tenths and hundredths before you introduce this concept within the context of time.) There may be confusion about why we can divide seconds into 10 or a hundred equal parts when we divide minutes and hours into 60 equal parts. It is important that pupils have the opportunity to explore this before they are asked to measure or calculate using tenths of a second. An interesting discussion would be to introduce the fact that the Egyptians were the first to divide the day into 24 equal parts around 1500 BCE. The second is an SI (metric) unit and is subdivided by powers of 10 into units such as deciseconds and milliseconds. Minutes, hours and days are not SI units and retain their original 60 or 24 divisions.

Resources

- Analogue clock faces
- Sand timers
- Stopwatch
- Counters
- Spinner
- Blank timetable
- Cubes

Key vocabulary

- **Actual**
- **Analogue**
- **Anticlockwise**
- **Clockwise**
- **Convert**
- **Destination**
- **Digital**
- **Durations**
- **Elapsed**
- **Estimate**
- **Formula**
- **Hourglass**
- **Instrument**
- **Interval**
- **Investigating**
- **Quarter**
- **Sundial**

Misconceptions

Potential misconception	Addressing the misconception
Units 8.2, 8.4 and 8.6 Pupils may think that time can only be measured using a clock and in seconds, minutes and hours.	Through questioning and quality discussion ensure that pupils are familiar with the many different **instruments of measure** that are used – clock, watch, stopwatch, sundial, calendar, diary… Also, ensure that pupils are familiar with the different **units of measure** that are used – milliseconds, seconds, minutes, hours, days, weeks, years, decades…
Unit 8.1 Pupils may think that the hands go around an analogue clock once in a day.	As there are only the digits 1–12 on the clock face, it is important to make pupils aware that the hands will rotate around the clock twice in 1 day/24 hours.
Units 8.1, 8.2 and 8.3 Using formal addition or subtraction would not be advised.	Using formal addition or subtraction calculations will not be accurate as you are working within 60, but 'chunking' along a number line will make it easier to visualise and calculate.

Potential misconception	Addressing the misconception
Units 8.5 and 8.6 There may be confusion about why we can divide seconds into 10 or a hundred equal parts when we divide minutes and hours into 60 equal parts.	It is important that pupils have the opportunity to explore this before they are asked to measure or calculate using tenths of a second.

Teaching ideas

Unit 8.1 Investigating how long a journey will take

- Ask the pupils to consider why it is important to know how long a journey will take. There is no right answer here, but possible responses might include, 'to know what time to leave…', 'to know what time you will arrive…', 'to plan another activity or event…', 'to be able to meet someone…'.
- With pupils, consider the consequences of not knowing how long an activity or journey might take Again, this may elicit a number of different answers, for example 'you might miss the bus…', 'you could keep someone waiting…', 'the dinner might burn…'.
- Tell the group they are going to plan a journey. The activity can be introduced in a number of ways – some pupils went to the cinema, to the shops, on a school trip…
- Ask pupils to create an itinerary for their activity using a bus or train timetable (check online to provide a timetable for the pupils to use).
- Ask questions to challenge pupils:
    - What information do you need and how will you get it?
    - When is the best time of day to travel? Why?
    - It is better to travel during the week or at the weekend?
    - How much will it cost?
    - How do you calculate the cost of the journeys?
    - Is it quicker/cheaper to travel by train or bus?
    - How will you present this information to the class?
- Each group will deliver a presentation to the class, sharing their learning and any challenges they have faced.

Unit 8.2 Calculating the duration of activities and events

- Ask pupils if they can tell you the duration of their school day. Ask them the following questions: **How do they know this? Can they check this information? Why is it important to know this information?**
- Use post-its to write down the different things that pupils do within the school day and get them to order them in sequential order, emphasising the timeline from morning to night.

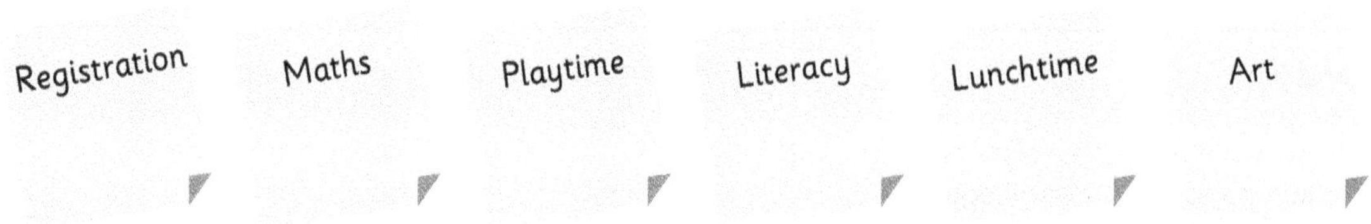

Next, attribute a time to each of them…

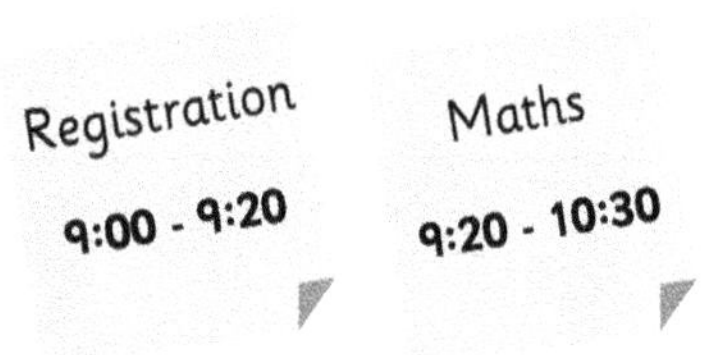

- Once you have established a structure for the day, you can then ask pupils: **How long do you spend at Registration?** This will reinforce their understanding of the passing of time and length of time they will spend doing a special activity or event.
- Tell pupils that playtime is 15 minutes and ask them to tell you what time Literacy will start. Guide them to the information that they already know, e.g. Maths finishes at 10:30, and encourage them to use their existing knowledge to solve the problem.
- Challenge pupils to work out what time Art will start by telling them that Literacy lasts for the same time as Maths and that Lunchtime is 1 hour long.

Unit 8.3 Investigating ways speed, time and distance can be measured

- Ask pupils to write their own definitions for speed, distance and time and explain them to each other, their group or the class.
- Show pupils the speed, distance, time triangle and ask them to explain their understanding of what it is.

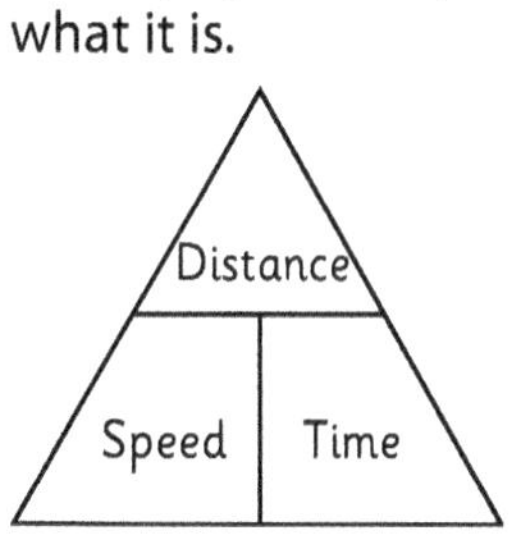

- Explore the way the triangle is set out and ask them why distance is on top.
- Ask pupils to identify the relationship between distance, speed and time.
- Once you have established an understanding, ask pupils what a formula is and if they can create a formula (or rule) to calculate distance, speed and time. The pupils can then use these rules to help them solve the problems in the unit.

Unit 8.4 Calculating time accurately using a stopwatch

- Establish with pupils their understanding of what a stopwatch is and what it is used for. Ask the following questions: **What is a stopwatch? When would you use a stopwatch? Why would you use a stopwatch and not a watch? Can you name activities that you would time with a stopwatch?**
- Establish an understanding of the units of measure that a stopwatch will display. Clearly explain the minutes, seconds and milliseconds to ensure that pupils are clear what they are recording prior to starting.

Unit 8.5 Converting between units of time

- Introduce this Unit by asking pupils to name as many units of time as they can. You are looking for them to have the basic units – seconds, minutes, hours, days, hours, days, weeks, months and years.
- For each unit of time that they list, ask them to give an example of what they would use this unit of time to measure to clarify their understanding of each one.
- Ask the pupils to create a poster to display the following facts – they can research online if they have internet access:

  60 seconds = 1 minute

  60 minutes = 1 hour

  24 hours = 1 day

  7 days = 1 week

  365 days = 1 year

  52 weeks = 1 year

  10 years = 1 decade

9 Measurement

Measurement 2A

Section overview

Textbook 2A units	Review units	Prior learning	Benchmarks/Es&Os
Unit 9.1 Using familiar objects to estimate length, mass, area and capacity	**1B** Units 8.1, 8.2, 8.3 and 8.4	• Estimate, measure and record length, mass and capacity using standard units. • Estimate, measure and record area using squares.	**MNU 2-11a** **MNU 2-11b** **MNU 2-11c** • Uses the comparative size of familiar objects to make reasonable estimations of length, mass, area and capacity. • Estimates to the nearest appropriate unit, then measures accurately: length, height and distance in millimetres (mm), centimetres (cm), metres (m) and kilometres (km); mass in grams (g) and kilograms (kg); and capacity in millilitres (ml) and litres (l). • Calculates the perimeter of simple straight-sided 2D shapes in millimetres (mm), centimetres (cm) and metres (m). • Calculates the area of squares, rectangles and right-angled triangles in square millimetres (mm^2), square centimetres (cm^2) and square metres (m^2). • Calculates the volume of cubes and cuboids in cubic centimetres (cm^3) and cubic metres (m^3). • Converts between common units of measurement using decimal notation; for example, 550 cm = 5·5 m; 3·009 kg = 3009 g. • Chooses the most appropriate measuring device for a given task and carries out the required calculation, recording results in the correct unit. • Reads a variety of scales accurately.
Unit 9.2 Estimating and measuring length	**1B** Units 8.1 and 8.8 **1C** Unit 8.1	• Estimate, measure and record length / height using standard units. • Read scales on measuring devices: length (whole numbers). • Read and record measurements of length to simple fractions.	
Unit 9.3 Estimating and measuring mass	**1B** Units 8.2 and 8.9 **1C** Unit 8.4	• Estimate, measure and record mass using standard units. • Read scales on measuring devices: mass (whole numbers). • Read and record measurements of mass to simple fractions.	
Unit 9.4 Converting units of length	**1B** Unit 8.5 **1C** Unit 8.3 **2A** Unit 6.5	• Convert simple standard units of length / height (cm, m). • Convert standard units of length (mm, cm). • Write decimal equivalents to tenths.	
Unit 9.5 Calculating the perimeter of simple shapes	**1C** Unit 8.1 **2A** Unit 9.2	• Estimate, measure and record length / height using standard units. • Read scales on measuring devices: length (whole numbers).	
Unit 9.6 Finding the area of regular shapes in square cm/m	**1B** Unit 8.3 **1C** Units 8.7 and 8.9 **2A** Unit 9.2	• Estimate, measure and record area using squares. • Read and record measurements of length. • Link area of a rectangle / square to arrays.	

Unit 9.6 Finding the area of regular shapes in square cm/m	**1B** Unit 8.3 **1C** Units 8.7 and 8.9 **2A** Unit 9.2	• Estimate, measure and record area using squares. • Read and record measurements of length. • Link area of a rectangle / square to arrays.	
Unit 9.7 Finding the volume of cubes and cuboids by counting cubes	**1C** Unit 8.8 **2A** Units 9.2 and 9.6	• Create shapes with a specified area. • Read and record measurements of length. • Find the area of regular shapes in square cm/m.	
Unit 9.8 Estimating and measuring capacity	**1B** Units 8.4 and 8.10 **1C** Unit 8.5	• Estimate, measure and record capacity using standard units. • Read scales on measuring devices: capacity (whole numbers). • Read and record measurements of capacity to simple fractions.	

Conceptual context

The Advisory Committee on Mathematics Education, ACME (2011) describe measurement as, 'a method for describing quantities. It requires: appreciation of conservation; understanding the nature and dimensionality of the quantities to be measured; understanding continuous number; understanding units and their iterative use; and knowing how to use the relevant tools.'

The activities presented in this book continue to build on pupils' understanding of standard metric units and their ability to convert between units. It is important that pupils develop an understanding of single units represented on a scale through their use of measuring devices, such as metre sticks, measuring jugs and scales for measuring mass. When measuring length, key concepts are developed, such as the requirement to leave no gaps between repeated units, ensuring that objects are properly aligned with the start of the scale (zero), and the understanding that the distance doesn't change when using alternative units (e.g. centimetres or millimetres). As pupils develop an understanding of the relationship between units of measurement, they will begin to appreciate when specific units are appropriate and when it is necessary to estimate or take a more accurate measurement.

As confidence of standard units increases, pupils can start to combine and compare lengths using calculations, e.g. finding the perimeter of a shape by adding the lengths of all the sides. As pupils' understanding of one-dimensional measurement increases, they will begin to see the links to two- and three-dimensional measurements, i.e. area and volume. Length, being the measurement of one point to another; area, being the measurement of surface space (measured in squares); and volume, being the measurement of three-dimensional space (measured in cubes for solid objects and litres for liquids).

Throughout this process, pupils will store benchmarks in their memory that will enable them to *estimate* measurements. Familiar objects become internal benchmarks, such as a bag of sugar is a kilogram or a finger-width is about a centimetre.

Pupils need many opportunities to practise estimating; they also need to know when it is right and proper to *only* estimate and when we need to measure *accurately*. Try to avoid always emphasising 'estimate then check' during a measurement task. Often, the emphasis should be that we sometimes measure to get better at estimating. For example, when decorating a room, I *estimate* the amount of paint I need by imaging the number of square metres to paint. I then check the paint tin instructions and buy enough to cover my estimate. When it comes to the carpet, I *accurately* measure the length and width to ensure a good fit. Pupils need to develop these skills and be able to justify their answer and say how confident they are. They should also start to justify whether an accurate answer is required or an approximate answer is satisfactory.

Resources

- Rulers
- Metre sticks
- Tape measures
- Trundle wheels
- Scales – balance, kitchen, hanging, bathroom
- Weights (one kilogram and 100 grams)
- Base 10 materials
- Measuring jugs
- Funnels
- Plastic bottles

Key vocabulary

- **area**
- **capacity**
- **centimetre**
- **dimension**
- **estimate**
- **gram**
- **height**
- **length**
- **kilogram**
- **mass**
- **metre**
- **millimetre**
- **metre stick**
- **perimeter**
- **volume**

Misconceptions

Potential misconception	Addressing the misconception
Units 9.2 and 9.4 Pupils fail to understand that when measuring length, units are measured from the start of the first unit.	• Use the broken ruler activity (see Teaching Ideas, Unit 9.2) to diagnose and address misconceptions.
Unit 9.5 Pupils fail to understand that perimeter is the measurement of length of all the sides; for example, they may only include the lengths of sides that are displayed in a diagram: 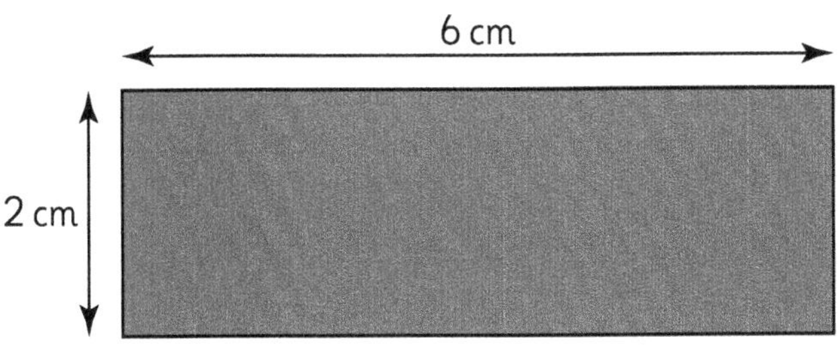 Pupils incorrectly calculate the perimeter to be 8 cm because they only include the lengths shown.	• Provide opportunities for pupils to measure the perimeter of objects in context (see Teaching Ideas, Unit 9.5).
Unit 9.6 Pupils believe, for example, that doubling the length of the sides of a square will result in the area being doubled.	• Provide centimetre squares for pupils to check the area of shapes (see Teaching Ideas, Unit 9.6).
Unit 9.6 Pupils confusing the area of a shape with the perimeter; for example, believing that a square with edges of 1 cm has an area of 4 cm^2 because they count up the length of all four sides.	• Pupils must have experience of using squares to measure area (see Textbook 1B, Unit 8.3 and 1C, Unit 8.7) in order to appreciate that area is a measure of surface space. • These concepts are introduced separately in the book to develop an appreciation of the differences between both. However, the links between area and perimeter should be investigated.
Unit 9.7 Pupils believe, for example, that doubling the length of the sides of a cube will result in the volume being doubled.	• Provide cubic centimetres for pupils to create 3D shapes (see Teaching Ideas, Unit 9.7).
Unit 9.7 Pupils struggle to measure volume where 2D representations are given for 3D objects (i.e. they are required to 'count' cubes that are not visible).	• Provide cubic centimetres for pupils to recreate 3D equivalents of 2D representations (see Teaching Ideas, Unit 9.7).
Unit 9.7 Pupils confusing the volume of an object with the area; for example, believing that a cube with edges of 1 cm has a volume of 6 cm^3 because they count up the number of sides.	• Provide cubic centimetres for pupils to recreate 3D equivalents of 2D representations (see Teaching Ideas, Unit 9.7). • Ask questions such as: **How many cubes are needed to make this shape?**
Units 9.3 and 9.8 Pupils fail to read the scale on a measuring jug or scales correctly because they are unable to identify the value of each increment.	• Provide opportunities for pupils to work out increments on number lines and scales (see Teaching Ideas, Unit 9.7).

Teaching ideas

Unit 9.1 Using familiar objects to estimate length, mass, area and capacity

- This topic could be introduced by reading the book, *Counting on Frank* by Rod Clement.
- Set up four measuring stations in your class. At each station provide a baseline (a familiar object) with which to estimate measurements of other objects. These 'baselines' should be used to compare and estimate the measurements of other objects:
    1. **Length:** base 10 'ten' (baseline: 10 cm) and various objects to estimate, e.g. pencil, jotter, table, etc.
    2. **Mass:** scales, pencil case (baseline: use scales to measure mass accurately) and various objects to estimate, e.g. water bottle, lunchbox, school bag, etc.
    3. **Capacity:** measuring jug, funnel, plastic cup of water (baseline: use measuring jug to measure capacity accurately) and various containers to estimate, e.g. water bottle, jug, cup, etc.
    4. **Area:** 100 cm^2 square paper (measuring 10 cm by 10 cm) and various objects to estimate the surface area, e.g. textbook, A4 paper, poster, etc.
- Organise pupils into four groups and challenge them to use the baseline at each table to estimate the measurements of the various items.
- Encourage groups to come up with statements for each item, e.g. *We think the jotter is two and a half times the length of the 'ten' so we think the jotter is 25 cm long. The plastic cup holds 100 ml and we think the water bottle will hold five times as much, so we think the capacity of the water bottle is 500 ml, etc.*
- Compare the results of each group. Ask questions such as: **Why do you think our estimates are different? When is it necessary to estimate measurements and when is it necessary to measure accurately?** *(e.g. when working out how much paint is needed to paint a wall, we can estimate how much paint is required; when working out how much medicine to give someone, we need an accurate measurement; etc.)*

Unit 9.2 Estimating and measuring length

- Make rulers with the pupils. This could be done using a 'ten' (base 10). Draw a straight line on a piece of paper and mark the intervals:

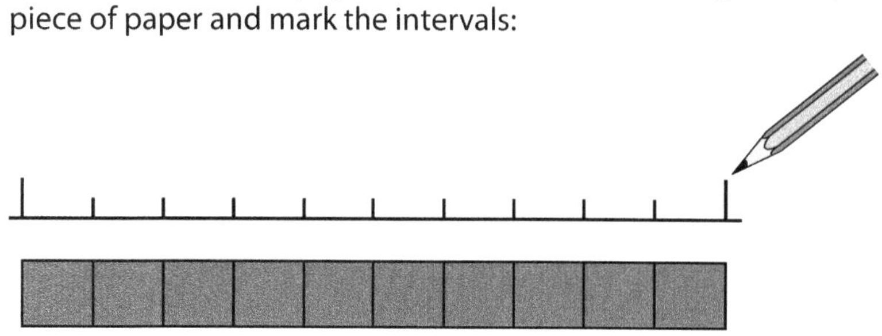

- The pupils must understand that the starting point (often zero but not always!) is the beginning of the first unit; they must count to the end of each unit:

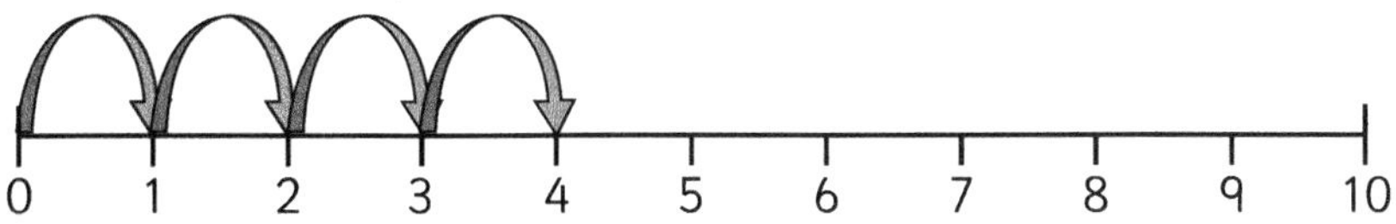

- Ask the pupils to measure the same object, e.g. an eraser or pencil, and compare their results.
- Ask the pupils questions such as: **Are our measurements the same? Why might they be different?**
- Provide pupils with 'broken rulers' to measure an object (to see if they understood the concept above and apply measurement without a zero):

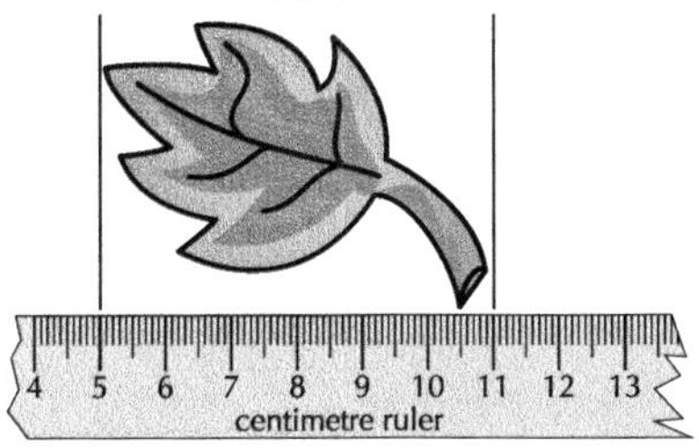

- An answer of 11 would suggest a lack of understanding of units.
- '7' suggests that the pupil counted 5, 6, 7, 8, 9, 10, 11 so 7.
- '6' is the correct answer; most will count the gaps but how many will find the difference of 11 and 5?

Unit 9.3 Estimating and measuring mass

- Set up a set of kitchen scales at the front of the class. You will also need a paper or plastic bag, 1 kg weights and 100 g weights.
- Organise pupils into pairs or groups and provide them with kitchen scales (ideally with increments of 100 g), kilogram weights and 100 gram weights.
- Place 1300 g inside the bag and put it on the scales (not allowing the pupils to see the contents of the bag). Challenge pupils to use the weights provided to match the weight displayed.
- Ask questions such as: **What weights have I used in the bag? What is the total mass of the bag of weights? What will happen if I put in another 100 g weight?**
- Discuss the value of each increment on the scale, i.e. that each increment represents 100 g.
- Repeat for the following weights:
    1. 600 g
    2. 1500 g
    3. 2100 g etc.
- Challenge pupils to display weights more or less than the teacher scales, e.g. **Measure out 300 grams more than I have on my scales.**

Unit 9.4 Converting units of length

- Organise pupils into pairs or groups and provide them with string, a metre stick (or measuring tape), a pen and scissors.
- Instruct pupils to cut two lengths of string measuring exactly 1 m each (use the pen to mark the string before cutting):

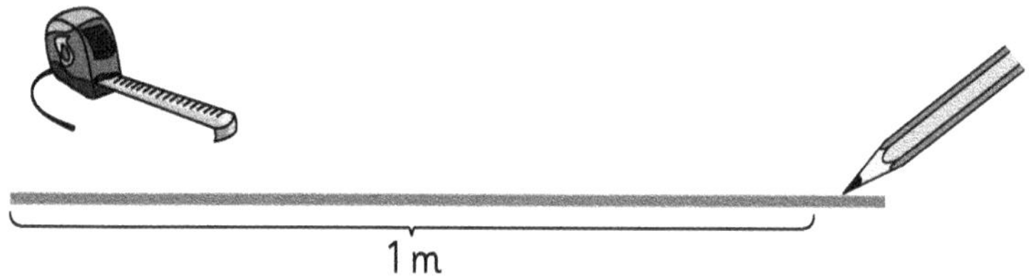

- Instruct pupils to cut one of the metre lengths of string into ten equal parts (use the pen to mark 10 cm intervals):

- **What fraction of a metre are each of these 10 cm lengths?** *(Tenths)*
- Instruct pupils to cut one of the 10 cm lengths of string into ten equal parts (use the pen to mark 1 cm intervals before cutting):

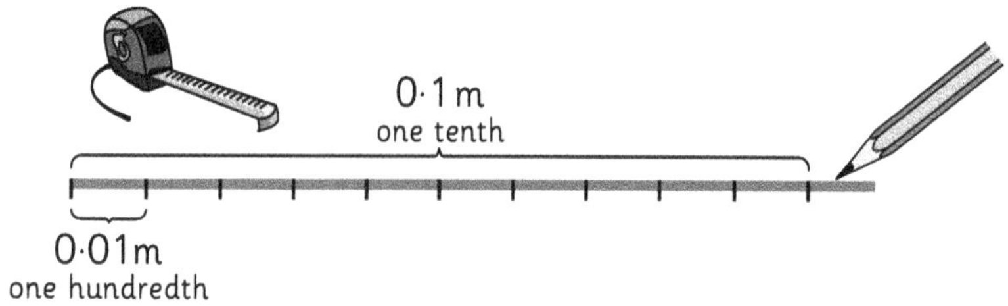

  What fraction of a metre are each of these 1 cm lengths? *(hundredths)*

  Note that, alternatively, a metre stick can be used as well as base 10 'tens' to represent tenths of a metre (0·1 m) and units to represent hundredths of a metre (0·01 m).

- Challenge pupils to measure each other's height (by lying on the floor) using the measuring tape (to measure in centimetres) and the lengths of string (to measure the decimal equivalent).

  Through discussion, ascertain that each of their heights can be expressed in centimetres and metres (using decimals), for example: *Nuria measures 124 cm. We needed 1 m length of string, 2 tenths and 4 hundredths. We can write this as 1·24 m.*

  Write each of your heights in both centimetres and metres.

- Provide pupils with rulers and ask: **How many millimetres are there in a centimetre?** *(10 mm)* **What fraction of a centimetre is one millimetre?** *(One tenth)* **Can you write one millimetre as a decimal fraction of a centimetre?** *(1 mm = 0·1 cm)*
- Provide pupils with objects to measure, e.g. sharpener, pencil, eraser, etc. Challenge pupils to measure the length of these objects in both millimetres and centimetres (using decimal notation).

Unit 9.5 Calculating the perimeter of simple shapes

- Organise pupils into pairs or groups and provide them with masking tape (or string).

  Set up various objects to measure perimeter, e.g. paint tray, whiteboard, table surface, gym mat, etc. Challenge pupils to measure out strips of tape (or string) that will wrap around the perimeter of each object (you may need to demonstrate wrapping the tape around the perimeter of one of the objects beforehand). Compare by measuring how long (or short) the pupils' attempts are.
- Provide pupils with rulers, metre sticks and measuring tapes. Repeat the challenge, this time allowing pupils to use the measuring devices to measure both the perimeters and masking tape. Ask questions such as: **What did you need to do to get an accurate measurement? What is the least number of sides you need to measure?**
- School grounds challenge: Provide pupils with string and suitable measuring devices (measuring tape, trundle wheels, etc.). Challenge pupils to measure the perimeter of various fixtures within the school grounds, e.g. flower bed, basketball court, picnic bench, etc. Ask questions such as: **Which measuring device was most suitable and why?**
- Provide pupils with playground chalk. Challenge pupils to draw shapes with a perimeter of:
  1. 6 metres
  2. 10 metres
  3. 1 metre, etc.

  Use the measuring devices to check the accuracy of their attempts.

Unit 9.6 Finding the area of regular shapes in square cm/m

- Organise pupils into pairs or groups and provide them with pencils, rulers and blank paper. Challenge pupils to draw a square or rectangle with an area of four square centimetres and compare results (some may draw a square 2 cm by 2 cm or a rectangle 1 cm by 4 cm or some may incorrectly draw a square 4 cm by 4 cm).

  How can we check which of these have an area of four square centimetres? *(We can check by cutting out square centimetres and seeing how many are needed to cover the shape.)*
- Provide pupils with one-centimetre squared paper and challenge them to prove which of their shapes are the correct area *(either by cutting out individual square centimetres or by cutting out four square centimetres in one go)*. Review results with pupils. Discuss the fact that we can create more than one shape with the same area (2 cm by 2 cm and 1 cm by 4 cm).
- Challenge pupils to draw (and compare / check results), further shapes with areas of for example:
  1. 6 square centimetres
  2. 9 square centimetres
  3. 10 square centimetres, etc.

Ask questions such as: **Is it necessary to check the area of these shapes using squares? Is there a more efficient way of working out the area?** Ascertain, through discussion, that each of their shapes is an array made up of square centimetres and that as such, the area can be calculated by multiplying the lengths of the two dimensions.

- Draw the following shapes:

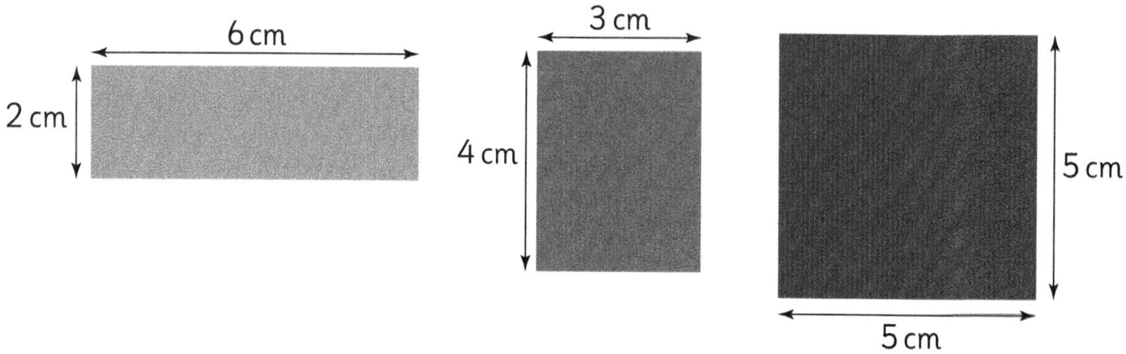

- Challenge pupils to calculate their areas and then check using 'cut out' squared paper. This activity can be repeated using measurements in metres (in the playground, gym hall, etc.) using metre sticks and chalk. It will be useful to cut out a square metre of fabric, paper, etc. so that pupils gain an appreciation of the size of a square metre, i.e. that it is not simply one hundred times the size of a square centimetre.

Unit 9.7 Finding the volume of cubes and cuboids by counting cubes

- Discuss pupils' understanding of the term 'volume'. *(Until now, volume has been associated with the amount of space taken up by a liquid.)* Show pupils a one centimetre cube (base 10) and challenge them to come up with statements to describe it: *Each of the faces are squares; it has six faces; the length of each edge is 1 cm; the area of each face is one square centimetre; the total surface area is 6 cm^2; etc.*

  Discuss the fact that volume measures the three-dimensional space taken up by something (whether it's a liquid, solid or gas) and that this is measured in cubes.

  What is the volume of this cube? Some pupils may give an answer of 6 cubic centimetres (or 6 square centimetres) because they relate it to the total surface area of all 6 sides.

  Draw a line measuring 1 cm. Discuss the fact that length measures one dimension (from one point to another) and only requires one measurement to be taken.

  Draw a square centimetre. Discuss the fact that area measures two dimensions (the space taken up by a 2D shape) and requires two measurements to be taken (length and breadth).

  Show the cubic centimetre again. Discuss the fact that volume measures three dimensions (the amount of three-dimensional space taken up by something) and requires three measurements to be taken (length, depth and height).

  Through discussion, ascertain the fact that the volume of the cube is one cubic centimetre (it may be helpful to show pupils that to create a volume of six cubic centimetres it would be necessary to use six of the cubes).

- Challenge pupils to make cubes or cuboids using the following volumes of cubic centimetres:
  1. 5 cubic centimetres
  2. 8 cubic centimetres
  3. 11 cubic centimetres
  4. 12 cubic centimetres
  5. 13 cubic centimetres

  Ask questions such as: **How many different cubes or cuboids can you make with each amount of cubes? Why do you think it's possible to make more with 12 cubes than it is with 13 cubes?** (13 is a prime number so only has factors of 1 and 13; 12 has 6 factors: 1, 2, 3, 4, 6 and 12, allowing for more alternatives.)

Unit 9.8 Estimating and measuring capacity

- Discuss pupils' understanding of the term 'volume'. *(Pupils should now understand that volume can relate to the space taken up by a liquid (measured in millilitres, litres, etc.) and the three-dimensional space taken up by a solid or gas (measured in cubic centimetres, metres, etc.)*

  You will need: one litre measuring jug, 'one thousand' cube and transparent cube (holds one volumetric litre). Fill the measuring jug with exactly one litre of water. **How many millilitres is in one litre?**
- Show them a base 10 one thousand block. **How many cubic centimetres is this?** Demonstrate that the thousand block fits exactly into the transparent cube. **What will happen when we pour the litre of water into the container?** Challenge pupils to discuss and justify their predictions. Discuss the fact that one thousand millilitres of liquid takes up the exact same space as one thousand cubic centimetres. Review the 'Let's learn' box (page 143).
- Display the following number lines (for example) and challenge pupils to identify the missing numbers:

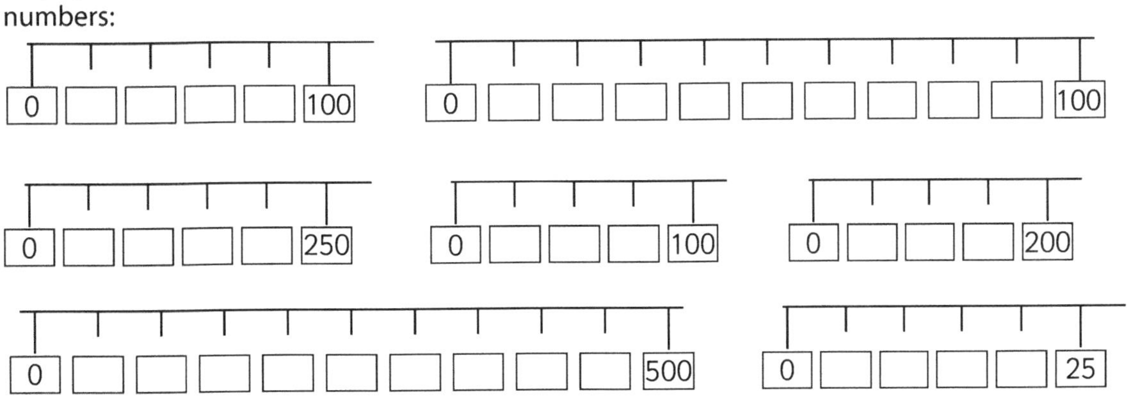

- Display images of various measuring jugs with different increments.

  Challenge pupils to identify the increments being used for each scale.

  Ask questions such as: **How did you work out the value of each increment?** See 'Let's Practise' question 1 (page 143) – review Nuria's strategy.

Measurement 2B

Section overview

Textbook 2B units	Review units	Prior learning	Benchmarks/Es&Os
Unit 9.1 Estimating and measuring length	**1C** Unit 8.3 **2A** Units 6.5, 9.2 and 9.4	• Estimate, measure and record length / height using standard units. • Read scales on measuring devices: length. • Write decimal equivalents to tenths. • Convert units of length.	**MNU 2-11a** **MNU 2-11b** **MNU 2-11c** • Uses the comparative size of familiar objects to make reasonable estimations of length, mass, area and capacity. • Estimates to the nearest appropriate unit, then measures accurately: length, height and distance in millimetres (mm), centimetres (cm), metres (m) and kilometres (km); mass in grams (g) and kilograms (kg); and capacity in millilitres (ml) and litres (l). • Calculates the perimeter of simple straight-sided 2D shapes in millimetres (mm), centimetres (cm) and metres (m). • Calculates the area of squares, rectangles and right-angled triangles in square millimetres (mm^2), square centimetres (cm^2) and square metres (m^2). • Calculates the volume of cubes and cuboids in cubic centimetres (cm^3) and cubic metres (m^3). • Converts between common units of measurement using decimal notation; for example, 550 cm = 5·5 m; 3·009 kg = 3009 g. • Chooses the most appropriate measuring device for a given task and carries out the required calculation, recording results in the correct unit.
Unit 9.2 Estimating and measuring mass	**1B** Unit 8.6 **1C** Unit 8.4 **2A** Units 6.5 and 9.3	• Estimate, measure and record mass using standard units. • Read scales on measuring devices: mass. • Write decimal equivalents to tenths. • Convert simple standard units of mass (g, kg).	
Unit 9.3 Estimating and measuring capacity	**1B** Unit 8.7 **1C** Unit 8.5 **2A** Units 6.5 and 9.8	• Estimate, measure and record capacity using standard units. • Read scales on measuring devices: capacity. • Write decimal equivalents to tenths. • Convert simple standard units of capacity (ml, l).	
Unit 9.4 Converting metric units	**1B** Unit 8.7 **2A** Unit 9.7 **2B** Unit 9.3	• Convert simple standard units of capacity (ml, l). • Find the volume of cubes and cuboids by counting cubes. • Estimate and measure capacity.	
Unit 9.5 Imperial measurement	**2A** Unit 9.4 **2B** Units 9.2 and 9.4	• Convert standard units of length, mass and capacity.	

Unit 9.6 Calculating perimeter	**2A** Units 9.4 and 9.5 **2B** Unit 9.1	• Estimate, measure and record length / height using standard units. • Convert units of length. • Calculate the perimeter of simple shapes.	• Reads a variety of scales accurately. • Draws squares and rectangles accurately with a given perimeter or area. • Demonstrates understanding of the conservation of measurement; for example, draw three different rectangles each with an area of 24 cm^2. • Shows awareness of imperial units used in everyday life; for example, miles or stones.
Unit 9.7 Calculating the area of regular shapes	**1C** Unit 8.9 **2A** Unit 9.6 **2B** Unit 9.1	• Estimate, measure and record length / height using standard units. • Link area of a rectangle / square to arrays. • Find the area of regular shapes in square cm/m.	
Unit 9.8 Volume of simple 3D shapes	**2A** Unit 9.7 **2B** Unit 9.7	• Calculate the area of regular shapes. • Find the volume of cubes and cuboids by counting cubes.	

Conceptual context

The activities outlined in Book 2A develop pupils' ability to take measurements with increasing accuracy and convert between standard units. It is important to note that the progression from non-standard to standard units is not linear. Pupils should not progress from using non-standard units exclusively to standard units exclusively but rather be given the opportunity to explore measurement through a combination of both. As such, Unit 9.1 continues to develop pupils' ability to estimate using non-standard units. Reference to this should be continued throughout the book as pupils' develop more accurate measurement strategies, i.e. we can *estimate* that the table is six times the length of the textbook. We can *measure accurately* with a tape measure that it is 1·8 metres.

Pupils will continue to use stored benchmarks in their memory that will enable them to *estimate* measurements. Familiar objects will be their internal benchmarks, such as a bag of sugar is a kilogram or a finger-width is about a centimetre. Pupils need many opportunities to practise estimating; they also need to know when it is right and proper to *only* estimate and when we need to measure *accurately*. Try to avoid always emphasising 'estimate then check' during a measurement task. Often the emphasis should be that we sometimes measure to get better at estimating.

Pupils will continue to develop their understanding of the differences between perimeter, area and volume and how they relate to each other. They should be given opportunities to explore concepts such as conservation of both perimeter and area and how these do not have a direct relationship.

For example, pupils may initially believe that doubling the length of one side of a rectangle will result in both the area and perimeter being doubled:

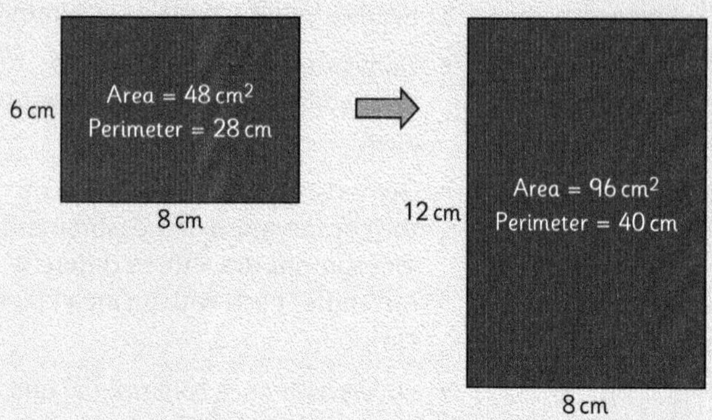

In this example, the effect of doubling the length results in the area being doubled. However, the perimeter has not increased by the same ratio.

Pupils need to experiment with the concept of conservation of area just as much as length and capacity. This can be practised using tessellation. Using a set amount of rectangles and/or triangles, pupils can explore conservation of area. Paper shapes can be cut up and compared; tangrams are an ancient version of this.

Area can also be challenging because we do not have an area 'ruler'. Pupils need to develop strategies for counting the area of a shape. Even spotting arrays in shapes will not be immediately obvious. They will need help to see arrays and then again to use multiplication to find the total. Unit 9.7 introduces formulae for calculating the areas of rectangles and triangles, allowing pupils to become more efficient with their calculation strategies. Thus, rather than having to rely on 'counting squares' to find the area of a rectangle, pupils can apply the formula *Area* = *Length* × *Width* by developing understanding of the conceptual links to arrays. Unit 9.8 further develops the understanding of calculating volume by identifying layers of cubes before progressing to the formula *Volume* = *Length* × *Breadth* × *Height*, explored in Book 2C.

Resources

- Rulers
- Metre sticks
- Tape measures
- Trundle wheels
- Scales – balance, kitchen, hanging, bathroom
- Weights (one kilogram and 100 grams)
- Base 10 materials
- Measuring jugs
- Funnels
- Plastic bottles

Key vocabulary

- area
- capacity
- centimetre
- estimate
- fluid ounce
- foot
- gallon
- gram
- height
- inch
- kilogram
- length
- mass
- metre
- metre stick
- mile
- millimetre
- ounce
- perimeter
- pint
- pound
- stone
- volume
- yard

Misconceptions

Potential misconception	Addressing the misconception
Units 9.1, 9.2 and 9.3 Pupils believe that using decimals changes the length/mass/capacity of the item being measured.	Review 'Let's learn' on pages 146–147. Provide opportunities for pupils to measure using estimates (e.g. the pencil is between 6 and 7 centimetres) and accurately (e.g. using a ruler displaying millimetres).
Units 9.2 and 9.3 Pupils fail to read the scale on a measuring jug or scales correctly because they are unable to identify the value of each increment.	Provide opportunities for pupils to work out increments on number lines and scales (see Teaching Ideas, Book 2A, Unit 9.7).
Unit 9.6 Pupils unable to identify the length of an unlabelled side in the following examples: 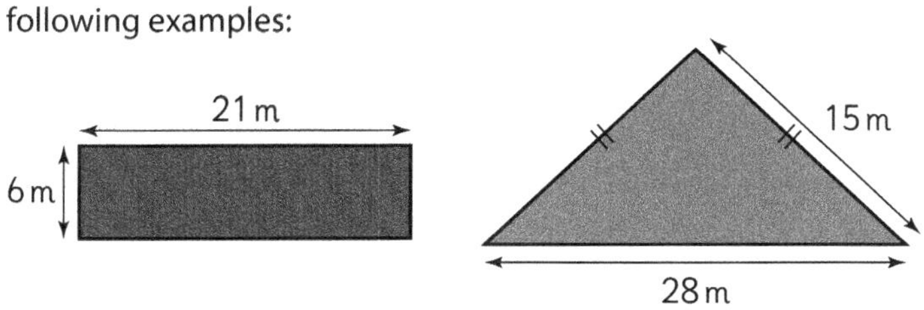	Provide opportunities for pupils to cut out the shapes and fold them in half to prove that opposite sides are equal in length.
Units 9.6 and 9.7 Pupils believing that conservation of area results in the conservation of perimeter (and vice versa). For example, when finding the area of a triangle by halving the area of a rectangle, the perimeter of the triangle is also half the perimeter of the rectangle.	See Teaching Ideas activities for Units 9.6 and 9.7. When perimeter and area are first introduced to the pupils, introduce perimeter (a measure of length) separately to area (a measure of a surface or plane). Continue to reinforce this difference in measurement when exploring area and perimeter of rectangles, i.e. perimeter is a length, area is a surface.

Potential misconception	Addressing the misconception
Unit 9.8 Pupils struggle to measure volume where 2D representations are given for 3D objects (i.e. they are required to 'count' cubes that are not visible).	Provide cubic centimetres for pupils to recreate 3D equivalents of 2D representations (see Teaching Ideas, Unit 9.8).

Teaching ideas

Unit 9.1 Estimating and measuring length

- Unit 9.1 consolidates the concepts explored in Book 2A, Unit 9.4, *Converting units of length* and as such, the activities outlined in the Teaching Ideas can be revisited.

  Provide pupils with rulers (that don't have millimetres marked) or use the pupil-made rulers outlined in Book 2A, Unit 9.4. Provide a selection of objects to measure, e.g. post-it note, tens frame, paint brush, etc.

  Challenge the pupils to measure the length of each object using the ruler and estimate to the nearest millimetre. Pupils should write estimates in both millimetres and centimetres (to one decimal place).

  Provide pupils with rulers (that have millimetres marked) to check their estimates. Note that pupils could be challenged to add up the total of the differences between their estimates and actual measurements to see who gave the most accurate estimates overall.

- Challenge pupils to draw a line of a specified length using estimation. For example: **Draw a line that you estimate to be 4·3 cm** (it may be useful for pupils to use a straight edge, e.g. a whiteboard, to help draw their lines).

  Pupils can then check how accurate their estimations were by measuring with a ruler.

  Repeat for several lengths:

  1. 3·6 cm
  2. 6·1 cm
  3. 8·7 cm

  This activity could be repeated in the playground using chalk and measurements in metres, e.g. **Draw a line that you estimate to be 2·1 m** (provide pupils with measuring tapes to check their accuracy). This can also be extended to link with the story the class are reading or with a history topic, e.g. **Draw a tiger to actual size, length 2·5 metres, tail 1 metre and height 1·2 metres** or **Draw a Viking ship, length 15 metres and beam 4·6 metres.**

Unit 9.2 Estimating and measuring mass

- Organise pupils into pairs or groups and provide them with balancing scales, 2 or 3 items to weigh (e.g. lunch box, pile of textbooks, etc.) and kilogram weights.

  Challenge pupils to weigh each of the items and come up with statements about each: *The lunchbox weighs more than one kilogram but less than two kilograms. The pile of books weighs between two and three kilograms.*

  Ask pupils questions such as: **How accurate are these measurements? Why could you not get the scales to balance? How could we measure the weight with greater accuracy?**

  Through discussion, ascertain the need for smaller weights to help make the scales balance.

  Introduce 100 g weights. Demonstrate (or challenge the pupils to find that) ten 100 g weights balance exactly with 1 kg.

  What fraction of 1 kg are each of the 100 g weights? *(One tenth of a kilogram or 0·1 kg.)*

  Challenge pupils to weigh the items again and come up with statements about each: *The lunchbox weighs more than 1·2 kg but less than 1·3 kg. The pile of books weighs about 2·6 kg.*

  Ask pupils questions such as: **How accurate are these measurements? Why could you not get the scales to balance? What did you notice when you added the 100 g weights?** *(Pupils may notice that the scales move very quickly when adding denominations of 1 kg but slow down when adding denominations of 100 g.)* **What does this tell us about the accuracy of the measurement?**

  Through discussion, ascertain that the introduction of smaller denominations has allowed for greater accuracy when describing the weight of each item.

  Originally, we said the lunchbox weighed between one and two kilograms. We can now say it weighs between 1·2 kg and 1·3 kg. The weight has remained exactly the same, but we can measure it with greater accuracy.

  Challenge pupils to measure out, for example, 0·4 kg of sand, cubes, marbles, etc. This could be done using balance scales or kitchen scales (see Challenge question on page 151).

  Repeat using different measurements: 0·7 kg, 1·4 kg, 2·1 kg, etc.

Unit 9.3 Estimating and measuring capacity

- Organise pupils into pairs or groups and provide them with a measuring jug (showing denominations of 1 l or 500 ml) and a large bottle of water (not completely filled).

  Challenge pupils to measure the volume of water by pouring it into the measuring jug and come up with a statement: *Our bottle contained less than a litre of water. Our bottle contained between one and a half and two litres, etc.*

  Ask pupils questions such as: **How accurate are these measurements? How could we measure the volume of water with greater accuracy?**

  Through discussion, ascertain the need for smaller denominations to help measure with greater accuracy.

Introduce 100 ml measuring jugs. Demonstrate (or challenge the pupils to find) that ten 100 ml of water measures exactly 1 l.

What fraction of 1 l is 100 ml? *(One tenth of a litre or 0·1 l.)*

Provide measuring jugs showing denominations of 100 ml. Challenge pupils to measure the volumes of water again and come up with a statement: *Our bottle contained between 0·8 l and 0·9 l of water. Our bottle contained more than 1·6 l but less than 1·7 l, etc.*

Ask pupils questions such as: **How accurate are these measurements? Is the volume of water any different than before?**

Through discussion, ascertain that the introduction of smaller denominations has allowed for greater accuracy when describing the volume of water.

Originally, we said, *for example*, **there was between one and a half litres and two litres. We can now say the volume is between 1·6 l and 1·7 l. The volume has remained exactly the same, but we can measure it with greater accuracy.**

How could we measure the volume of water with even greater accuracy? *(By introducing smaller denominations, i.e. hundredths of a litre.)*

Challenge pupils to measure out, for example, exactly 1·4 l of water.

We can write this as 1·4 l or 1400 ml.

Repeat using different measurements: 0·3 l, 1·8 l, 2·4 l, etc.

Unit 9.4 Converting metric units

- Unit 9.4 consolidates the concepts explored in Book 2A, Unit 9.8, *Estimating and measuring capacity* and as such it may be useful to revisit the activity in the Teaching Ideas section proving that 1000 cubic centimetres is equal to 1 litre.

Organise pupils into pairs or groups and provide them with base 10 materials.

Part fill a bottle with water.

Challenge pupils to estimate the volume of water using the base 10 materials.

Pour the water into a measuring jug to check the exact volume and compare with their estimations.

Provide pupils with a bottle, measuring jug and funnel. Challenge pupils to repeat the activity for different volumes of water.

Display 175 cubic centimetres using base 10 materials.

Challenge pupils to fill their bottles to approximately the equivalent volume of water. Pupils can then check by pouring the water into the measuring jug to compare their results.

Repeat with different volumes:

1. 250 cm^3
2. 400 cm^3
3. 525 cm^3

Unit 9.5 Imperial measurement

- Organise pupils into pairs or groups and provide them with access to the internet.

  Challenge pupils to investigate the imperial system of measurement. Start the activity by agreeing how the information will be presented and what it will contain. This will establish the success criteria for each piece of research. Some suggestions are given below.

  How will your research be presented?

  1. Poster
  2. Talk
  3. Written piece
  4. PowerPoint

  What information will be included? They may wish to use the following headings:

  1. History / Origins
  2. Units
  3. Current use
  4. Metric equivalents

  Provide opportunities for pupils to present their findings to the class.

- Display the following:

Length	**Mass**	**Capacity**
1 foot is about 30 centimetres	1 pound is about 450 grams	1 pint is about half a litre
1 foot = 12 inches	1 pound = 16 ounces	1 pint = 20 fluid ounces
1 yard = 3 feet	1 stone = 14 pounds	1 gallon = 8 pints
1 mile = 1760 yards		

  Set up the following resources at stations for pupils to create imperial measurements:

  Length: Rulers, measuring tapes, string, scissors.

  Challenge pupils to create lengths of string that measure:

  1. 1 inch
  2. 1 foot
  3. 1 yard

  Mass: modelling clay, scales.

  Challenge pupils to measure out the following masses of modelling clay:

  1. 1 pound
  2. 1 ounce
  3. 1 stone

  Capacity: measuring jugs, large plastic bottles, plastic cups, large basin.

  Challenge pupils to measure out the following volumes of water:

1. 1 pint (bottle)
2. 1 fluid ounce (plastic cup)
3. 1 gallon (basin)

Ask questions such as: **How did you measure out an ounce?** For example, some pupils may have worked out the metric equivalent (28 g) and measured using this equivalent. An alternative method could be to measure out one pound and then find a sixteenth by halving it, halving one half to make quarters, halving a quarter to make eighths then halving an eighth to make a sixteenth. **How did you measure out one fluid ounce?** It will be much easier to calculate the metric equivalent (by dividing 500 ml by 20 to get 25 ml) than attempting to split one pint into twenty equal parts.

Unit 9.6 Calculating perimeter

- This unit extends the learning developed in Book 2B, Unit 9.5 by progressing on to more accurate measurements of perimeter involving decimal numbers. As such, the activity outlined in the Teaching Ideas section can be extended to include decimal numbers to gain a more accurate measurement of perimeter.

  The concept of conservation of both area and perimeter can be explored at this point.

  Organise pupils into pairs or groups and provide them with a rectangular piece of paper.

  Challenge pupils to create a shape that has exactly the same area but a larger perimeter.

  One possible solution is to draw a straight line diagonally from one corner to the opposite corner.

  Cut along the line and reposition the two parts to form a parallelogram.

  Ask pupils: **What has happened to the area of the shape?** *(It has remained the same.)* **What has happened to the perimeter of the shape?** *(It has increased.)*

- Provide pupils with centimetre squared paper.

  Challenge pupils to create three shapes each with an area of 12 cm^2 that have a perimeter of:

  1. exactly 16 cm
  2. less than 16 cm
  3. more than 16 cm

  Possible solutions may include:

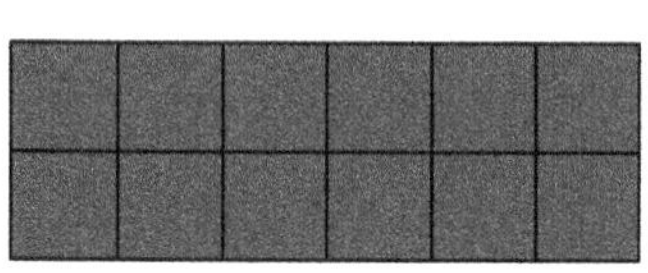

Perimeter = 16 cm

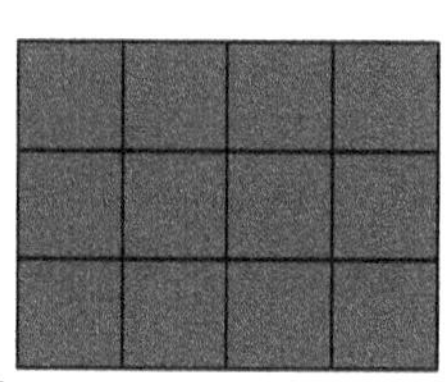

Perimeter = 14 cm

Perimeter = 26 cm

Discuss the fact that changing the shape conserves the area but does not conserve the perimeter.

Challenge pupils to draw three shapes, each with a perimeter of 18 cm that have an area of:

1. Exactly 14 cm^2
2. Less than 14 cm^2
3. More than 14 cm^2

Discuss the fact that conserving the perimeter does not result in the area being conserved.

Unit 9.7 Calculating the area of regular shapes

- Draw or display the following array:

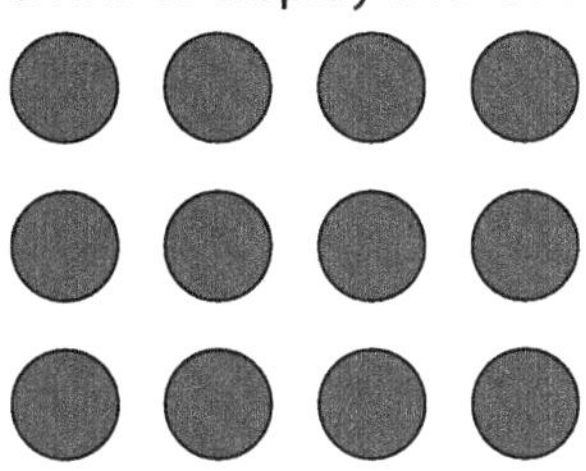

Ask pupils: **What multiplication fact does this array represent?** *($3 \times 4 = 12$)*

Draw or display the following rectangle:

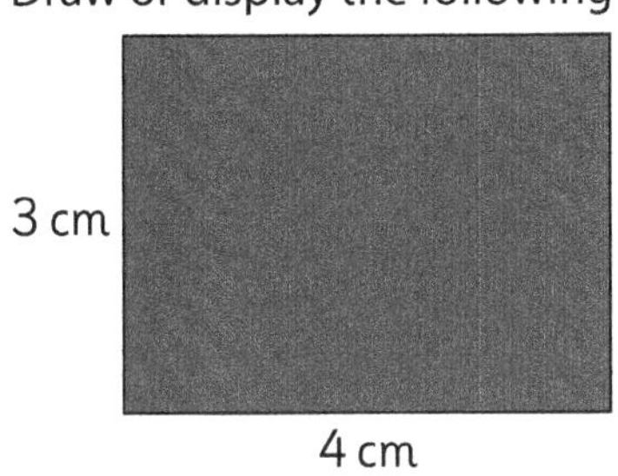

Ask pupils: **How could we work out the area of this rectangle?** *(We could work out how many square centimetres it would take to fill the rectangle.)*

Discuss with pupils whether it is necessary to draw the squares to work out the area. Draw attention to the fact that the area can be calculated in the same way as the total of the array, i.e. $3\,cm \times 4\,cm = 12\,cm^2$

Provide pupils with blank paper and challenge them to draw (or sketch) as many rectangles as they can with areas of:

1. 16 cm^2
2. 20 cm^2
3. 24 cm^2

Draw or display the same rectangle as before and discuss how the area was calculated.

Draw a straight line from one corner to the corner diagonally opposite in order to create two triangles and ask pupils: **What fraction of the rectangle is each triangle?** *(Half)* **What is the area of each triangle?** *(6 cm²)* **How did you work it out?** *(The area of the triangle is half the area of the rectangle.)* **Is the perimeter of the triangle half of the perimeter of the rectangle?** *(No)*

Discuss the formula for the area a triangle, i.e. area $= \frac{1}{2} \times$ base $\times$ height.

Provide pupils with blank paper and challenge them to draw (or sketch) triangles with areas of:

1. $16\,cm^2$
2. $20\,cm^2$
3. $24\,cm^2$

Unit 9.8 Volume of simple 3D shapes

- Build and hide the following cuboid using base 10 'ones':

  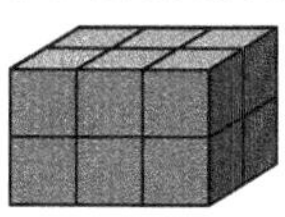

  Tell the pupils you have built a cuboid and you have hidden it. Challenge pupils to come up with questions to help them work out the volume of the cuboid. They may ask questions such as: **How many rows are there? How many cubes in each row? How many layers does the cuboid have? Etc.**

  Repeat with other cuboids.

  Organise pupils into teams and provide cubes. Instruct them to build a cuboid and hide it.

  Pair up teams and challenge them to work out the volume of each others' cuboid by taking turns to ask one question at a time.

- Provide pupils with a dice. Roll the dice three times to get the dimensions of a cuboid, i.e. first roll: number of rows; second roll: number in each row; third roll: number of layers.

  Challenge pupils to work out the volume, then check by building the cuboid using the cubes. (This could be done as a whole class activity by organising the pupils into teams and the teacher rolling the dice.)

Measurement 2C

Section overview

Textbook 2C units	Review units	Prior learning	Benchmarks/Es&Os
Unit 9.1 Estimating and measuring length	**2B** Units 6.4 and 9.1	• Estimate, measure and record length / height using standard units. • Read scales on measuring devices: length. • Write decimal equivalents to hundredths. • Convert units of length.	**MNU 2-11a** **MNU 2-11b** **MNU 2-11c** • Uses the comparative size of familiar objects to make reasonable estimations of length, mass, area and capacity. • Estimates to the nearest appropriate unit, then measures accurately: length, height and distance in millimetres (mm), centimetres (cm), metres (m) and kilometres (km); mass in grams (g) and kilograms (kg); and capacity in millilitres (ml) and litres (l). • Calculates the perimeter of simple straight-sided 2D shapes in millimetres (mm), centimetres (cm) and metres (m). • Calculates the area of squares, rectangles and right-angled triangles in square millimetres (mm^2), square centimetres (cm^2) and square metres (m^2). • Calculates the volume of cubes and cuboids in cubic centimetres (cm^3) and cubic metres (m^3). • Converts between common units of measurement using decimal notation; for example, 550 cm = 5·5 m; 3·009 kg = 3009 g.
Unit 9.2 Estimating and measuring mass	**2B** Units 6.4 and 9.2	• Estimate, measure and record mass using standard units. • Read scales on measuring devices: mass. • Write decimal equivalents to hundredths. • Convert simple standard units of mass (g, kg).	
Unit 9.3 Estimating and measuring area	**2B** Units 6.4 and 9.7	• Estimate, measure and record length / height using standard units. • Link area of a rectangle / square to arrays. • Calculate the area of regular shapes in square cm/m. • Write decimal equivalents to hundredths.	
Unit 9.4 Estimating and measuring capacity	**2B** Units 6.4 and 9.3	• Estimate, measure and record capacity using standard units. • Read scales on measuring devices: capacity. • Write decimal equivalents to hundredths. • Convert simple standard units of capacity (ml, l).	
Unit 9.5 Estimating imperial measurements	**2B** Unit 9.5 **2C** Units 9.1, 9.2 and 9.4	• Estimate, measure and record length, mass and capacity using standard units. • Identify and use imperial units.	
Unit 9.6 Converting imperial measurements	**2B** Units 9.4 and 9.5 **2C** Units 9.1, 9.2 and 9.4	• Convert units of length, mass and capacity. • Identify and use imperial units.	

Unit 9.7 Calculating perimeter (rectilinear shapes)	**2B** Unit 9.6 **2C** Unit 9.1	• Estimate, measure and record length / height using standard units. • Calculate the perimeter of simple shapes.	• Chooses the most appropriate measuring device for a given task and carries out the required calculation, recording results in the correct unit. • Reads a variety of scales accurately. • Draws squares and rectangles accurately with a given perimeter or area. • Demonstrates understanding of the conservation of measurement; for example, draw three different rectangles each with an area of 24 cm^2. • Shows awareness of imperial units used in everyday life; for example, miles or stones.
Unit 9.8 Calculating area: triangles	**2B** Unit 9.7	• Calculate the area of regular shapes.	
Unit 9.9 Measuring area (composite shapes)	**2B** Unit 9.7 **2C** Units 9.1 and 9.8	• Calculate the area of regular shapes. • Estimate and measure length. • Calculate the area of a triangle.	
Unit 9.10 Calculating volume	**2B** Units 9.7 and 9.8	• Calculate the area of regular shapes. • Find the volume of a cuboid.	
Unit 9.11 Volume: composite shapes	**2C** Units 9.9 and 9.10	• Calculate the area of composite shapes. • Calculate the volume of a cuboid.	
Unit 9.12 Capacity problems	**2B** Unit 9.4 **2C** Unit 9.10	• Convert units of capacity (cm^3, ml). • Calculate the volume of a cuboid.	

Conceptual context

The learning outcomes set out in Book 2C continue to develop the concepts explored in the preceding books. The first four units challenge pupils to measure length, mass, area and capacity with increasing accuracy and convert from one unit to another using decimal notation to two decimal places. Once again, it is important that this is complemented with an increasing awareness of where and when to apply more accurate measurements.

Pupils will continue to use stored benchmarks in their memory that will enable them to *estimate* measurements. Familiar objects will be their internal benchmarks, such as a bag of sugar is a kilogram, or a finger-width is about a centimetre.

Pupils need many opportunities to practise estimating and also need to know when it is right and proper to *only* estimate and when we need to measure *accurately*. Try to avoid always emphasising, 'estimate then check' during a measurement task. Often the emphasis should be that we sometimes measure to get better at estimating.

Units 9.5 and 9.6 continue to develop pupils' understanding of imperial measurements commonly used in this country. It is important that pupils are able to compare imperial and metric measurements in a society where they are so prominent, whether it be distances on road signs, measurements on a sports' field or ingredients for a recipe.

Unit 9.7 focuses on calculating the perimeter of rectilinear shapes. Very often misconceptions arise relating to the links between perimeter and area. Pupils should be given the opportunity to explore relationships through investigation. For example, by investigating questions such as: **Do all shapes with the same perimeters have the same areas?** Pupils should be given opportunities to investigate shapes like the following in order to draw comparisons:

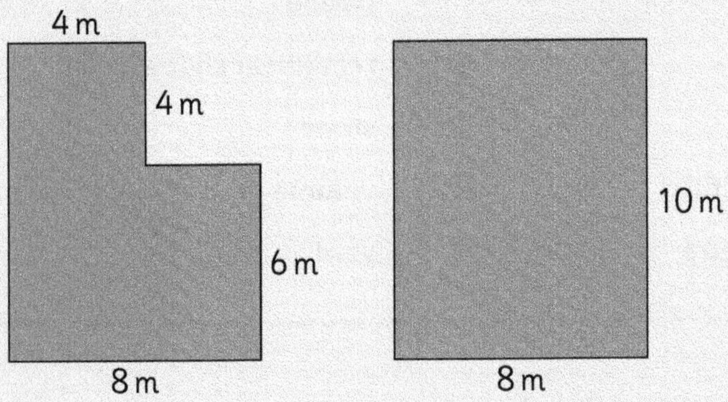

Pupils need to experiment with the concept of conservation of area just as much as length and capacity. This can be practised using tessellation. Using a set number of rectangles and/or triangles, pupils can explore conservation of area. Paper shapes can be cut up and compared. Tangrams are an ancient version of this. To be good at breaking up compound shapes, pupils will need time to practise creating simple shapes from complex shapes and understanding that the area is conserved. Units 9.8 and 9.9 explore calculating the area of triangles and composite shapes respectively. Through investigation, pupils should develop their ability to partition increasingly complex shapes into regular shapes in order to calculate area.

Units 9.10 to 9.12 explore concepts relating to capacity and volume. Once again, pupils will develop partitioning strategies to calculate the volumes of composite 3D shapes. Pupils can often find it difficult to distinguish between volume and capacity and as such, these units explore the commonalities and differences between both.

Resources

- Rulers
- Metre sticks
- Tape measures
- Trundle wheels
- Scales – balance, kitchen, hanging, bathroom
- Weights (one kilogram and 100 grams)
- Base 10 materials
- Measuring jugs
- Funnels
- Plastic bottles
- String
- Elastic bands
- Pegboards
- Cuisenaire rods

Key vocabulary

- area
- capacity
- centimetre
- composite shape
- dimension
- estimate
- fluid ounce
- foot
- gallon
- gram
- height
- inch
- kilogram
- length
- mass
- metre
- metre stick
- millimetre
- mile
- ounce
- perimeter
- pint
- pound
- rectilinear shape
- stone
- volume
- yard

Misconceptions

Potential misconception	Addressing the misconception
Units 9.1, 9.2, 9.3 and 9.4 Pupils believe that, for example, 1·8 km is a shorter distance than 1·43 km because it has fewer digits.	• Give pupils the opportunity to identify and use tenths and hundredths in the context of length, mass, area and capacity (see Teaching Ideas for Units 9.1 to 9.4). • Use bar models to demonstrate scale visually.
Unit 9.6 When converting from one unit to another, for example, from inches to centimetres, pupils are unsure whether to multiply or divide, i.e. *I have to convert 6 inches into centimetres. Do I multiply or divide by 2·54?*	• Use bar models to represent the conversion visually (see Teaching Ideas for Unit 9.6).
Unit 9.7 Pupils believing that when combing two or more regular shapes to form a composite shape, the resultant perimeter is calculated by adding the perimeters of the regular shapes.	• See Teaching Ideas activities for Unit 9.7.
Unit 9.10 Pupils believe that, for example, when doubling each of the dimensions of a cuboid the resultant volume is also doubled.	• Provide opportunities for pupils to build and investigate 3D shapes. Ask questions such as: **What happens when we double the length, width and height?** • Provide opportunities for pupils to work out increments on number lines and scales (see Teaching Ideas, Book 2A, Unit 9.7).

Potential misconception	Addressing the misconception
Units 9.9 and 9.11 Pupils failing to partition a composite 2D or 3D shape (in order to calculate area or volume, respectively) and therefore double-counting parts. For example, calculating the area of this shape by calculating $(6\,m \times 4\,m) + (7\,m \times 3\,m) = 45\,m^2$. 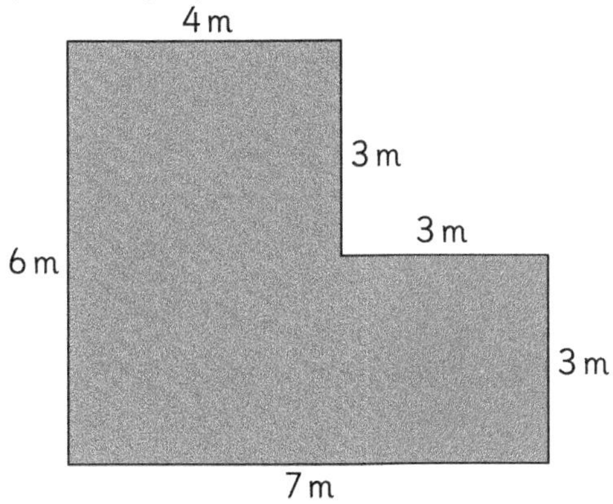	• Provide opportunities for pupils to sketch 2D composite shapes and construct 3D composite shapes and partition both. • See Teaching Ideas for Units 9.9 and 9.11.

Teaching ideas

Unit 9.1 Estimating and measuring length

- This unit extends the learning developed in Book 2A, Unit 9.4, *Converting units of length* and Book 2B, Unit 9.1, *Estimating and measuring length* and as such, the activities within the Teaching Ideas section can be revisited.
- Draw a bar and tell the pupils that the length represents 1 km. Ask pupils, **how many metres is 1 km?**
- Ask pupils: **What will be the value of each part if I split the bar into 10 equal parts?** *(0·1 km or 100 m)*
- Ask pupils: **What will be the value of each part if I split one tenth of the bar into 10 equal parts?** *(0·01 km or 10 m)*
- **So we can say that:**
  1. **1 km = 1000 m**
  2. **0·1 km = 100 m**
  3. **0·01 km = 10 m**
- Challenge pupils to convert the following measurements into metres:
  1. 1·25 km
  2. 2·07 km
  3. 3·90 km
  4. 10·46 km

- ...and the following into kilometres:
    1. 2730 m
    2. 3060 m
    3. 6800 m
    4. 10 940 m

Unit 9.2 Estimating and measuring mass

- Book 2B, Unit 9.2, *Estimating and measuring mass* introduced tenths of a kilogram (0·1 kg or 100 g) in order to increase the accuracy of our measurements. This unit extends the accuracy by which we can measure mass by introducing hundredths of a kilogram (0·01 kg or 10 kg).
- Organise pupils into pairs or groups and provide them with balancing scales, 1 kg weights, 100 g weights and 10 g weights.
- Ask pupils: **How many 100 g weights will balance with a 1 kg weight?** (*10 × 100 g*) **What fraction of 1 kg is 100 g?** (*one tenth of a kilogram*) **How do we write this as a decimal?** (*0·1 kg*) Instruct pupils to check using the weights.
- Ask pupils: **How many 10 g weights will balance with a 100 g weight?** (*10 × 10 g*) **How many 10 g weights will balance with a 1 kg weight?** (*100 × 10 g*) **What fraction of 1 kg is 10 g?** (*One hundredth of a kilogram*) **How do we write this as a decimal?** (*0·01 kg*) Pupils could check this using 9 × 100 g weights and 10 × 10 g weights to balance with 1 kg.
- Provide pupils with a suitable material to be weighed, e.g. sand or modelling clay, and a bag or container.
- Challenge pupils to measure out the following weights of the material, then write each weight in kilograms (using decimals):
    1. 60 g
    2. 130 g
    3. 250 g
    4. 1 kg 70 g
- Convert the following weights into grams:
    1. 0·45 kg
    2. 0·08 kg
    3. 1·3 kg
    4. 2·09 kg

Unit 9.3 Estimating and measuring area

- Show pupils the following shape and discuss how they would calculate the area, i.e. *area = length × width so 3 cm × 2 cm = 6 cm²*:

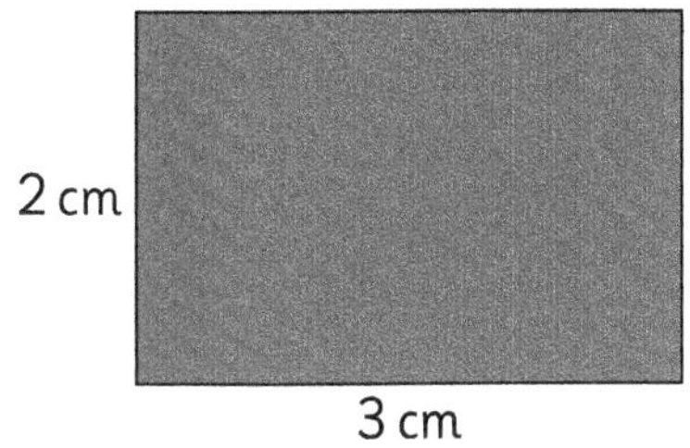

- Now show pupils the following rectangle and challenge them to find the area:

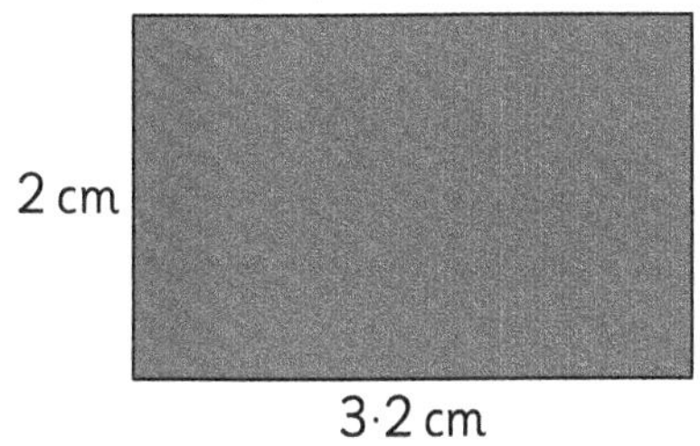

- Note that some pupils may convert the measurements into millimetres and perform the following calculation: 20 mm × 32 mm = 640 mm² (which is the correct area).
- When pupils attempt to convert this into square centimetres, they may incorrectly assume that they divide 640 mm² by ten (because there are 10 mm in 1 cm) giving them an answer of 64 cm². If this is the case, discuss with pupils whether they think this is an appropriate estimate of the area. **How many square millimetres are there in a square centimetre?** *One hundred.*
- Provide pupils with base 10 materials and challenge them to replicate the rectangle (where a one hundred 'flat' represents one square centimetre):

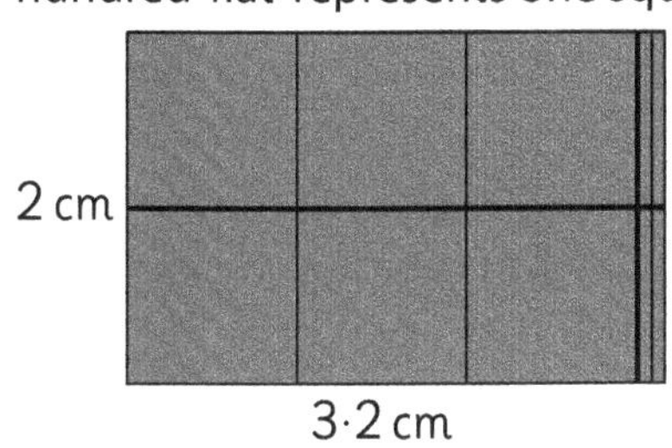

- Ask pupils: **How many whole square centimetres are there?** *(6 cm²)* **What fraction of a square centimetre does each 'ten stick' represent?** *(One tenth)* **How many tenths are needed to complete the area?** *(4 tenths)* **What is the area of the whole shape?** *(6·4 cm²)*
- Provide pupils with centimetre squared paper and a ruler and challenge them to draw the rectangle to scale showing the partitioned tenths.

- Challenge pupils to draw and calculate the area of following rectangles:

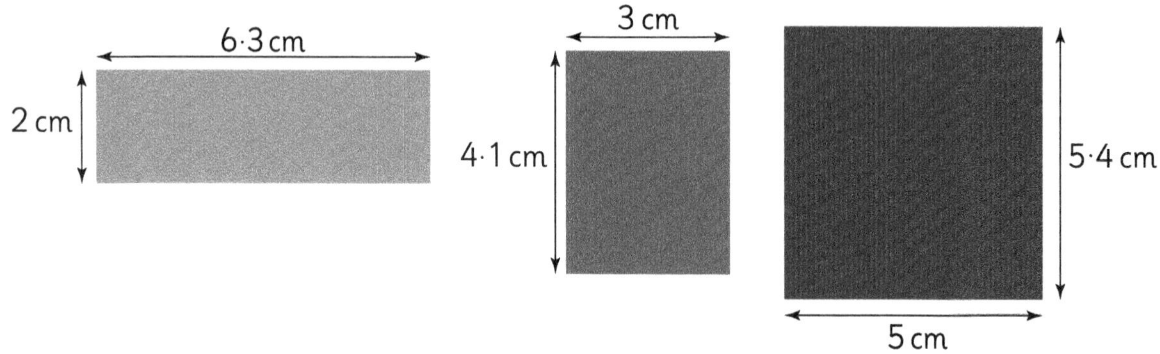

Unit 9.4 Estimating and measuring capacity

- Book 2B, Unit 9.4, *Estimating and measuring capacity* introduced tenths of a litre (0·1 l) in order to increase the accuracy of our measurements. This unit extends the accuracy by which we can measure capacity by introducing hundredths of a litre (0·01 l).
- Organise pupils into pairs or groups and provide them with a measuring jug (showing denominations of 100 ml or 0·1 l) and a large bottle of water (not filled completely).
- Challenge pupils to measure the volume of water by pouring it into the measuring jug and come up with a statement: *Our bottle contained less than 0.7 l of water. Our bottle contained between 0·3 l and 0·4 l, etc.*
- Ask pupils questions such as: **How accurate are these measurements? When do we need to make really accurate measurements?** *(For example, when measuring out medicine.)* **How could we measure the volume of water with greater accuracy?**
- Through discussion, ascertain the need for smaller denominations to help measure with greater accuracy.
- Show pupils a tablespoon and/or a 10 ml syringe (usually provided alongside children's medicine bottles).
- Fill a measuring jug with water to, for example, exactly 0·4 litres.
- Ask pupils: **How many tablespoons (or 'syringe fulls') will we have to add so that there are exactly 0·5 litres of water?**
- Demonstrate (or challenge the pupils to find) that ten 10 ml of water must be added to increase the volume of water from 0·4 to 0·5 l.
- **What fraction of 1 l is 10 ml?** *(One hundredth of a litre or 0·01 l.)*
- Provide measuring jugs and tablespoons (or syringes). Challenge pupils to measure out 0·24 litres of water and write this measurement in both litres (0·24 l) and millilitres (240 ml). Pupils may achieve this by filling the measuring jug to exactly 0·2 litres then adding 4 tablespoons of water.
- Repeat using different measurements: 0·36 l, 0·78 l, 1·15 l, etc.

Unit 9.5 Estimating imperial measurements

- This unit extends the learning developed in Book 2B, Unit 9.5, where imperial measurement was introduced. It will be useful to recreate the following imperial measurements as per the activity outlined in the Teaching Guides section in Book 2B:
    1. Length: 1 foot of string
    2. Mass: 1 pound of sugar
- Display the following:
    1. Length: 1 foot is about 30 cm
    2. Mass: 1 pound is about 450 grams
    3. Capacity: 1 pint is about half a litre
- Set up the following resources at stations for pupils to create imperial measurements and use them to estimate the measurements of various objects. Length: rulers, measuring tapes, string, scissors.
- On a length of string, mark increments with a pen one foot apart (up to about 10 to 12 feet).
- Challenge pupils to estimate and then measure (using the marked string) the length /height (in feet) of the following:
    1. themselves
    2. table
    3. classroom
    4. how far they can jump
- Mass: modelling clay (or one-pound weights), scales.
- Use the scales to measure out blocks of one pound of modelling clay (or use one-pound weights if you have them).
- Challenge pupils to estimate and then measure the mass of the following in pounds:
    1. school bag
    2. pile of jotters / textbooks
    3. paint bottle(s)
- After measuring, the pupils should check how accurate their estimates were.
- Capacity: measuring jugs, funnel.
- Use the measuring jug to measure out one pint of water.
- Challenge pupils to estimate and then measure the capacity of the following in pints:
    1. cup
    2. large plastic bottle
    3. basin or bucket
- After measuring, the pupils should check how accurate their estimates were.

Unit 9.6 Converting imperial measurements

- Display the following:

Length	Mass	Capacity
1 inch = 2·54 cm	1 ounce = 28·3 g	1 fluid ounce = 28·4 ml
1 foot = 30·5 cm	1 pound = 453·6 g	1 cup = 240 ml
1 yard = 91·4 cm	1 kilogram = 2·2 pounds	1 litre = 1·76 pints
1 mile = 1·61 km	1 stone = 6·35 kg	1 gallon = 4·55 litres

- Draw the following bar on the board to represent 1 inch equalling 2·54 cm:

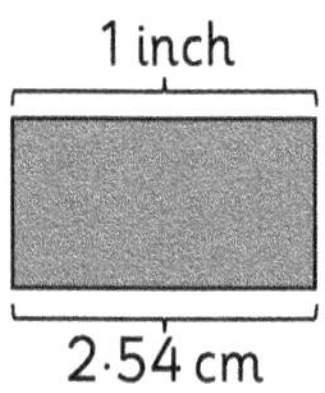

- Challenge pupils to draw a bar representing 6 inches.
- Ask pupils: **How would we convert 6 inches into centimetres?** (*By multiplying: 6 × 2·54 cm = 15·24 cm*)
- Demonstrate using a calculator (or instruct pupils) to perform the calculation.
- Ask pupils: **How would we convert 20 centimetres into inches?** (*By dividing: 20 cm ÷ 2·54 = 7·87 inches*)
- Challenge pupils to draw bars to represent converting the following:
    1. 5 feet into centimetres
    2. 8 kg into pounds
    3. 10 litres into pints
- Ask pupils: **How did you convert each measurement?** (*By multiplying the total number of units by the conversion.*)
- Challenge pupils to calculate the following:
    1. 20 km into miles
    2. 30 kg into stones
    3. 50 litres into gallons
- Ask pupils: **How did you convert each measurement?** (*By dividing the total number of units by the conversion.*)
- This activity could be extended by challenging pupils to convert their height from imperial to metric or vice versa.

Unit 9.7 Calculating perimeter (rectilinear shapes)

- Draw or display the following rectangles (the pupils may benefit from having some scrap paper and scissors handy):

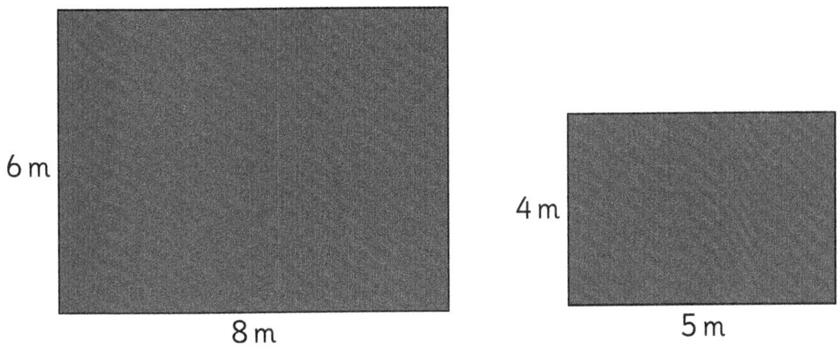

- Explain to pupils that a family want to have these sections of lawn within their garden and that they want to surround them with fencing.
- Ask pupils: **How much fencing would they need to surround each section?** (*6 m + 6 m + 8 m + 8 m = 28 m for the first section and 4 m + 4 m + 5 m +5 m = 18 m for the second section, so 46 m of fencing altogether.*)
- Challenge pupils to find a way of reducing the amount of fencing required without changing the shape of each section, e.g. by combining the sections to form a composite shape.
- Ask pupils: **What is the perimeter of this design?** (*6 m + 8 m + 5 m + 4 m + 5 m + 2 m + 8 m = 38 m*) **Can you find an alternative design that would require even less fencing?**

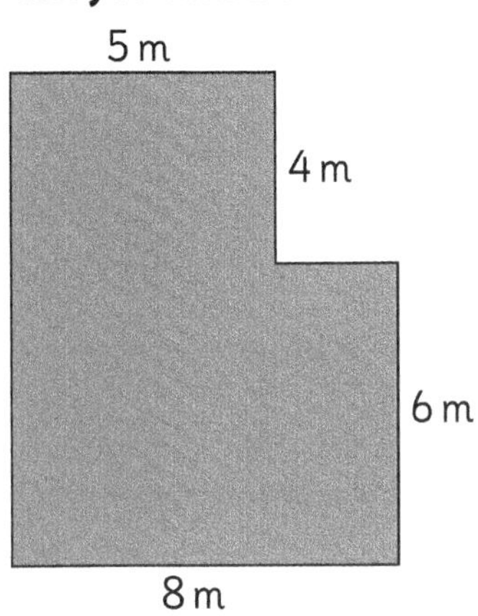

- Ask pupils: **What is the perimeter of this design?** (*5 m + 4 m + 3 m + 6 m + 8 m + 10 m = 36 m*)
- Ask questions such as: **What has happened to the total area of lawn?** (*It has stayed the same.*) **What has happened to the perimeter of the lawned section?** (*It has reduced.*) **Why has the perimeter not remained the same?** (*Part of the boundaries of each of the original sections no longer lie on the boundary of the composite shape.*)

- Display the following rectangles for pupils:

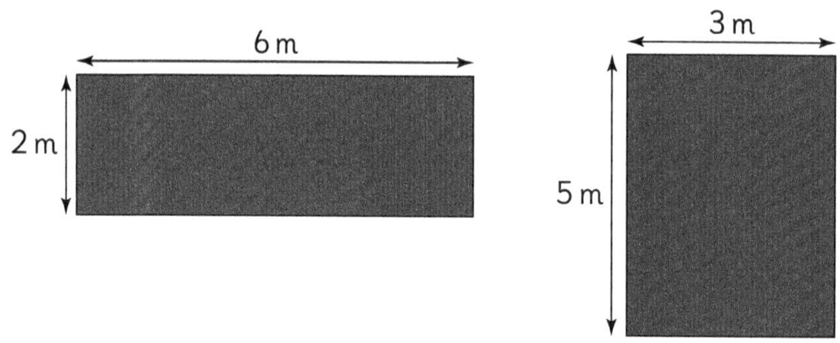

- Challenge them to combine the rectangles to form composite shapes with different perimeters (but conserving the area). **How many different perimeters can you create?**
- Note that, by partially overlapping the edges of each shape, there are multiple possibilities.
- Ask pupils to compare these two shapes by calculating their perimeter:

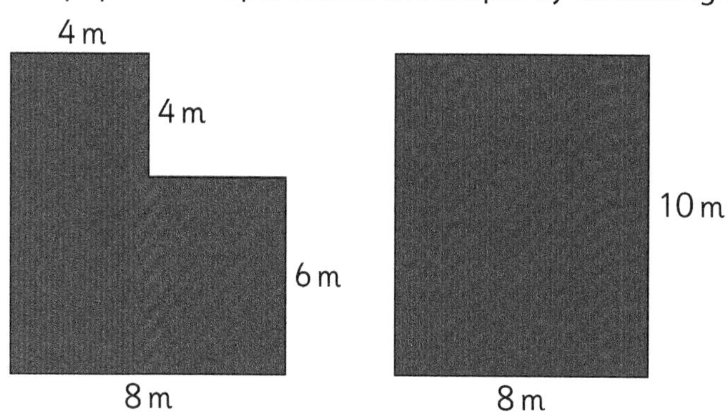

- Ask questions such as: **Do all shapes with the same perimeter have the same area? How might we calculate the area of the first shape?**
- Challenge pupils to find as many shapes as they can with the same perimeter but different areas.

Unit 9.8 Calculating area: triangles

- Draw or display the following right-angled triangle and challenge pupils to calculate the area:

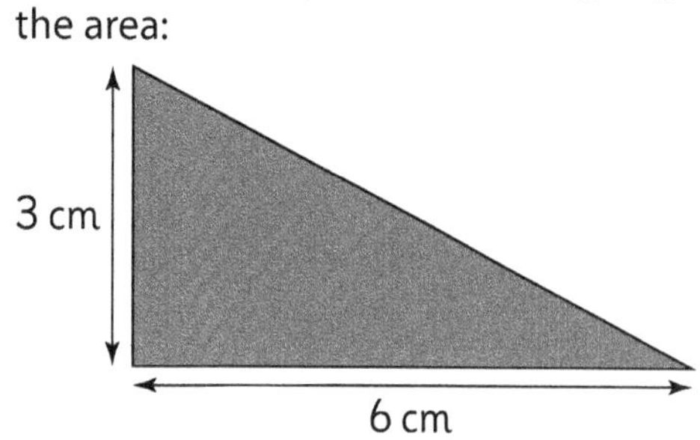

- Ask pupils: **What formula did you use to calculate the area of the triangle?** *(Area* $= \frac{1}{2} \times$ *length* $\times$ *width* $= \frac{1}{2} \times 6\,cm \times 3\,cm = 9\,cm^2$*)*
- Draw or display the following triangle and challenge pupils to calculate the area:

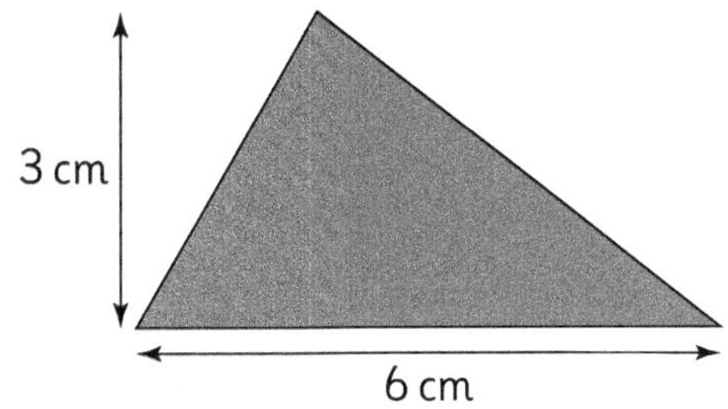

- Demonstrate (or elicit through discussion) that the triangle can be split into two right-angled triangles.
- Two imaginary triangles can then be drawn to create a rectangle:

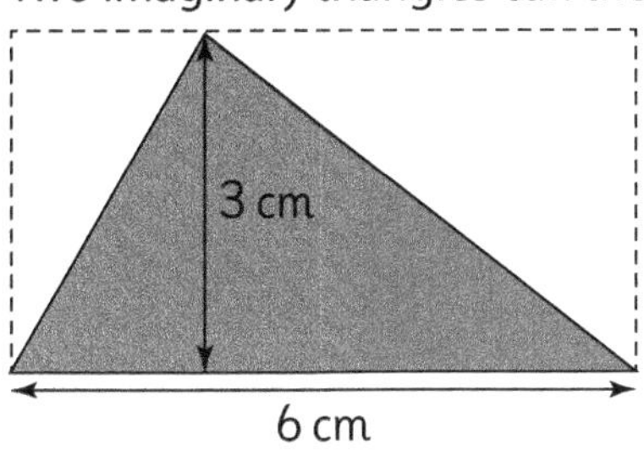

- Ask pupils: **What fraction of the rectangle on the left is the triangle on the left?** (*Half*) **What fraction of the rectangle on the right is the triangle on the right?** (*Half*) **What fraction of the whole rectangle is the whole triangle?** (*Half*)
- **Therefore, the area of this triangle can still be calculated using the formula: Area =** $\frac{1}{2} \times$ **length × height.**
- Draw or display a rectangle 6 cm long, 4 cm tall:
- Challenge pupils to draw the rectangle, use it to create a triangle of half the area and repeat. **How many different triangles can you create with an area half that of the rectangle?**
- Note that pupils can create multiple triangles, for example:

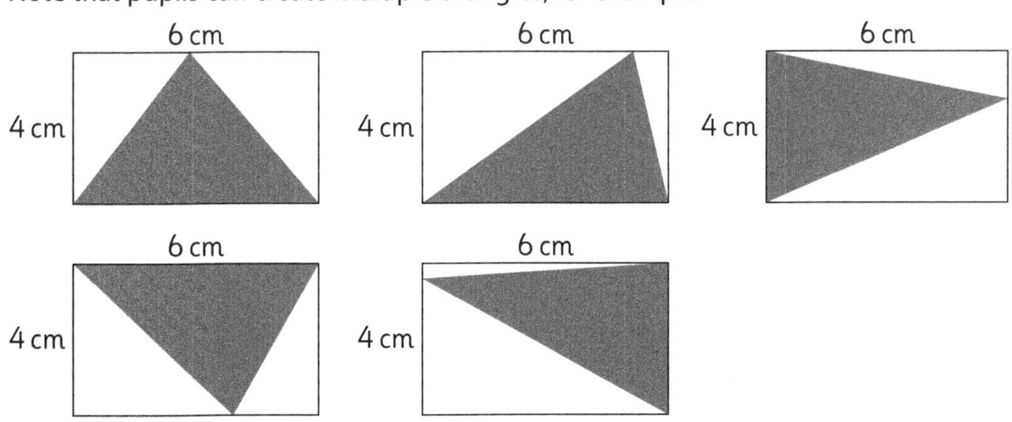

- Show pupils the following patterns and challenge them to identify those that are half shaded:

Unit 9.9 Measuring area (composite shapes)

- Provide pupils with squared paper and instruct them to draw and cut out the following rectangles:

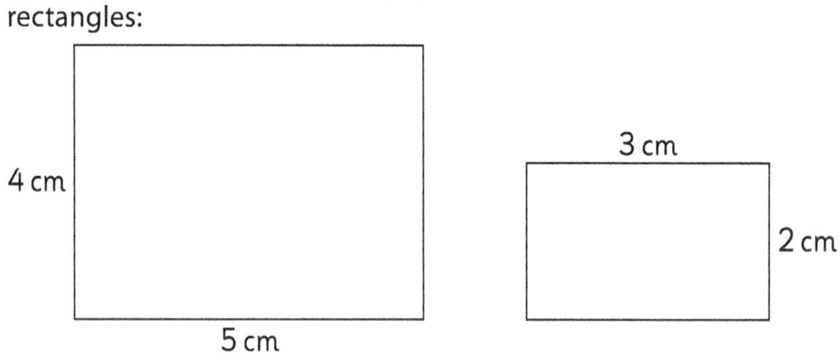

- Challenge pupils to use the two rectangles to create composite shapes with the same overall area. **How many different composite shapes can you create?**

- Draw or display the following composite shape:

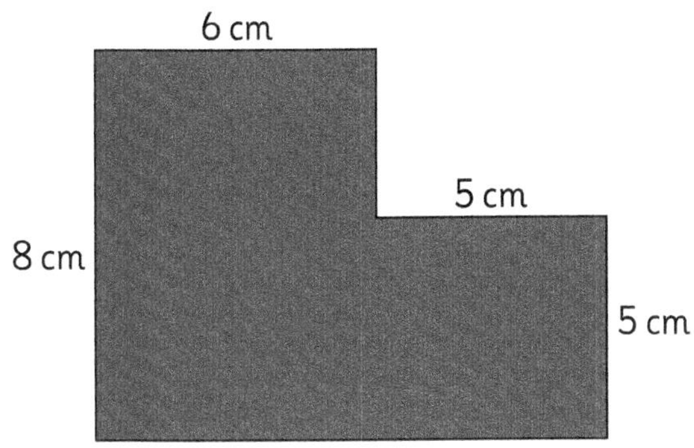

- Provide pupils with blank paper. Challenge them to draw the shape and calculate the area.
- Possible solutions:

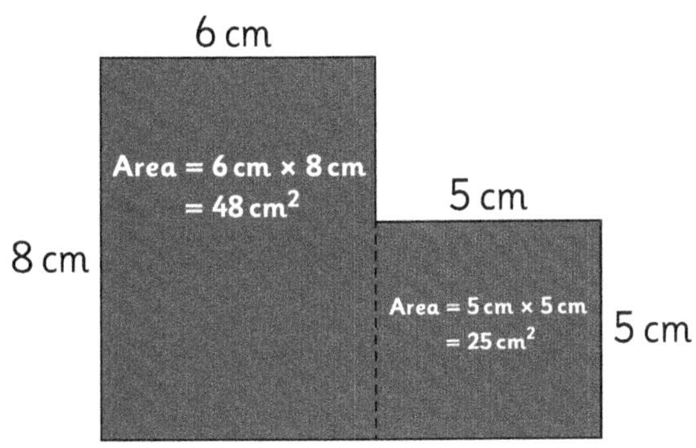

Total Area = $48\ cm^2 + 25\ cm^2 = 73\ cm^2$

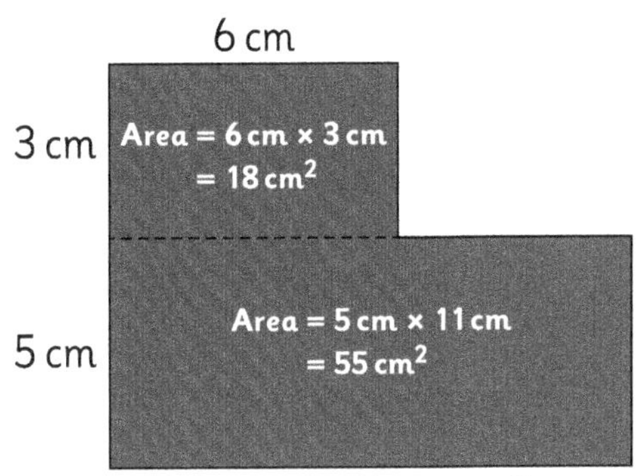

Total Area = $18\ cm^2 + 55\ cm^2 = 73\ cm^2$

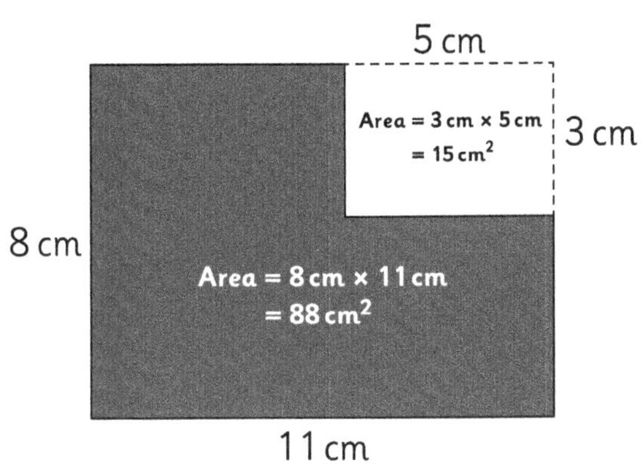

Total Area = $88\ cm^2 - 15\ cm^2 = 73\ cm^2$

- Discuss the fact that any composite shape can be split into smaller regular shapes and that the total area can be calculated using the area of the regular shapes.
- Challenge pupils to create composite shapes with the same area as the example above. Note that, rather than splitting the shape into two parts and adjusting their positions, pupils might split it into three or more parts to create more complex composite shapes.

Unit 9.10 Calculating volume

- This unit extends the learning developed in Book 2B, Unit 9.8, *Volume of simple 3D shapes;* therefore, the activities outlined in the Teaching Ideas section can be revisited.
- Draw or display the following cuboid:

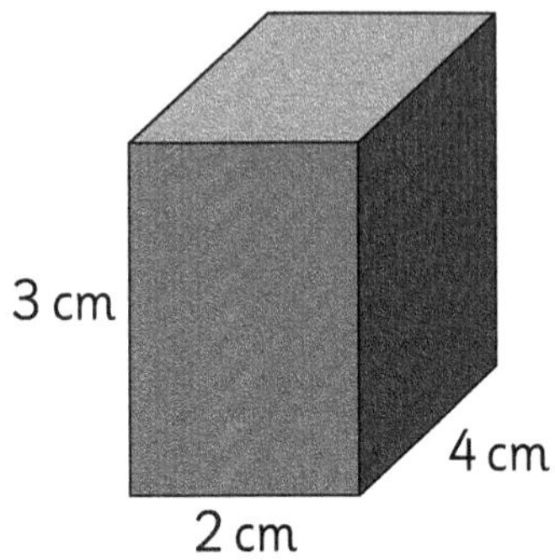

- Challenge pupils to work out the volume without using cubes to build it. Note that pupils may choose to sketch the cuboid and mark the cubic centimetres.
- Ask questions such as: **How could you work out how many cubes there are on the bottom layer without drawing it? How many cubes are there in each layer? How many layers are there? How could we work out the total number of cubes? Etc.**
- Through discussion, elicit that the number of cubes in a layer can be found by multiplying the length by the breadth i.e. $4 \times 2 = 8$ cubes, and the total volume can be found by multiplying the height by the number of cubes in a layer i.e. $3 \times 8 = 24$ cubes.
- **Can we come up with a formula to find the volume of any cuboid?** i.e. volume = length × breadth × height
- Provide pupils with cubes and challenge them to check their predictions by building the cuboid.
- Organise pupils into pairs. Challenge them to sketch a cuboid (ensuring they label the dimensions) and swap with their partner to calculate the volume.
- Draw or display the following cube and box:

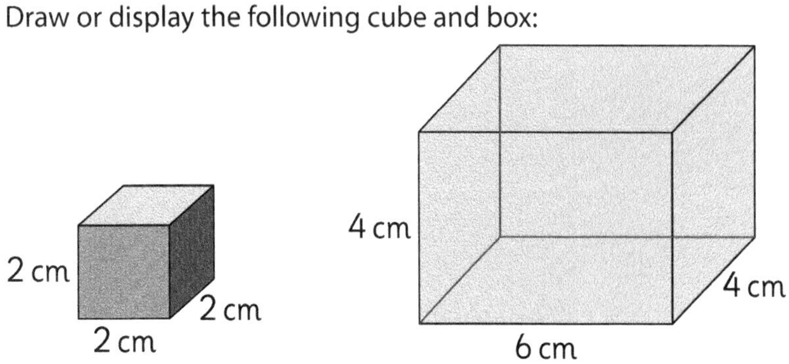

- Challenge pupils to calculate the volume of each and work out how many cubes could fit in the box.

Unit 9.11 Volume: composite shapes

- Draw or display the following composite shape:

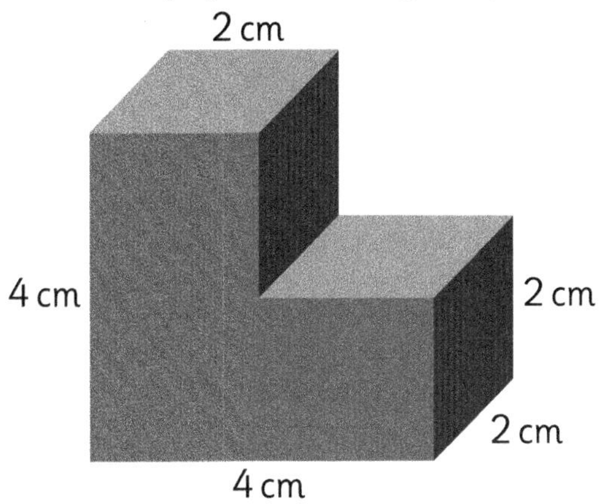

- Challenge pupils to work out the volume without using cubes. Ask questions such as: **What shapes can you see?** (*A cuboid and a cube or three cubes.*) **How can you work out the volume of these shapes?** (*volume = length × breadth × height*) **How might this help you?**
- Provide pupils with cubes and challenge them to build the shape and check their answers.
- Organise pupils into pairs or groups and instruct them to build each of the following cuboids:

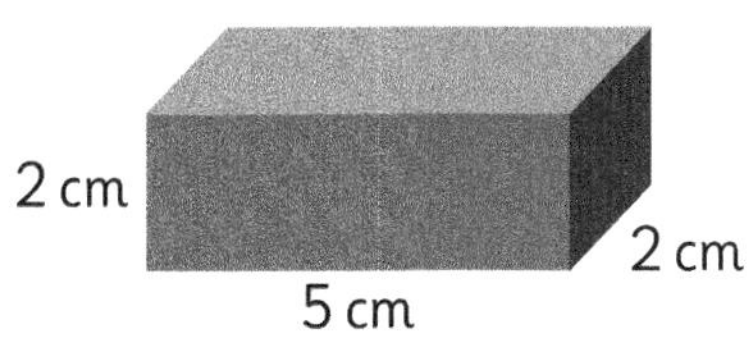

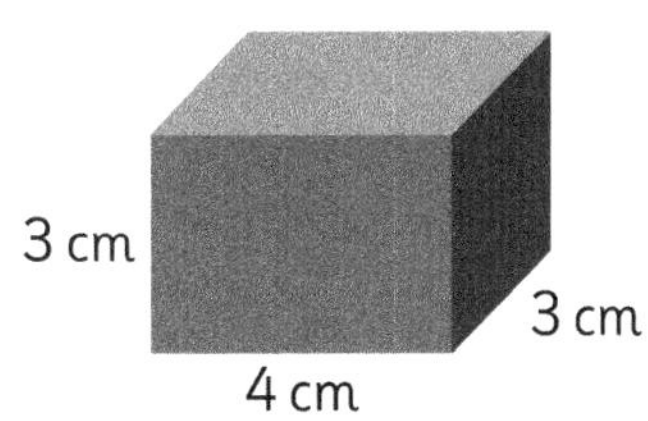

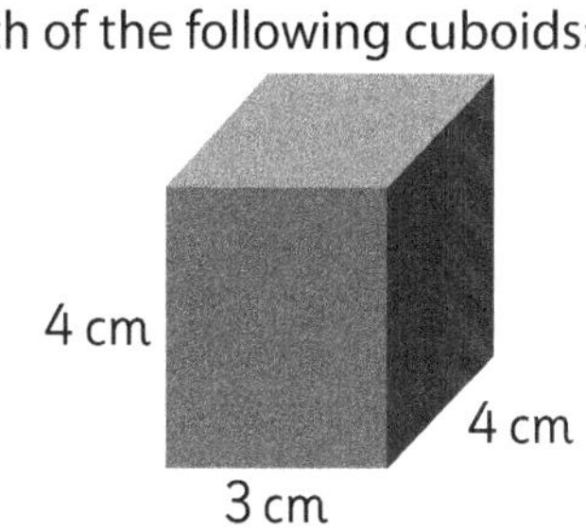

- Ask questions such as: **What is the volume of each cuboid?** (*20 cm³; 36 cm³; 48 cm³*) **How did you calculate this?** (*V = L × B × H.*)
- Challenge pupils to make different composite shapes using a combination of the cuboids and calculate their volume.
- Ask questions such as: **How did you work out the volume of each composite shape you made?** (*Add the volumes of the cuboids used.*) **What volumes are possible using these cuboids?** (*There are four possible volumes of composite shapes: 56 cm³, 68 cm³, 84 cm³ and 104 cm³.*) **How many composite shapes can be made?** (*Despite there being only four possible total volumes, there are multiple possibilities of shapes depending on how the cuboids are positioned in relation to one another.*)
- Draw or display the following composite shape:

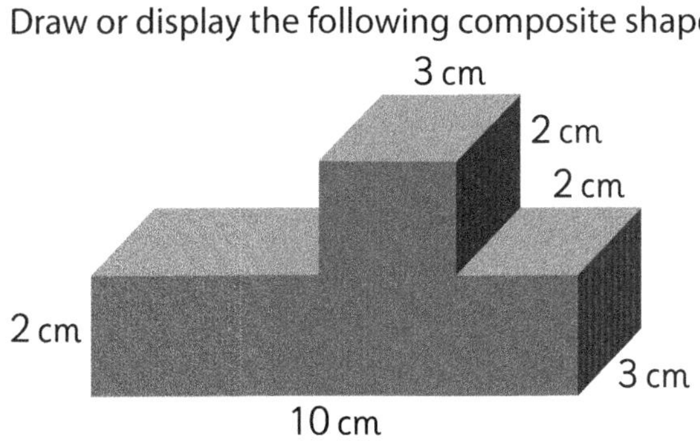

- Discuss how the volume of this shape might be calculated. Challenge pupils to find different ways to partition the shape, for example:

- Ask questions such as: **Which strategy is the most efficient and why?** (*The third example because it has been partitioned into the least number of parts so there are fewer calculations involved.*)
- Challenge pupils to work out the volume of the shape (*78 cm³*).
- Challenge pupils to create composite shapes with the same volume. Note that it may be easier if pupils use Cuisenaire rods to create their shapes.

Unit 9.12 Capacity problems

- This unit extends and combines the learning developed in Book 2B, Unit 9.4, *Converting metric units* and Book 2C, Unit 9.10, *Calculating volume*. As such, the activities outlined in both sets of Teaching Guides can be revisited.
- Show pupils a base 10 'one thousand block' and ask: **What volume is this cube?** (*1000 cm³*) **What are the dimensions of the cube?** (*10 cm by 10 cm by 10 cm*)
- Draw or display the following container and ask: **What is the capacity of this container?** (*1000 ml or 1 litre*):

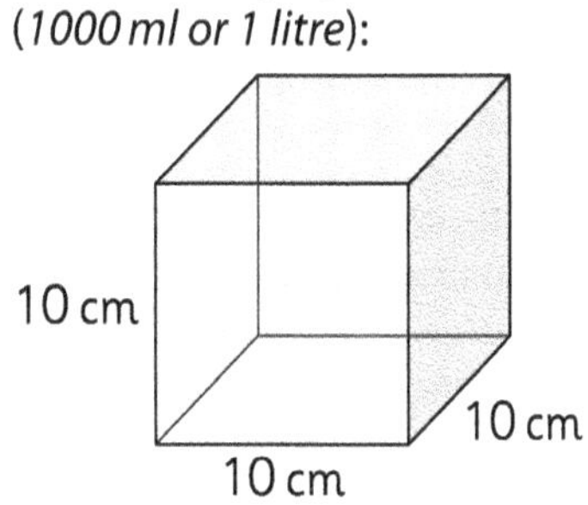

- Draw or display the following container and ask: **What is the capacity of this container?** (*8000 ml or 8 litres.*) **How do we calculate the capacity?** (*20 cm × 20 cm × 20 cm = 8000 cm³ which can be converted into millilitres or litres.*)

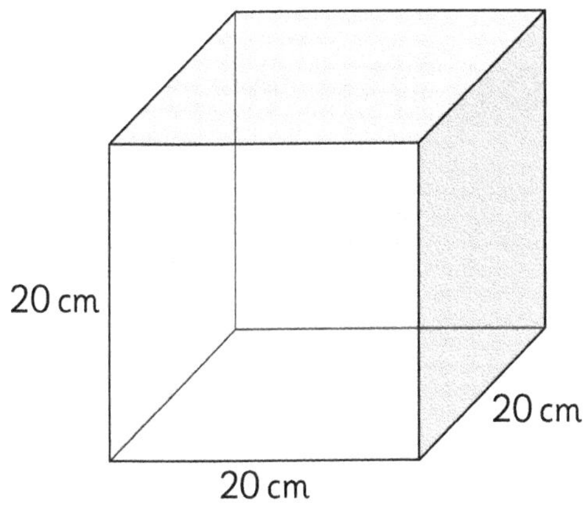

- It may be necessary to construct this using base 10 'one thousand blocks' as some pupils may incorrectly presume that the volume doubles because the dimensions have doubled.
- Organise pupils into pairs and challenge them to sketch containers (cubes or cuboids) with the following capacities:
  1. 48 ml
  2. 96 ml
  3. 250 ml
  4. 480 ml
- Ask questions such as: **How many different containers can you find for each capacity?, What strategy did you use to determine the dimensions?** Note that some pupils may employ 'guess, check and improve' whereas a more efficient strategy would be to use factors, e.g. 4 is a factor of 48 so the length could be 4 cm. If we divide 48 by 4 we are left with 12. $2 \times 6 = 12$ so the dimensions could be $4\,cm \times 2\,cm \times 6\,cm = 48\,cm^3$.

10 Mathematics, its impact on the world, past, present and future

Mathematics, its impact on the world, past, present and future 2A

Section overview

Textbook 2A units	Review units	Prior learning	Benchmarks/Es&Os
Unit 10.1 Mathematical inventions and different number systems	**1C** Unit 10.1	• I can describe a variety of ways that number is used in real-life contexts. • I can name different number systems – Roman numerals, Egyptian hieroglyphics, etc. • I have investigated the work of Scottish mathematicians.	**MTH 2-12a** • Researches and presents examples of the impact mathematics has in the world of life and work. • Contributes to discussions and activities on the role of mathematics in the creation of important inventions, now and in the past.

Conceptual context

The study of mathematics has led to great inventions that have raised standards of living and created the modern society that we live in today. Famous mathematicians litter history and their lives and discoveries are still discussed today. People such as Lovelace, Einstein, Aryabhatta, Nightingale, Pythagoras and Turing have all contributed to the development of human thinking and their ideas are celebrated in the mathematics community.

Interestingly, the study of mathematics was often undertaken alongside other disciplines or not even thought of as mathematics. Many historical figures that we recognise for mathematics often considered themselves as philosophers. For example, Isaac Newton (1642 – 1727) published his famous laws of motion in the book entitled, *Philosophiæ Naturalis Principia Mathematica* (Mathematical Principles of Natural Philosophy).

Engaging with the history of mathematics and the colourful figures from the past (and present) that have led to our current understanding of mathematics can help to build motivation in pupils and identify cross-curricular links.

The textbook unit presents a section on mathematical number systems and ends with a challenge to research a famous historical figure. By learning about different number systems, pupils can start to recognise that mathematics is a worldwide phenomenon and that mathematics is a universal language. Number systems have been used for years to communicate amounts of things and these have evolved over the years to the modern base 10 system we use today. A number system is a way of recording and expressing numbers using digits, letters or symbols in a consistent way. From simple tally marks on a stick to complex written number systems that could be used to perform many types of calculations, people in the ancient world developed and used a wide variety of number systems. Around 3000 BCE, the ancient Egyptians used a system of numerical notation or 'hieroglyphs' based on a combination of shapes and marks.

When challenged to research a historical figure, pupils can also develop skills in literacy, history and ICT. By creating a presentation, pupils develop their ability to communicate and share ideas.

Finally, by studying the impact of mathematics on the world, pupils will begin to see how much of our lives are connected with mathematics. Every financial purchase or transaction relies on a mathematical analysis to make a good choice. Everyday risk is informed by chance and uncertainty and the way information is presented and used requires a sound statistical grounding, to ensure that facts are separated from falsehoods.

Resources

- Roman Numeral Cards (Resource sheet 10a)
- Number Cards (Resource sheet 10b)

Key vocabulary

- **Civilisation**
- **Invention**
- **Mathematician**
- **Numerical notation**
- **Research**

Misconceptions

Potential misconception	Addressing the misconception
Unit 10.1 Pupils may think that number systems must contain digits.	Pupils might think that because of the word 'number,' a number system should contain a digit. Explain that a number system measures a quantity that can be represented using letters, tally marks, diagrams, symbols…, etc.

Teaching ideas

Unit 10.1 Mathematics, inventions and different number systems

- Before starting this unit, a broad discussion about what pupils think mathematics actually is might be interesting. Ask the following questions:

  What is mathematics?

  Why is mathematics important in your daily life?

  What do you use mathematics for outside of your school day?

Through discussion you can explore pupils' knowledge and understanding of how they perceive mathematics in relation to their lives and daily routine.

- Ask pupils to think of as many different ways they know of to record numbers or quantities. These could include: tally marks, diagram, pictures, Roman numerals, etc.
- Ask them how they know what each of the marks mean? How can they check or clarify this?
- Either in pairs or groups, give pupils a copy of the Roman Numeral Cards and the Number Cards.
- Explain to pupils that the Romans used a system of numerical notation based on a combination of letters. The following letters were used in various combinations to record quantities:

  I V X L C D M

- Ask pupils if they can guess what each of these letters might stand for. Go over the following values:

I = 1	C = 100
V = 5	D = 500
X = 10	M = 1000
L = 50	

- Using this information, ask pupils to work with a partner or with their group to match the Roman Numeral Cards with the correct Number Cards.
- Discuss with pupils if they still see Roman numerals in society today. Ask them to identify the different places that they might be used – clock face, watch, door numbers…
- The Challenge section asks pupils to research John Napier. This type of activity can be presented as a whole-class activity. Start the activity be agreeing how the information will be presented and what it will contain. This will establish the success criteria for each piece of research. Some suggestions are given below.
- **How will your research be presented?**
  - o Poster
  - o Talk
  - o Written piece
  - o PowerPoint

- **What information will be included?**
    - o Date of birth (and death)
    - o Nationality
    - o Famous for?
    - o Interesting fact
    - o A picture
    - o Are they still relevant today?
    - o Did they invent anything?
    - o Did they have famous friends?
- Suggested famous mathematicians around the time of John Napier:
    - o John Napier (1550 – 1617) invented Napier's bones
    - o Blaise Pascal (1623 – 1662) invented the first adding machine in 1642
    - o Galileo Galilei (1564 – 1642) astronomer and maker of telescopes
    - o Isaac Newton (1642 – 1727) inventor of his laws of motion and constructed his own reflecting telescope.
- Split pupils into smaller groups and ask them to use the internet to search for information or use an encyclopaedia in the library. Pupils could also find or create an object similar to that used by their famous historical mathematician.

Mathematics, its impact on the world, past, present and future 2B

Section overview

Textbook 2B units	Review units	Prior learning	Benchmarks/Es&Os
Unit 10.1 Mathematical inventions and different number systems	**2A** Unit 10.1	• I can describe ways that numbers are used in real-life contexts. • I can name different number systems – Roman numerals, Egyptian hieroglyphics, etc. • I have investigated the work of Scottish mathematicians. • I can discuss famous mathematicians and their work.	**MTH 2-12a** • Researches and presents examples of the impact mathematics has in the world of life and work. • Contributes to discussions and activities on the role of mathematics in the creation of important inventions, now and in the past.

Conceptual context

The study of mathematics has led to great inventions that have raised standards of living and created the modern society that we live in today. Famous mathematicians litter history and their lives and discoveries are still discussed today. People such as Lovelace, Einstein, Aryabhatta, Nightingale, Pythagoras and Turing, have all contributed to the development of human thinking and their ideas are celebrated in the mathematics community.

Interestingly, the study of mathematics was often undertaken alongside other disciplines or not even thought of as mathematics. Many historical figures that we recognise for mathematics often considered themselves as philosophers. For example, Isaac Newton (1642 – 1727) published his famous laws of motion in the book entitled, *Philosophiæ Naturalis Principia Mathematica* (Mathematical Principles of Natural Philosophy).

Engaging with the history of mathematics and the colourful figures from the past (and present) that have led to our current understanding of mathematics can help to build motivation in pupils and identify cross-curricular links.

The textbook unit presents a section on mathematical number systems and ends with a challenge to research a famous historical figure. By learning about different number systems, pupils can start to recognise that mathematics is a worldwide phenomenon and that mathematics is a universal language. Number systems have been used for years to communicate amounts of things and these have evolved over the years to the modern base 10 system we use today. A number system is a way of recording and expressing numbers using digits, letters or symbols in a consistent way. From simple tally marks on a stick to complex written number systems that could be used to perform many types of calculations, people in the ancient world developed and used a wide variety of number systems. Around 3000 BCE, the ancient Egyptians used a system of numerical notation or 'hieroglyphs' based on a combination of shapes and marks.

When challenged to research a historical figure, pupils can also develop skills in literacy, history and ICT. By creating a presentation, pupils develop their ability to communicate and share ideas.

Finally, by studying the impact of mathematics on the world, pupils will begin to see how much of our lives are connected with mathematics. Every financial purchase or transaction relies on a mathematical analysis to make a good choice. Everyday risk is informed by chance and uncertainty and the way information is presented and used requires a sound statistical grounding, to ensure that facts are separated from falsehoods.

Resources

- Roman Numeral Cards (Resource sheet 10a)
- Number Cards (Resource sheet 10b)
- Egyptian Hieroglyphic Number Cards (Resource sheet 10c)
- Number Cards (Resource sheet 10d)
- Scots Gaelic Number Cards (Resource sheet 10e)

Key vocabulary

- **Civilisation**
- **Invention**
- **Mathematician**
- **Numerical notation**
- **Research**

Misconceptions

Potential misconception	Addressing the misconception
Unit 10.1 Pupils may think that all number systems must contain digits.	Pupils might think that due to the word 'number' that a number system should contain a digit. Explain that a number system measures a quantity which can be represented using letters, tally marks, diagrams, symbols, etc.

Teaching ideas

Unit 10.1 Mathematical inventions and different number systems

- Learning to count in Gaelic is fairly straightforward and was traditionally based on twenties and not the decimal system. Up to twenty, it is identical in both systems. Unlike numbers between 11 and 20 which in English have irregular forms like 'eleven' and 'twelve', the Gaelic number system mirrors other languages like French by adding 'deug' to the appropriate numeral. In the case of twelve, it is dhà dheug, with the deug in its lenited form.

neoini	zero		
aon	one	aon deug	eleven
dhà	two	dhà dheug	twelve
trì	three	trì deug	thirteen
ceithir	four	ceithir deug	fourteen
còig	five	còig deug	fifteen
sia	six	sia deug	sixteen
seachd	seven	seachd deug	seventeen
ochd	eight	ochd deug	eighteen
naoi	nine	naoi deug	nineteen
deich	ten	fichead	twenty

- Using the Scots Gaelic Number Cards (Resource sheet 10e) ask pupils to match them to the Number Cards (Resource sheet 10b). Discuss with pupils any observations they have when looking at the numbers in comparison.
- The Challenge section in the textbook asks pupils to research James Gregory. This type of activity can be presented as a whole-class activity. Start the activity by agreeing how the information will be presented and what it will contain. This will establish the success criteria for each piece of research. Some suggestions are given below.
- **How will your research be presented?**
  - o Poster
  - o Talk
  - o Written piece
  - o PowerPoint
- **What information will be included?**
  - o Date of birth (and death)
  - o Nationality
  - o Famous for?
  - o Interesting fact
  - o A picture

  - o Are they still relevant today?
  - o Did they invent anything?
  - o Did they have famous friends?
- Suggested famous Scottish mathematicians:
  - o James Gregory (1638 – 1685) astronomer and maker of telescopes
  - o Mary Somerville (1780 – 1872) the first female member of the Royal Astronomical Society (jointly with Caroline Herschel)
  - o Flora Phillip (1865 – 1943) the first female member of the Edinburgh Mathematical Society
  - o Colin Maclaurin (1698 – 1746) mathematician who had a special series named after him
- Pupils can split into smaller groups and use the internet to search for information or use an encyclopaedia in the library. Pupils could also find or create an object similar to that used by their famous historical mathematician.

Mathematics, its impact on the world, past, present and future 2C

Section overview

Textbook 2C units	Review units	Prior learning	Benchmarks/Es&Os
Unit 10.1 Mathematical inventions and different number systems	**2B** Unit 10.1	• I can explain ways that maths is used in real-life contexts. • I can name different number systems – Roman numerals, Egyptian hieroglyphics, etc. • I have investigated the work of famous mathematicians. • I can discuss famous mathematicians and their work.	**MTH 2-12a** • Researches and presents examples of the impact mathematics has in the world of life and work. • Contributes to discussions and activities on the role of mathematics in the creation of important inventions, now and in the past.

Conceptual context

The study of mathematics has led to great inventions that have raised standards of living and created the modern society that we live in today. Famous mathematicians litter history and their lives and discoveries are still discussed today. People such as Lovelace, Einstein, Aryabhatta, Nightingale, Pythagoras and Turing, have all contributed to the development of human thinking and their ideas are celebrated in the mathematics community.

Interestingly, the study of mathematics was often undertaken alongside other disciplines or not even thought of as mathematics. Many historical figures that we recognise for mathematics often considered themselves as philosophers. For example, Isaac Newton (1642 – 1727) published his famous laws of motion in the book entitled, *Philosophiæ Naturalis Principia Mathematica* (Mathematical Principles of Natural Philosophy).

By engaging with the history of mathematics and the colourful figures from the past (and present) that have led to our current understanding of mathematics can help to build motivation in pupils and identify cross-curricular links.

The textbook unit presents a section on mathematical number systems and ends with a challenge to research a famous historical figure. By learning about different number systems, pupils can start to recognise that mathematics is a worldwide phenomenon and that mathematics is a universal language. Number systems have been used for years to communicate amounts of things and these have evolved over the years to the modern base 10 system we use today. A number system is a way of recording and expressing numbers using digits, letters or symbols in a consistent way. From simple tally marks on a stick to complex written number systems that could be used to perform many types of calculations, people in the ancient world developed and used a wide variety of number systems. Around 3000 BCE, the ancient Egyptians used a system of numerical notation or 'hieroglyphs' based on a combination of shapes and marks.

When challenged to research a historical figure, pupils can also develop skills in literacy, history and ICT. By creating a presentation, pupils develop their ability to communicate and share ideas.

Finally, by studying the impact of mathematics on the world, pupils will begin to see how much of our lives are connected with mathematics. Every financial purchase or transaction relies on a mathematical analysis to make a good choice. Everyday risk is informed by chance and uncertainty and the way information is presented and used requires a sound statistical grounding, to ensure that facts are separated from falsehoods.

Resources

- Roman Numeral Cards (Resource sheet 10a)
- Number Cards (Resource sheet 10b)
- Egyptian Hieroglyphic Number Cards (Resource sheet 10c)
- Number Cards (Resource sheet 10d)

Key vocabulary

- **Civilisation**
- **Invention**
- **Mathematician**
- **Numerical notation**
- **Research**

Misconceptions

Potential misconception	Addressing the misconception
Unit 10.1 Pupils may think that number systems must contain digits	Pupils might think that due to the word 'number' that a number system should contain a digit. Explain that a number system measures a quantity which can be represented using letters, tally marks, diagrams, symbols, etc.

Teaching ideas

Unit 10.1 Mathematical inventions and different number systems

- Shepherds in the Lake District, the north of England and in parts of Scotland used a traditional counting system, 'Yan Tan Tethera', to keep a total of the sheep in their flock. Keeping an accurate count was important to the livelihood of the shepherd. The system was also used to keep a count when knitting. This system is quite rare now but fun to learn. The topic can be introduced as a link to agriculture or through exploring songs or stories about sheep or as a link to the Scots language.

Number	Scots	Number	Scots
1	Yan	11	Yanadik
2	Tyan	12	Tyanadik
3	Tethera	13	Tetheradik
4	Methera	14	Metheradik
5	Pimp	15	Bumfitt
6	Sethera	16	Yanabumfitt
7	Lethera	17	Tyanabumfitt
8	Hovera	18	Tetherabumfitt
9	Dovera	19	Metherabumfitt
10	Dik	20	Giggot

- Introduce the first five numbers using soft toys or blocks to count, you can display *Yan, Tan, Tethera, Methera and Pimp* on the board as an aide memoir. Ask: **Why did the shepherds use these funny words? Why didn't they write down the count?** You can explore with pupils that the shepherds are working outside, in all weathers, often miles from any settlement. Once pupils get the hang of the words, introduce the words up to 10. For 11 to 20 there is a pattern to 14. Ask: **Can you see the pattern for 11 to 14?** i.e. one and ten, two and ten etc. Then introduce 15-20 and again challenge pupils to spot the pattern of 1 and 16, 2 and 17. Yan Tan Tethera is based on the Celtic language which is base 20 Celtic (vigesimal). You can compare this number system to our own decimal base 10 system to explore the differences.
- You can extend the task by completing all the counts, like lunches and line ups for the day/week in Yan Tan Tethera. There are other languages that can also be explored, such as Oksapmin from New Guinea. This language uses body parts to keep count to 27.
- The Challenge section asks pupils to research a famous mathematician. This type of activity can be presented as a whole-class activity to focus on ancient mathematicians to fit with the Babylonian number system presented in the Let's Learn section. Start the activity by agreeing how the information will be presented and what it will contain. This will establish the success criteria for each piece of research. Some suggestions are given below.

- **How will your research be presented?**
    - o Poster
    - o Talk
    - o Written piece
    - o PowerPoint
- **What information will be included?**
    - o Date of birth (and death)
    - o Nationality
    - o Famous for?
    - o Interesting fact
    - o A picture
    - o Are they still relevant today?
    - o Did they invent anything?
    - o Did they have famous friends?
- Suggested famous mathematicians from ancient history:
    - o Pythagoras of Samos (about 569 – 475 BCE) philosopher, mathematician, astronomer and musical theorist.
    - o Thales of Miletus (about 624 – 547 BCE) philosopher and mathematician, famous for geometry.
    - o Eratosthenes (276 – 194 BCE) chief librarian and first to calculate the circumference of the Earth. Introduced the sieve of Eratosthenes.
    - o Euclid (around 300 BCE) described as the founder of geometry.
    - o Liu Hsin (50 BCE – 23 CE) Chinese philosopher and politician. Used decimal fractions to approximate pi.
- Pupils can split into smaller groups and use the internet to search for information or use an encyclopaedia in the library. Pupils could also find or create an object similar to that used by their famous historical mathematician.

11 Patterns and relationships

Patterns and relationships 2A

Section overview

Textbook 2A units	Review units	Prior learning	Benchmarks/Es&Os
Unit 11.1 Exploring and extending number sequences	**1C** Unit 11.1	I can describe a pattern or sequence. • I can continue a pattern or sequence. • I can recognise and continue of a pattern or sequence.	**MTH 2-13a** • Explains and uses a rule to extend well-known number sequences including square numbers, triangular numbers and Fibonacci sequence. • Applies knowledge of multiples, square numbers and triangular numbers to generate number patterns.

Conceptual context

Within maths, there are patterns of numbers and patterns of shapes. Some patterns are repeating, and some are sequential. Patterns and relationships provide the basis for a number of the strategies pupils use to solve number calculations. According to research, the ability to spot mathematical patterns can help to predict future mathematical success. In simple terms, patterns are relationships that exist between various elements. A sequential pattern needs to have a starting point and a rule to explain the consistent growth of the pattern. It is important that pupils are challenged to identify patterns and relationships. Pupils can use different strategies to record patterns and relationships including in pictures, in diagrams and in tables.

Previously, pupils have looked at how patterns and sequences can be made by using the four operations. Within this unit, pupils' understanding is challenged by asking them to identify and explain the pattern or sequence. Pupils may recognise familiar numbers and sequences without fully understanding, mathematically, the relationship between the digits. For example, if pupils have rote-learned their 3 times table (3, 6, 9, 12…) they will recognise a sequence that uses multiples of 3 without being able to articulate that it is increasing by 3 each time; thus, in a similar pattern that starts at 162, (165, 168…) this might be problematic. Providing pupils with the correct vocabulary and framework to articulate the relationship within a pattern or sequence is a vital part of deepening their understanding.

Resources

- Unifix cubes (different colours)

Key vocabulary

- **Elements**
- **Exploring**
- **Extending**
- **Pattern**
- **Recur**
- **Relationship**
- **Repeating**
- **Sequence**
- **Term**

Misconceptions

Potential misconception	Addressing the misconception
Unit 11.1 Pupils may think that patterns or sequences only contain numbers.	Provide pupils with lots of examples of patterns or sequences that contain letters, shapes, colours and numbers to highlight that it is not only numerical.
Unit 11.1 Pupils may think that patterns or sequences cannot have two of the same thing next to one another – ABB, Blue Red Red…	Introduce pupils to a wide variety of repeating patterns to highlight how they can contain duplicate letters, numbers, shapes and so on and still form a pattern – A, B, B, C, D, D…

Teaching ideas

Unit 11.1 Exploring and extending number sequences

- Ask pupils to sing the song *Head, Shoulders, Knees and Toes, Knees and Toes*. Discuss the sequence of the actions and ask if they can identify a pattern.

  H S K T K T H S K T K T

- Introduce pupils to a variety of repeating patterns, progressing from ABC to ABB to ABBC, to emphasise that you can have two of the same thing next to one another in a pattern. Encourage pupils to discuss the relationship between the letters and how this creates a pattern. By developing this dialogue, pupils will be able to deepen their understanding and then apply this to shapes, colours and numbers and soon, using coloured cubes; for example, a pattern such as:

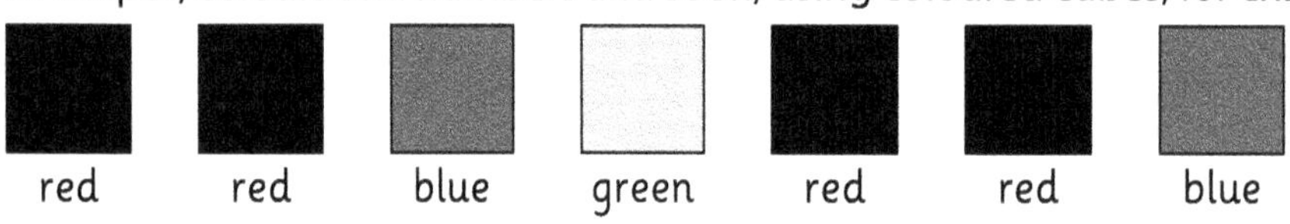

- Ask pupils to copy this pattern. It is important that pupils can both copy the original pattern and compare their construction with it to demonstrate their pattern awareness. This can then be extended by asking them what colour of cube they would add next and why. Some pupils will copy a pattern accurately but are unable to continue it. Understanding the 'unit of repeat' leads to multiplicative thinking and this involves counting groups of items and not just single items, which is important when understanding number composition.
- Looking at the Let's learn section with pupils, emphasise the fact that patterns are created as a result of the relationship between a variety of different things including shapes, pictures, letters, numbers or actions. Highlight the importance of being able to discuss the relationship as the key mathematical thinking point.

Patterns and relationships 2B

Section overview

Textbook 2B units	Review units	Prior learning	Benchmarks/Es&Os
Unit 11.1 Exploring and extending number sequences	**2A** Unit 11.1	• I can continue a sequence using a rule explained in words. • I can describe a simple sequence using words. • I can write the rule to a simple sequence. • I can find the missing number in a simple sequence.	**MTH 2-13a** • Explains and uses a rule to extend well-known number sequences including square numbers, triangular numbers and Fibonacci sequence. • Applies knowledge of multiples, square numbers and triangular numbers to generate number patterns.

Conceptual context

Patterns and relationships exist in numbers, shapes, letters and in the world around us. Pupils that can recognise, generalise and use these are better problem-solvers, have deeper understanding of mathematics, and are more able to engage in algebraic thinking, involving recognising mathematical features, identifying the relationship between elements, and noting regularities. Developing pattern-based thinking or using patterns to analyse and solve problems is very important and leads to a greater appreciation of how functions are used to describe relationships. The ability to recognise, construct and extend patterns, in addition to categorising and classifying objects and identifying rules, are key features of pattern-based thinking.

Pupils should be given the opportunity to recognise, construct and extend patterns using blocks, sticks, cubes and other concrete materials. Through this they will develop the tools to become better learners of mathematics and better problem-solvers, making connections between maths and other curricular areas.

Resources

- Buttons
- Hundred square
- Unifix cubes
- Whiteboards
- Whiteboard pens

Key vocabulary

- Element
- Exploring
- Extending
- Formula
- Pattern
- Recur
- Relationship
- Repeating
- Sequence
- Symbol
- Term

Misconceptions

Potential misconception	Addressing the misconception
Pupils expect patterns to always increase in size.	Ensure that pupils encounter a range of different patterns that have increasing and decreasing rules.
Pupils always expect patterns to change by the same amount each time.	Ensure that the pupils make patterns created by all four operations and patterns that are created according to shape (square and triangular numbers) or patterns that have a multi-step rule, such as the Fibonacci sequence.
Pupils do not realise that they can use strategies to work out the rule for generating the pattern, instead they expect the 'rule' to pop into their head.	Ensure that you explain your reasoning for continuing a pattern and encourage them to do the same. Show the pupils that you look for the gaps between numbers, you use your number facts, like multiplication facts and your knowledge of special sequences like square and triangular numbers and the Fibonacci sequence.

Teaching ideas

Unit 11.1 Exploring and extending number sequences

- Ask pupils to look around the classroom and identify any patterns they can see, encouraging them to be as creative as they can. Record their ideas on a whiteboard. It could be the order they are sitting – boy, girl, boy, girl, or it could be that they sit in groups of 4. Once they have had time to record their observations, give pupils time to discuss their ideas, encouraging them to think about the relationship between the patterns and sequences they have found.
- Using a hundred square, ask pupils to create patterns and explain the relationships to each other.

1	2	●	4	5	●	7	8	●	10
11	●	13	14	●	16	17	●	19	20
●	22	23	●	25	26	●	28	29	●
31	32	●	34	35	●	37	38	●	40
41	●	43	44	●	46	47	●	49	50
●	52	53	●	55	56	●	58	59	●
61	62	63	64	65	66	67	68	69	70
71	72	73	74	75	76	77	78	79	80
81	82	83	84	85	86	87	88	89	90
91	92	93	94	95	96	97	98	99	100

In this example, pupils have marked the multiples of 3 up to 60. It could also be described as adding 3 each time.

- Using a hole punch that makes 2 holes in each piece of paper, ask pupils to consider how the number of holes changes each time another piece of paper is added. Ask them if they can think of a way to record this – using number sentences, diagrams, pictures or a table.

Patterns and relationships 2C

Section overview

Textbook 2C units	Review units	Prior learning	Benchmarks/Es&Os
Unit 11.1 Applying knowledge of multiples, square numbers and triangular numbers to generate number patterns.	**2B** Unit 11.1	• I can use a table or diagram to identify a number pattern. • I can describe a number pattern. • I can use a formula to work out a number pattern. • I can find the missing numbers in a number pattern.	**MTH 2-13a** • Explains and uses a rule to extend well-known number sequences including square numbers, triangular numbers and Fibonacci sequence. • Applies knowledge of multiples, square numbers and triangular numbers to generate number patterns.

Conceptual context

Finding patterns in number, shapes, colour and the many other examples that can be found in our lives will support students to recognise, describe and continue sequences. We should realise that patterns can be used to analyse and predict real-life situations and through this process we will make links with algebraic concepts such as formulae, gradient and the equations for a straight line.

Investigating patterns in an organised and systematic way develops pattern-based thinking which, in turn, encourages problem-solving. This leads to an appreciation of how functions can be used to describe relationships between number and number patterns. Pupils should use manipulatives to investigate what happens in a pattern as the number of terms is extended or as the beginning of a number is changed. As pupils look at square numbers and triangular numbers, they will study patterns involving multiplication and division and, as categories become more complex, more than one attribute is used to categorise and classify objects. Pupils will work to organise their study of patterns more carefully and systematically by learning to use tables more effectively to record, identify and extend their findings. When pupils start to collect data and look for a pattern to solve a problem, they are often unsure what they are looking for. By organising this data into a chart or a table, patterns can begin to appear that will help pupils to determine a solution.

Resources

- Dotted paper (see Resource sheet 11a)
- Squared paper (see Resource sheet 11b)

Key vocabulary

- **Element**
- **Exploring**
- **Extending**
- **Formula**
- **Multiple**
- **Pattern**
- **Recur**
- **Relationship**
- **Repeating**
- **Sequence**
- **Square number**
- **Symbol**
- **Term**
- **Triangular number**

Misconceptions

Potential misconception	Addressing the misconception
Unit 11.1 Pupils may think that a multiple of one number cannot be a multiple of another.	When discussing the multiples of 3 within the Let's Learn section, take time to discuss multiples of other numbers and that pupils understand that there will be an overlap.
Pupils expect patterns to always increase in size.	Ensure that pupils encounter a range of different patterns that have increasing and decreasing rules.
Pupils always expect patterns to change by the same amount each time.	Ensure that the pupils meet patterns created by all four operations and patterns that are created according to shape (square and triangular numbers) or patterns that have a multi-step rule, like the Fibonacci sequence.
Pupils do not realise that they can use strategies to work out the rule for generating the pattern, instead they expect the 'rule' to pop into their head.	Ensure that you explain your reasoning for continuing a pattern and encourage them to do the same. Show the pupils that you look for the gaps between numbers, you use your number facts, like multiplication facts and your knowledge of special sequences like square and triangular numbers and the Fibonacci sequence.

Teaching ideas

Unit 11.1 Applying knowledge of multiples, square numbers and triangular numbers to generate number patterns

- Introduce this unit by establishing prior knowledge of key terms that will be used – multiple, square number and triangular number.
- Ask pupils to share their knowledge and understanding of multiples, square numbers and triangular numbers to establish a baseline. Using whiteboards, ask pupils to write down a definition for each of the terms and to provide an example of each one.
- Using squared paper, demonstrate the first 4 square numbers to illustrate why they are called square numbers:

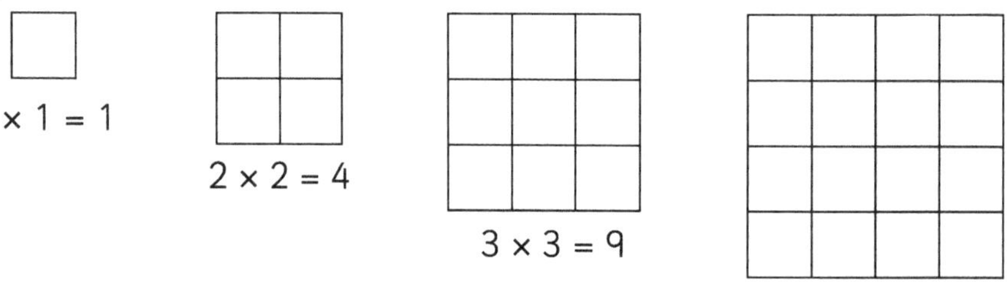

$1 \times 1 = 1$ $2 \times 2 = 4$ $3 \times 3 = 9$ $4 \times 4 = 16$

- Ask pupils to draw the next 3 square numbers to consolidate their understanding.
- Using dotted paper, demonstrate the first 4 triangular numbers to illustrate why they are called triangular numbers:

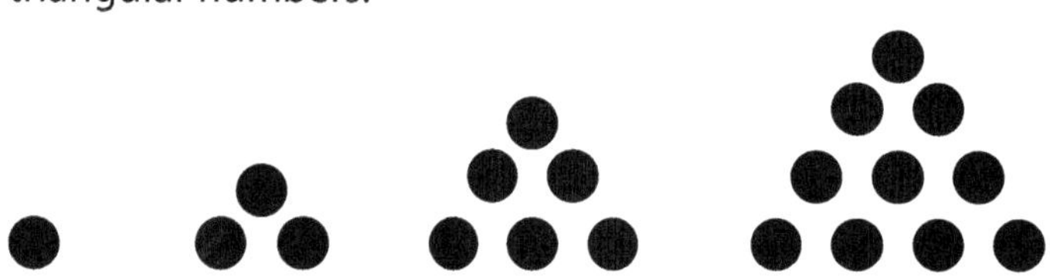

1 $1 + 2 = 3$ $3 + 3 = 6$ $6 + 4 = 10$

- Ask pupils to draw the next 3 triangular numbers to consolidate their understanding.
- Introduce the Lucas series, which is similar to the Fibonacci numbers. Edouard Lucas (1842 – 1891) was a French mathematician who gave the name to the series of numbers 0, 1, 1, 2, 3, 5, 8, 13… the Fibonacci Numbers. Each Lucas number is defined to be the sum of its two immediate previous terms, thereby forming a Fibonacci integer sequence.

12 Expressions and equations

Expressions and equations 2A

Section overview

Textbook 2A units	Review units	Prior learning	Benchmarks/Es&Os
Unit 12.1 Solving simple equations using known number facts	**1C** Unit 12.1	• I understand and can use the terms 'equal to' and 'not equal to' including their related symbols ($=$, $\neq$). • I understand and can use the terms 'less than' and 'greater than' including their related symbols ($<$, $>$). • I can apply my knowledge of number facts to solve simple expressions and equations.	**MTH 2-15a** • Solves simple algebraic equations with one variable; for example, $a - 30 = 40$ and $4b = 20$.

Conceptual context

A sound understanding of a range of part-whole calculation strategies, where a pupil can not only carry out arithmetical operations but can also understand the meaning of the operations and choose the most efficient method, is an important precursor to success with algebra. But this will only support a pupil so far; algebra is more than 'generalised arithmetic' and research tells us that pupils need teaching and scaffolding to move from arithmetical approaches to adopt algebraic conventions. This will take multiple experiences over time.

For most pupils at primary school, learning how to use the commutative, associative and distributive properties of number operations will support their initial algebraic understanding. These concepts are introduced through the number units.

Pupils also have to understand the concept of 'inverse' as well as operations. Addition and subtraction are inverse to each other; and multiplication and division are inverse to each other. This understanding will help pupils to solve hidden number problems where the answer is not a simple number bond. For example, $6 + g = 13$ is a straightforward addition problem, but $f + 73 = 219$ is easier to solve if the inverse subtraction is used.

In addition to these number conventions, primary-aged pupils will be introduced to algebraic symbols from an early age. The equals sign (=) is one of the first algebraic symbols used by pupils. However, in number work it usually means 'calculate'; in algebra, it more often means 'is equal to' or 'is equivalent to'. A simple step is to ensure that pupils use the equals sign in a range of contexts.

In mathematics, the terms 'expression' and 'equation' are used frequently, but people often misunderstand an expression for an equation. These two mathematical terms are not the same; the main difference is in how they are arranged. The simplest way to identify an expression or equation is that if it contains an *equal to* sign (=) it is an equation. However, if it does not contain an *equal to* (=) sign, it is an expression.

Pupils have simplified expressions and solved equations when they were asked to add and subtract. They probably didn't know that 5 + 3 is an expression and that 7 + 2 = 9 is an equation. It is important for pupils to understand these terms to clearly understand the similarities and differences between equations and expressions. They will begin to use algebraic expressions that feature variables as well. By employing their addition and subtraction skills, they will be simplifying expressions and evaluating algebraic expressions. Stress the progression from an expression to an equation by demonstrating how equal expressions are found on either side of an equals sign.

With the introduction of solving missing number problems, there are several potential problems and misconceptions that come under the banner of *'Fruit salad'* algebra. Questions should be carefully constructed to avoid the misconceptions highlighted below.

Resources

- Coloured card/paper squares or mosaic pieces
- Whiteboards and pens

Key vocabulary

- **Equation**
- **Expression**
- **Statement**
- **Symbols**
- **Terms**
- **Values**

Misconceptions

Potential misconception	Addressing the misconception
Unit 12.1 The pupil thinks that an expression and an equation is the same thing.	Make sure that pupils are clear about the difference between an expression and an equation from the start as both terms are used throughout the unit.
The pupil thinks that the missing values are related to the alphabet, a = 1, b = 2, x + 2 = z, etc.	Try to avoid this misconception by planning questions that avoid this rule.
Pupils may think that letters stand for objects, e.g. 'a' for apples.	Try to avoid using letters that match objects. In the textbook, a paint splodge is often used so that the pupil can think, 'Oh, someone has painted over the number. I need to work out what number that could be'. There are exceptions to this misconception, e.g. units of measurement, like 'm' for metres.

Potential misconception	Addressing the misconception
Pupils may think that letters all have different values.	In algebra, if a letter or symbol is used to represent a missing value it can have the same value as a different symbol or letter. Practice and experience are required, alongside carefully crafted questions. Try to avoid always using the same letter or symbol, such as '?' or 'x' to represent the missing value.
Pupils may think that a symbol or letter has the same value in different problems.	Unless the questions are directly related, each new question should be solved independently of the last question. Sometimes, a whole exercise may have the unknown represented by a '?'. Each question will have a different solution even though the same symbol is used.

Teaching ideas

Unit 12.1 Solving simple equations using known number facts

- Write 6 + 3 and 6 + 3 = 9 on the board and then ask a pupil to read aloud the expression 6 + 3 and the equation 6 + 3 = 9.
- Discuss with pupils what the difference is between 6 + 3 and 6 + 3 = 9. Pupils may respond by stating that one has an equals sign and the other does not.
- Clarify that 6 + 3 is called an *expression* and explain that an expression consists of a number or a combination of numbers joined by one of the operation symbols: +, −, ×, ÷. Tell pupils that an expression **does not** have an equals sign.
- Ask pupils to look at 6 + 3 = 9 and to consider what is on the left side of the equals sign and what is on the right side.
- Guide pupils to recognise that, in 6 + 3 = 9, an expression or value is on both sides of the equals sign by encouraging them to describe the numbers as well as the operation. The left side is an expression, 6 + 3, that is equal to the value of the right side, 9.
- Explain that 6 + 3 = 9 is called an equation and that one side of an equation equals the other side.
- Use more examples, including using two numbers on each side, making sure both sides of these equations are equal.

  $9 - 3 = 2 \times 3$

  $20 = 4 \times 5$
- Explain that both sides have the same total, therefore they balance.
- Give pupils small squares of coloured card/plastic tiles (18 each/mixed colours) and ask them how many different ways can they be arranged to create an array that can then be explained using an equation?
- Get pupils to share their ideas.

Expressions and equations 2B

Section overview

Textbook 2B units	Review units	Prior learning	Benchmarks/Es&Os
Unit 12.1 Solving equations using mathematical rules	**2A** Unit 12.1	• I can use my knowledge of Fact Families to find missing numbers where symbols are used. • I can create numbers statements using <, > and =. • I understand the meaning of 'not equal to'. • I understand that an equals sign is a balance within a statement.	**MTH 2-15a** • Solves simple algebraic equations with one variable; for example, a – 30 = 40 and 4b = 20.

Conceptual context

Commutative, associative and distributive properties of number operations continue to be an important concept to support algebraic understanding. Pupils also have to understand 'inverse' as well as operations. Addition and subtraction are inverse to each other; and multiplication and division are inverse to each other. This understanding will help pupils to solve hidden number problems where the answer is not a simple number bond. For example, 6 + g = 13 is a straightforward addition problem, but f + 73 = 219 is easier to solve if the inverse subtraction is used.

Pupils who are new to solving equations will cling onto calculation strategies or their own ad hoc number-specific methods and will need scaffolding to help them move from concrete models to abstract modes of solution. Research suggests that the 'function machine' approach leads to a better understanding of what an equation is and the variable nature of x; but it is only suitable for 'one-sided' equations. Balance methods of solving appeal to pupils' wish for concrete meaning. This method works for linear equations but does not work for negative signs.

As outlined in section 2A, the terms 'expression' and 'equation' are used frequently but people often misunderstand an expression for an equation. However, these two mathematical terms are not same, and the main difference is in how they are arranged. The simplest way to identify an expression or equation is that if it contains an equal-to sign (=) it is an equation. However, if it does not contain an equal-to sign (=), then it is an expression.

Pupils have simplified expressions and solved equations when they were asked to add and subtract. They probably didn't know that 5 + 3 is an expression and that 7 + 2 = 9 is an equation. It is important for pupils to understand these terms to clearly understand the similarities and differences between equations and expressions.

Key vocabulary

- **Balance**
- **Equation**
- **Expression**
- **Statement**
- **Symbols**
- **Terms**
- **Values**

Misconceptions

Potential misconception	Addressing the misconception
Unit 12.1 The pupil thinks that an expression and an equation are the same thing.	Make sure that pupils are clear about the difference between an expression and an equation from the start, as both terms are used throughout the unit.
The pupil thinks that the missing values are related to the alphabet, a = 1, b = 2, x + 2 = z, etc.	Try to avoid this misconception by planning questions that avoid this rule.
Pupils may think that letters stand for objects, e.g. 'a' for apples.	Try to avoid using letters that match objects. In the textbook, a paint splodge is often used so that the pupil can think, 'Oh, someone has painted over the number. I need to work out what number that could be'. There are exceptions to this misconception, e.g. units of measurement, like 'm' for metres.
Pupils may think that letters all have different values.	In algebra, if a letter or symbol is used to represent a missing value it can have the same value as a different symbol or letter. Practice and experience are required, alongside carefully crafted questions. Try to avoid always using the same letter or symbol, such as '?' or 'x' to represent the missing value.
Pupils may think that a symbol or letter has the same value in different problems.	Unless the questions are directly related, each new question should be solved independently of the last questions. Sometimes a whole exercise may have the unknown represented by a '?'. Each question will have a different solution even though the same symbol is used.

Teaching ideas

Unit 12.1 Solving equations using mathematical rules

- Write 10 × 2 and 10 × 2 = 4 × 5 on the board and then ask a pupil to read aloud the expression and the equation.
- Discuss with pupils: **What is the difference between 10 × 2 and 10 × 2 = 4 × 5?**
- Pupils may respond by stating that one has an equals sign and the other does not.

- Reinforce that 10×2 is called an expression and explain that an expression consists of a number or a combination of numbers joined by one of the operation symbols: +, −, ×, ÷. Tell pupils that an expression **does not** have an equals sign.
- Ask pupils to look at $10 \times 2 = 4 \times 5$ and consider what is on the left side of the equals sign and what is on the right side.
- Guide pupils to recognise that an expression or value is on both sides of the equals sign in $10 \times 2 = 4 \times 5$ by encouraging them to describe the numbers as well as the operation. The left side is an expression, 10×2, that is equal to the value of the right side 4×5.
- Explain that $10 \times 2 = 4 \times 5$ is called an equation and that the left side of an equation equals the right side.
- Use additional examples, including using two numbers on each side, making sure both sides of these equations are equal.
- Show pupils the following expressions:

25×4	$36 \div 4$
$28 \div 7$	30×3
$99 \div 11$	$20 \div 5$
10×9	20×5

  Ask pupils to match the expressions so that if they add an = in between each one they will have an equation with both sides having the same/equal answer.
- Now pupils are familiar with the concept that each side of the = sign must match, introduce the concept of balancing the equation. Ask pupils what happens when they sit on a seesaw.

- Discuss the idea that the heavier/bigger person causes the seesaw to go down at one side, therefore it is not balanced. Think about having an adult on one side and 2 children on the other to balance the weight.

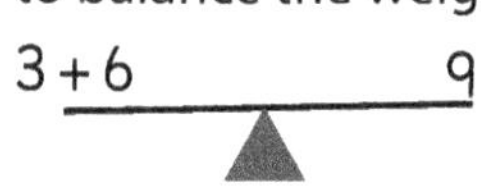

- What number is missing to balance the equation?

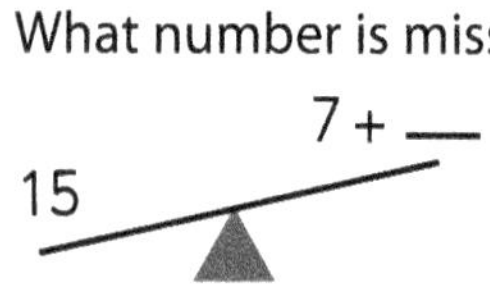

- Ask pupils to identify what you need to add to 7 to make 15.

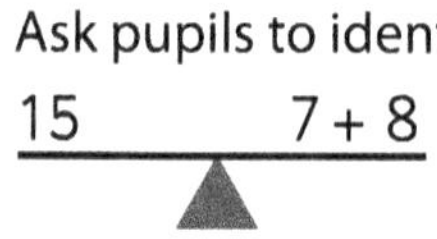

- Go over different examples with pupils, including ones using −, × and ÷ to ensure that they understand that they can balance the equation using any of the 4 operations.

  $14 \div$ ___ $= 5 + 2$

  $24 = 8 \times$ ____

  $100 -$ ___ $= 50$

Expressions and equations 2C

Section overview

Textbook 2C units	Review units	Prior learning	Benchmarks/Es&Os
Unit 12.1 Solving equations with inequalities	**2B** Unit 12.1	• I can create and solve number statements using <, >, = and ≠. • I understand the meaning of 'not equal to', 'greater than', 'less than'. • I understand that an equals sign is a balance within a statement.	**MTH 2-15a** • Solves simple algebraic equations with one variable, for example, a – 30 = 40 and 4b = 20.

Conceptual context

Inequality signs may have been introduced at an early age to support and record the relationship between two objects. For example, if there are two containers of liquid, we can represent the quantities using letters to represent relationships such as a > b or e = f. This can be applied to weight, size, length, etc.

Equations and inequalities are both mathematical sentences created when two expressions are related to one another. In an equation, the two expressions are equal and this is shown by the '=' symbol, e.g. 5 = 3 + 2, or 5 is equal to 3 add 2.

Mathematics is not always about 'equals', sometimes we only know that something is greater or less than something else. An inequality though is where the two expressions are not necessarily equal, which is shown by the symbols: > or <.

9 > 7   9 is greater than 7

5 < 10   5 is less than 10

An equation or an inequality that contains at least one variable is called an 'open sentence'.

When you substitute the variable in an open sentence with a number, the resulting statement is either true or false. If the statement is true, the number is a solution to the equation or inequality. When you solve inequalities it is the same as solving equations and we do most of the same things but we must also pay attention to the direction of the inequality.

The following things do not affect the direction of the inequality:

- Adding or subtracting a number from both sides.
- Multiplying or dividing both sides by a number.
- Simplifying a side.

Resources

- Wooden shapes
- Coloured paper shapes

Key vocabulary

- **Balance**
- **Equation**
- **Expression**
- **Inequality**
- **Operators**
- **Statement**
- **Symbols**
- **Terms**
- **Values**

Misconceptions

Potential misconception	Addressing the misconception
Unit 12.1 The pupil thinks that an expression and an equation are the same thing.	Make sure that pupils are clear about the difference between an expression and an equation from the start as both terms are used throughout the unit.
The pupil thinks that the missing values are related to the alphabet, a = 1, b = 2, x + 2 = z, etc.	Try to avoid this misconception by planning questions that avoid this rule.
Pupils may think that letters stand for objects, e.g. 'a' for apples.	Try to avoid using letters that match objects. In the textbook, a paint splodge is often used so that the pupil can think, 'Oh, someone has painted over the number. I need to work out what number that could be'. There are exceptions to this misconception, e.g. units of measurement, like 'm' for metres.
Pupils may think that letters all have different values.	In algebra, if a letter or symbol is used to represent a missing value, it can have the same value as a different symbol or letter. Practice and experience are required, alongside carefully crafted questions. Try to avoid always using the same letter or symbol, such as '?' or 'x' to represent the missing value.
Pupils may think that a symbol or letter has the same value in different problems.	Unless the questions are directly related, each new question should be solve independently of the last questions. Sometimes a whole exercise may have the unknown represented by a '?'. Each question will have a different solution even though the same symbol is used.

Teaching ideas

Unit 12.1 Solving equations with inequalities

- Introduce the concept by asking pupils what they would do if they had a bag of 25 lollipops and 5 friends that they wanted to share them between. Discuss the different ways that they might write this down to work it out.
- Look at both options and explain that they are both number sentences that need to be balanced and this is something that they are doing all the time. Tell pupils that these are inequalities that need to be solved by balancing each side of the number sentence.
- Ask pupil for other examples that they can think of, such as:
- Sharing out pens/pencils
- Dividing up a birthday cake
- Giving out textbooks
- Making teams when playing a game
- In the same way as the lollipop scenario, go through a few examples on the board to reinforce their understanding.
- Share 10 textbooks between 20 pupils.

  How many pupils will need to share a textbook?

  $20 \div 10 = 2$ there is 1 textbook between every 2 pupils
- Share 20 football cards between 4 friends.

  How many cards will each person receive?

  $20 \div 4 = 5$ each one will receive 5 football cards
- Now, introduce the concept of different things having the same value:

3 gummy bears are 10p 1 chew is 10p 1 lollipop is 10p

- Therefore, if 3 pupils had 10p to spend, they could all buy something different but spend the same amount. Also, if they had 30p they could get as many different combinations of things that would all add up to 30p.

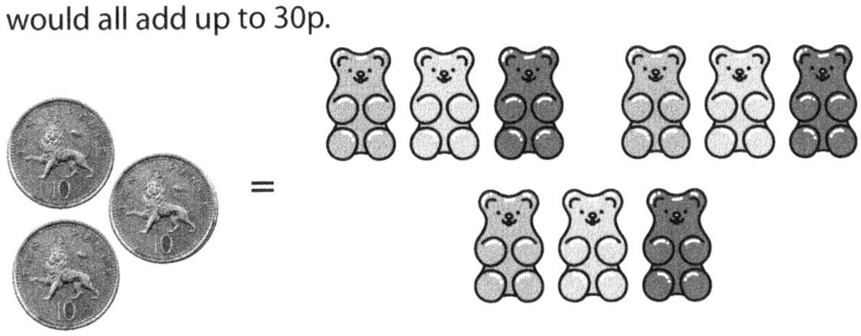

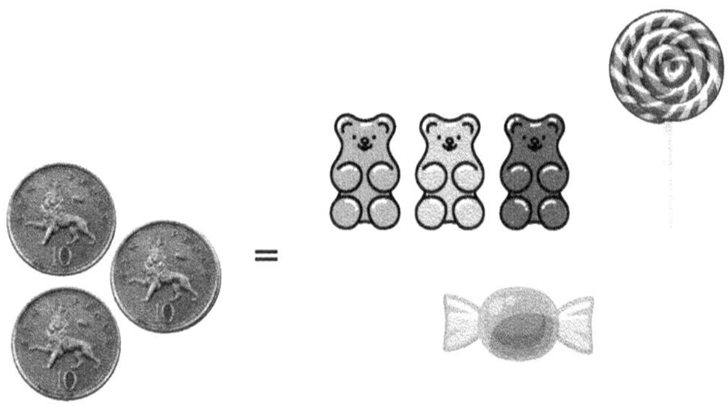

- They could also spend some of the money and keep the rest:

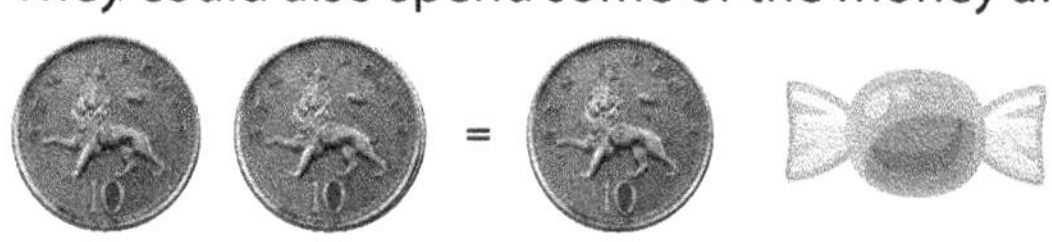

- All of the above are examples of number sentences that have been balanced as there is an equal value on each side of the = sign.
- On the board, draw a square, a triangle and a circle.
- Now tell them that the square has the value 6, the triangle has the value 2 and the circle has the same value as 2 triangles.
- Ask them to tell you what the value of the circle is.
- Once you have established that the circle has the value 4, ask pupils to imagine they had to balance a set of scales. Challenge them to come up with as many different combinations of the shapes as possible that will all balance the scales. The following are a few possible examples:

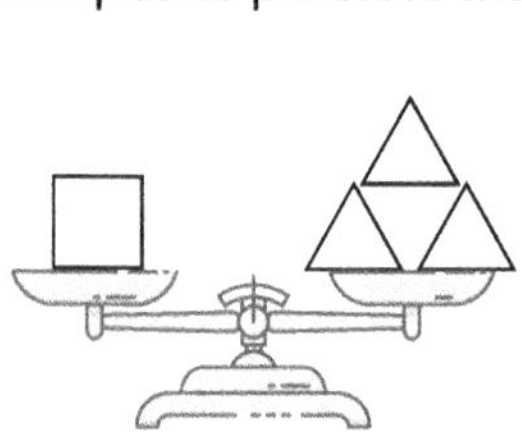
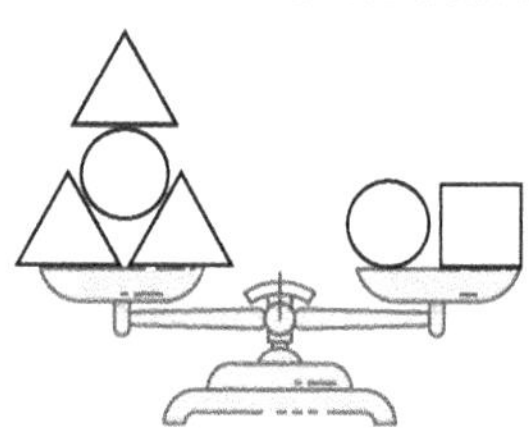

13 2D shapes and 3D objects

2D shapes and 3D objects 2A

Section overview

Textbook 2A units	Review units	Prior learning	Benchmarks/ Es&Os
Unit 13.1 Drawing 2D shapes	**1A** Units 12.1 and 12.2 **1B** Unit 12.1	• I know that vertex is another word for corner. • I can describe shapes by saying how many sides and how many vertices. • I know that a right angle is 90 degrees and can find right angles in my classroom. • I can identify a square, rectangle, pentagon, hexagon, triangle and circle.	**MTH 2-16a** **MTH 2-16b** **MTH 2-16c** • Describes 3D objects and 2D shapes using specific vocabulary including regular, irregular, diagonal, radius, diameter and circumference. Applies this knowledge to demonstrate understanding of the relationship between 3D objects and their nets. • Identifies and describes 3D objects and 2D shapes within the environment and explains why their properties match their function. • Knows that the radius is half of the diameter. • Uses digital technologies and mathematical instruments to draw 2D shapes and make representations of 3D objects, understanding that not all parts of the 3D object can be seen.
Unit 13.2 Naming and sorting 2D shapes	**1A** Units 12.1 and 12.2 **1B** Unit 12.1	• I can use a ruler to draw a straight line. • I can use a ruler to measure accurately. • I can use a protractor to measure 90 degrees. • I know the properties of squares and rectangles. • I can find the area of a rectangle.	
Unit 13.3 Drawing 2D shapes and 3D objects	**1A** Unit 12.3	• I can identify a cube, cuboid, cylinder, square-based pyramid. • I can identify a square, rectangle, triangle, circle, pentagon, hexagon, octagon. • I know that vertex is another word for corner.	
Unit 13.4 Describing and sorting prisms and pyramids	**1B** Unit 12.1	• I can identify a cube and a cuboid. • I can describe shapes by saying how many sides and how many vertices. • I know that a right angle is 90 degrees and can find right angles in my classroom.	

Conceptual context

Pupils should be able to visualise, draw and model shapes and to think and reason geometrically. This will initially involve investigating, exploring and experimenting with everyday objects and manipulatives. Before naming an object or idea, pupils should experience 2D shapes, 3D objects and their representations in many different forms and orientations. This supports visualisation, an essential skill that pupils need to develop for geometrical understanding. Pupils will then see figures, 2D shapes and 3D objects as collections of properties. They will be able to recognise and name properties of geometric figures, 2D shapes and 3D objects, seeing the relationships between these properties. It is vital, at this stage, that pupils discuss the names and characteristics of 2D shapes and 3D objects to clarify their mathematical understanding.

Without understanding concepts such as edges, sides, corners and faces, the definitions of shapes are meaningless. When describing and naming quadrilaterals, pupils should identify and discuss their similarities and differences using properties such as length of sides, size of angles, whether they have parallel lines, etc. When describing, sorting and comparing prisms and pyramids, discuss and highlight the similarities and differences between their properties.

Finally, when developing geometrical understanding, it is important to make it clear to pupils when a diagram should be drawn accurately and when a sketch is acceptable.

Resources

- Straight-sided 2D shapes: rhombus, trapezium, kite, v-kite, square, rectangle, parallelogram and other non-quadrilateral polygons. Both regular and irregular polygons.
- Isometric paper
- Multilink cubes
- Examples of prisms and pyramids: cube, square-based pyramid, rectangular prism, triangular prism, hexagonal prism, pentagonal-based pyramid, triangular-based pyramid, pentagonal prism and non-examples including cylinders and cones.
- Hula hoops

Key vocabulary

- Acute
- Base line
- Curved
- Edge(s)
- Faces
- Flat
- Irregular shape
- Parallel
- Polygon
- Prism
- Protractor
- Pyramid
- Quadrilaterals
- Regular shape
- Right angle
- Scale
- Side(s)
- Straight
- Vertex (vertices)

Misconceptions

Potential misconception	Addressing the misconception
Units 13.1, 13.2, 13.3 and 13.4 Pupils experience only typical, classic visual images of shapes, which limits their understanding of shapes, e.g. always seeing regular pentagons, shapes orientated with their base flat along the horizontal.	• Teachers must expose pupils to shapes of varying sizes, orientations and type (both regular and irregular).
Units 13.1, 13.2, 13.3 and 13.4 Pupils may struggle with inclusivity and exclusivity of definitions. For example, classifying squares and rectangles as different, discrete shapes.	• When defining quadrilaterals, investigate the properties of four-sided shapes, paying attention to the similarities and differences; ensure that the pupils understand that squares are regular rectangles; and squares, rectangles and rhombi (rhombuses) are all parallelograms.
Units 13.1, 13.2, 13.3 and 13.4 Pupils may be confused by the variety of ways that shapes can be defined, e.g. length of side; number of sides; parallel or adjacent sides; the size of interior angles; the ability to tessellate; the order of rotation and lines of symmetry.	• Whilst it is important for pupils to know, describe and name shapes according to their properties, they must be given ample opportunity to discover and understand the relationships within and between shapes.
Units 13.1, 13.2, 13.3 and 13.4 Pupils may define a polygon as any closed shape, whereas a polygon is any closed, plane shape made up of straight lines.	• Pupils need to understand that the word 'polygon' only relates to straight-sided shapes. A circle is not a polygon.

Teaching ideas

Unit 13.1 Drawing 2D shapes

- Pupils need to:
    - know the properties of squares and rectangles and their similarities and differences.
    - be able to use a ruler to measure and draw straight lines accurately.
    - be able to use a protractor to draw a 90-degree angle.
- Ask the pupils to draw squares and rectangles of different side lengths on 1-cm squared paper using a ruler. Now challenge them to do the same on blank paper.
- Ask pupils: **How could we draw a square or a rectangle on blank paper?** They may suggest tracing the squared paper or using it as a guide. They might suggest drawing around a commercially produced shape. You may want them to try both and to discuss the accuracy of their results. **What other piece of equipment might help us to be more accurate?**

Unit 13.2 Rounding whole numbers to the nearest 100

- Pupils should work with physical representations of 2D shapes, and be encouraged to notice and discuss similarities, differences and interrelationships whilst investigating their properties.
- Have a variety of pre-prepared polygons (some with right angles, some without, and all providing a variety of different side lengths). For example, right angled triangles, isosceles triangles; quadrilaterals (square, rectangle, parallelogram, rhombus, trapezium); regular and irregular pentagons; and regular and irregular hexagons. Make sure the polygons are big enough so that the pupils can easily check the size of angles using a protractor. Do not use commercially produced shapes for this activity as these often have slightly rounded corners.
- Ask the pupils to sort the shapes according to different properties: first sort by side length, then sort the shapes by the number of right angles. Next, sort by number of sides, number of vertices and finally by whether they have parallel sides. A photograph or a table can be used to capture how the shapes are sorted each time. After each sort ask: **What properties do these shapes have in common? Why are these shapes not included?**
- Encourage pupils to use the correct mathematical vocabulary. Recap vocabulary used in First Level, explaining that vertices can be used to describe corners in 3D objects and 2D shapes.

- Other activities you can use to develop key vocabulary and to support the pupils to identify the properties of shapes are:
    - Shape identification games or guess the object in a feely bag.
    - Ask the pupils to search pictures in order to locate similar shapes with different orientations.
    - Take the pupils on a walk to recognise shapes in the environment. Use a camera to capture the shapes and objects they find.
    - Examples and non-examples of shapes.
    - Shape reveal with an interactive whiteboard.
    - Shape treasure hunts – find a shape that has …
    - Geometric shape puzzles – cut shapes into pieces and pupils reassemble them.
    - Use Venn diagrams and Carroll diagrams for sorting 2D shapes.

Unit 13.3 Drawing 2D shapes and 3D objects

- Have a large cuboid for the pupils to look at and ask: **How can we represent this and other 3D objects on paper?** Pupils may give answers that involve shading or drawing lines to show the outline of the 3D object.
- Ask: **What shapes is this cuboid made from? What shapes do you see when I place it on the table?** Using a whiteboard pen, highlight all the edges of the cuboid so that they stand out.
- Ask pupils to work in pairs to discuss their observations of the cuboid and how this might be represented on the isometric paper. Do they recognise that opposite lines are parallel? Do they see that they can use the dots as guidelines to represent the cuboid?
- As a group, draw the 3D object on paper or an interactive whiteboard, ensuring that the key points above are highlighted to the pupils. Next, pupils can help to create cuboids with multilink cubes then move on to creating a range of shapes with 5 cubes and, in pairs, drawing these on the isometric paper.

Unit 13.4 Describing and sorting prisms and pyramids

- Sort a range of 3D objects (e.g. cube, square-based pyramid, rectangular prism, triangular prism, hexagonal prism, pentagonal-based pyramid, triangular-based pyramid, pentagonal prism and cylinders) into 'prism' and 'pyramid' categories. Ask: **What properties do the prisms and pyramids have? Can we create a rule for how we can tell whether a 3D object is a prism or a pyramid?**
- Play 'What am I?' with the above shapes in a feely bag. Encourage the pupils to pay close attention to the shape of the base, the shape of the sides and the number of vertices. Can they match the physical 3D object to the drawings of the 3D objects on page 162?
- Use hula hoops to create Venn diagrams to allow pupils to sort the given 3D objects in a variety of ways. Encourage the use of the properties of 'faces', 'vertices' and 'prisms/pyramids' or 'not a prism/pyramid' in their sorting.

2D shapes and 3D objects 2B

Section overview

Textbook 2B units	Review units	Prior learning	Benchmarks/Es&Os
Unit 13.1 Naming and sorting polygons	**1B** Unit 12.1	• I know that a pentagon has five sides and five vertices. • I know that a hexagon has six sides and six vertices. • I know and can use the terms sides, vertices and parallel. • I know that some shapes are regular and some shapes are irregular.	**MTH 2-16a** **MTH 2-16b** **MTH 2-16c** • Describes 3D objects and 2D shapes using specific vocabulary including regular, irregular, diagonal, radius, diameter and circumference. Applies this knowledge to demonstrate understanding of the relationship between 3D objects and their nets. • Identifies and describes 3D objects and 2D shapes within the environment and explains why their properties match their function. • Knows that the radius is half of the diameter. • Uses digital technologies and mathematical instruments to draw 2D shapes and make representations of 3D objects, understanding that not all parts of the 3D object can be seen.
Unit 13.2 Describing and drawing circles	**1B** Unit 12.1	• I can identify a circle. • I know that a circle has no vertices and one curved side. • I can use a ruler to measure accurately.	
Unit 13.3 Constructing 3D objects	**1B** Unit 12.3	• I can describe 3D objects by their properties. • I know and can use the terms face, vertex/vertices and edges. • I can identify a cube, cuboid, triangular-based pyramid, square-based pyramid, and a triangular prism.	
Unit 13.4 Constructing nets	**1B** Unit 12.2 **1C** Unit 12.2 **2A** Unit 13.4 **2B** Unit 13.3	• I can identify a cube, a cuboid and triangle- and square-based pyramids. • I can draw a square and a rectangle accurately. • I can describe the properties of cubes, cuboids and pyramids using vertex/vertices, edges and faces.	

Conceptual context

Pupils should be able to visualise, draw and model shapes, and to think and reason geometrically. This will initially involve investigating, exploring and experimenting with everyday objects and manipulatives. Before naming an object or idea, pupils should experience 2D shapes, 3D objects and their representations in many different forms and orientations. This supports visualisation, an essential skill that pupils need to develop for geometrical understanding (i.e. pupils will see figures, 2D shapes and 3D objects as collections of properties). They will be able to recognise and name properties of geometric figures, 2D shapes and 3D objects, seeing the relationships between these properties. It is vital, at this stage, that pupils discuss the names and characteristics of 2D shapes and 3D objects to clarify their mathematical understanding.

When introducing shape names, it can be helpful to discuss the origin of the word, ensuring that pupils understand that 'decagon' is from the Greek deka meaning 'ten' and gon meaning 'angles' (or corners). Pupils can then link this to other words they know either with the same beginning (for example decade, decimal) or ending (for example hexagon, pentagon).

When constructing and drawing 2D shapes and 3D objects, pupils must be encouraged to view shapes and objects from a variety of different perspectives, understanding how to orientate the component parts. Pupils should engage in practical activities (such as drawing nets of shapes), and observe what goes wrong when insufficient attention is paid to details of shape, size and placement of the component parts.

Depending on how one defines a 'side', a circle is not a polygon, it is a 2D shape with one curved side – see misconceptions below.

Resources

- Variety of 2D shapes
- Tacks, Blu Tack, string and a pencil
- Pairs of compasses
- A pair of board compasses or equivalent
- Rulers (cm/mm)
- Blank paper
- A variety of 3D objects (e.g. from a commercial pack and real-life objects, such as a cereal box and tennis ball)
- Straws and modelling clay (or something similar)
- Interlocking cubes
- 3D objects: cubes, cuboids, triangle-based pyramids, square-based pyramids
- Polydron or Clixi
- Squared paper

Key vocabulary

- Centre of a circle
- Circumference
- Diameter
- Radius (radii)
- Pair of compasses

Misconceptions

Potential misconception	Addressing the misconception
Units 13.1 and 13.2 Pupils may think that a circle has only one side.	• How many sides a circle has depends on how one defines a 'side'. If one defines a side as a straight-line segment, then a circle has no sides. If one defines a side as a curved boundary, a circle could have one curved side. One could also define a circle as having an infinite number of sides as a polygon with an infinite number of sides approximates a circle. • It is important to include the word 'curved' and more useful to describe a circle as having a curved edge rather than side. • This will provide a rich source of discussion.
Unit 13.2 Pupils may not understand that the diameter must pass through the centre of a circle and is not just a line that connects one side of the circumference of the circle to the opposite side.	• Make sure that when accurately drawing or sketching a circle, the centre is always identified.
Unit 13.2 Pupils may confuse the radius and diameter by suggesting that the radius is twice the length of the diameter.	• Ask the pupil to explain what they mean to help determine whether this is a simple confusion of the two terms or there is a deeper lack of understanding.
Unit 13.2 Pupils may struggle to draw a circle because, for instance, the pair of compasses keep slipping or they cannot keep the pencil on the paper.	• Ensure that pupils hold the pair of compasses at the top, not on the arms, and suggest that they keep the pair of compasses still and rotate the piece of paper instead.
Unit 13.4 Pupils may be unable to visualise which nets will form a cube or other 3D objects.	• Pupils need to experience opening and folding nets to see how faces and edges of 3D objects fit together.

Teaching ideas

Unit 13.1 Naming and sorting polygons

- Introduce 'right angles' to be another way of saying 'square vertices' as mentioned in the textbook for 2A.
- Display regular and irregular pentagons and hexagons. Ask pupils to sort these into 5- and 6-sided shapes. Discuss the link between the number of sides and the name of the shape. Ask pupils to label the sides and the vertices accurately and identify any parallel sides. Ensure that pupils make the connection between the sides being the same length and the angles being the same size, meaning that the shape is *regular*.
- In pairs, ask pupils to draw two heptagons, octagons, nonagons and decagons using a ruler but not a protractor. Get the pairs of pupils to join up with another pair and to sort their shapes into 7-sided, 8-sided, 9-sided and 10-sided groups. Provide regular heptagons, octagons, nonagons and decagons for the pupils to add to these groups (pupils will not be able to draw these independently at this stage).
- Shape Bingo/Who am I?

Unit 13.2 Describing and drawing circles

- Pupils will need to be able to work with a pair of compasses. This will require patience on the teacher's part and resilience from the pupils. To introduce the formal vocabulary for circles (radius, circumference and diameter), consider using a tack, a looped piece of string and a pencil to draw a circle *before* drawing circles with pairs of compasses.
- Give the pupils different lengths of looped string. Once the pupils have drawn a circle, remove the tack and draw several lines from the centre of the circle to its circumference. **Measure the length of each of these lines.** Discuss the results and define this length as the radius of the circle.
- Draw different lines that go from one side of the circle to the opposite side which all pass through the centre point. Define these longer lines as examples of diameters. Ask the pupils: **What do you notice about the length of the diameter compared with the length of the radius?**
- Collate information about the radii and diameters from different pupils to show that, despite circles being different sizes, the relationship between the radius and diameter is always diameter = 2 × radius.
- Move on to drawing circles using a pair of compasses whilst continuing to consolidate their circle vocabulary. Make sure that you check that the pairs of compasses are tightened and that the compass pencil-holders will grip the pencils.
- Model how to draw a circle using a pair of compasses with a board compass, using a visualiser or interactive whiteboard software. Next, challenge pupils to draw their own circles with a given radius or diameter. Once pupils become confident using a compass, challenge them to create a compass flower.

Unit 13.3 Constructing 3D objects

- Use commercial sets of 3D shapes with regular and irregular polyhedra (remembering to remove the objects with curved surfaces as these are not polyhedra) as well as 3D objects from real life, e.g. a cereal box, ball, etc. Ask a pupil to secretly choose one of the objects and describe it to the rest of the group. Encourage the use of *number of edges, number of edges meeting at a vertex* and *faces* in their descriptions. This will allow pupils to revisit and practice key vocabulary.
- Provide 12 straws of the same length and 8 blobs of modelling clay (or a similar material). Ask the pupils: **What shape can be made using all these materials? What does the modelling clay represent?**
- State: **I can make a skeleton of a square-based pyramid using 3 straws and 3 balls of modelling clay. Am I correct? Why/why not?** If pupils are finding it challenging to move from using straws and modelling clay to something larger in size, link making skeleton models out of interlinking cubes to buildings being built out of bricks or concrete blocks.

Unit 13.4 Constructing nets

- Pupils will need to be able to identify the 2D-shaped faces that make up the nets of the 3D object. Using a selection of differently sized cubes, cuboids, triangular-based pyramids and square-based pyramids, review the pupils' knowledge of faces, vertices and edges by doing a feely bag activity or playing shape bingo – these do not need to be commercially produced 3D objects; they can be, for instance, cardboard packaging. It is important that the pupils have the 3D objects in front of them.
- Ask: **What are the similarities and differences between a cube and a cuboid? What are the similarities and differences between a triangular-based pyramid and a square-based pyramid? What are the similarities and differences between a pyramid and a prism?** Concentrate on identifying the 2D shapes that make up the 3D objects' faces.
- In pairs, use Polydron or Clixi to build a cube, a cuboid, a triangular-based pyramid and a square-based pyramid. Use these 3D objects to complete the 'Before we start' activity on p.182. To ensure accurate counting of faces, edges and vertices, use Blu Tack to put blobs on the faces, edges and vertices, which are then removed as they are counted. This strategy will be used again when investigating Euler's formula in 2C, Unit 13.3. 'Flatten out' the Polydron/Clixi objects to reveal the nets. **Has everyone got the same nets; if not, why not?**
- 'Flatten out' cardboard containers to investigate how they are made, including the flaps. Discuss why flaps are needed and look at how they are designed.

2D shapes and 3D objects 2C

Section overview

Textbook 2C units	Review units	Prior learning	Benchmarks/Es&Os
Unit 13.1 Describing and sorting triangles	**2B** Unit 14.2	• I know that a triangle has three straight sides and three vertices. • I can distinguish between a regular and an irregular shape. • I can identify the line of symmetry in a 2D shape.	**MTH 2-16a** **MTH 2-16b** **MTH 2-16c** • Describes 3D objects and 2D shapes using specific vocabulary including regular, irregular, diagonal, radius, diameter and circumference. Applies this knowledge to demonstrate understanding of the relationship between 3D objects and their nets. • Identifies and describes 3D objects and 2D shapes within the environment and explains why their properties match their function. • Knows that the radius is half of the diameter. • Uses digital technologies and mathematical instruments to draw 2D shapes and make representations of 3D objects, understanding that not all parts of the 3D object can be seen.
Unit 13.2 Drawing 2D shapes	**2A** Unit 13.1	• I can identify a square, rectangle, parallelogram, kite, trapezium and rhombus. • I can identify a pentagon, hexagon, octagon. • I know and can use the terms parallel, vertex, vertices and sides.	
Unit 13.3 Making representations of 3D objects		• I can recognise cubes, cuboids and prisms. • I can describe the properties of cubes, cuboids and prisms.	
Unit 13.4 Nets of prisms	**1C** Unit 12.1 **1C** Unit 12.2 **2B** Unit 13.4	• I can identify nets of cubes, cuboids and pyramids. • I have investigated possible nets of pyramids and cubes. • I know that prisms have two equal faces.	

Conceptual context

Pupils should be able to visualise, draw and model shapes and think and reason geometrically. Initially, this will involve investigating, exploring and experimenting with everyday objects and manipulatives. Before naming an object or idea, pupils should experience 2D shapes, 3D objects and their representations in many different forms and orientations. This supports visualisation, an essential skill that pupils need to develop for geometrical understanding. Subsequently, pupils will see figures, 2D shapes and 3D objects as collections of properties. They will be able to recognise and name properties of geometric figures, 2D shapes and 3D objects, and see the relationships between these properties. It is vital, at this stage, that pupils discuss the names and characteristics of 2D shapes and 3D objects to clarify their mathematical understanding. An increased emphasis on using unambiguous, formal mathematical vocabulary in describing the characteristics, properties and relationships of 2D shapes and 3D objects is expected at this stage of the pupils' development.

When drawing 2D representations of 3D objects, pupils must be encouraged to view the objects from a variety of different perspectives, and to understand how to orientate the component parts. When drawing nets, through trial and error, pupils should learn to observe what goes wrong when insufficient attention is paid to details of shape, size and placement of the component 2D shapes.

Resources

- Protractors
- Rulers
- Blank paper
- A variety of congruent 2D shapes: triangles, rectangles, pentagons, hexagons, decagons
- 'Yes' 'No' 'Not sure' labels for Runaround/Corners game
- Modelling clay and a cutting tool
- Selection of prisms
- Nets of cubes
- A variety of 3D objects: a pentagonal-based pyramid, a cone, a triangular-based pyramid (tetrahedron), a square-based pyramid, a triangular prism and a hexagonal prism
- Cardboard packaging (e.g. cereal boxes, sweets/chocolate boxes, etc.)
- Polydron or Clixi
- Scissors

Key vocabulary

- **Angle size**
- **Congruent – identical shape and size**
- **Polyhedron – a solid object with flat faces**
- **Side length**

Misconceptions

Potential misconception	Addressing the misconception
Unit 13.1 Pupils can only identify prototypical orientations of triangles, i.e. where the base is along the horizontal.	• Teachers must ensure that pupils are exposed to a variety of sizes and different rotations of triangles.
Unit 13.2 Pupils may think that congruency is about only one shape property being the same.	• This is related to the misconception above. Ensure pupils understand that a shape can be orientated in different ways and still be the same shape, have the same length of sides and the same size of angles and is therefore congruent.
Unit 13.4 Pupils may think that a cylinder is a prism.	• A prism is a polyhedron, which means all its faces are flat. A cylinder is therefore not a prism because it has a curved side. A cylinder acts like a prism because it has the same cross-section and we calculate its volume in the same way we would for a prism.

Teaching ideas

Unit 13.1 Describing and sorting triangles

- Display various quadrilaterals (square, rectangle, parallelogram, kite, trapezium and rhombus) and three different types of triangle (equilateral, scalene and isosceles). Ask pupils to sort these into quadrilaterals and triangles. Recap on the properties of the quadrilaterals and their names. Give pupils a selection of equilateral, scalene and isosceles triangles, a ruler and a Carroll diagram with the labels 'sides all the same length', 'two sides the same length' and 'no equal sides'.
- Explain that 'equilateral' means equal-sided and comes from the Latin *aequus*, meaning 'even, or level' and *latus*, meaning 'side'. 'Isosceles' comes from the Greek word *isos*, meaning 'equal' and *Skelos*, meaning 'leg'. 'Scalene' also comes from the Greek word *skalenos*, which originally meant 'stirred up', which related to uneven ground.
- Pupils have already had experience of 'right-angled triangles' in Book 1B. Explain that right-angled triangles can also be scalene and isosceles. Challenge the pupils to draw a sketch of right-angled triangles that are scalene and then isosceles.

Unit 13.2 Drawing 2D shapes

- Play 'Runaround' or 'Corners' in an open space. Each pupil has a shape (one of triangle, rectangle, pentagon, hexagon, decagon). Put up the following labels in different corners of the space you are using: 'Yes', 'No' or 'Not Sure'. Call out clues to the properties of shapes named above, e.g. 'None of my angles are right angles'. The pupils then have to move to the correct corner that has 'Yes', 'No' or 'Not Sure'. Still in the same open space, ask the pupils to find others with the same shape. Ask: **What is the same and what is different about the shapes you have matched?**
- Introduce the word 'congruent' and explain that it means having the same shape, length of sides and size of angles.

  Ask: **Can a pentagon with side length 5 cm be congruent to a hexagon with side length 5 cm? Can a square and a rectangle, which both have four right angles, be congruent?**
- In pairs, get pupils to design a picture for Q1 of the 'Let's learn' section on p.179. Pupils should accurately draw congruent triangles and rectangles using rulers and protractors but should draw different sized pentagons, hexagons and decagons, cutting out the shapes to make their picture.

Unit 13.3 Making representations of 3D objects

- Distribute modelling clay (or similar), a cutting tool and various prisms with flat faces to allow pupils to explore whose statement is correct from the 'Before we start' activity on p.181. *Both are correct in their definition, though a cylinder is not a prism as a prism is a polyhedron. It therefore requires flat faces and the same cross-section when you cut it along its length.*
- Make cubes and hang them from different high points in the classroom to help pupils to appreciate that these are regular hexahedrons. Pupils often find it difficult to recognise that a cube is a special kind of prism.
- Use modelling clay (or similar) when investigating Euler's formula, putting blobs on the faces, edges and vertices, which are then removed when counting to ensure accurate counting of these.

Unit 13.4 Nets of prisms

- Provide a variety of 3D objects for the pupils to access. Make sure that the objects include a pentagonal-based pyramid, a cone, a triangular-based pyramid (tetrahedron), a square-based pyramid, a triangular prism and a hexagonal prism. Ask the pupils to describe the faces of the 3D objects and to discuss how these faces are connected to each other.
- Ask the pupils to build the 3D objects from Polydron or Clixi.
- Challenge the pupils to sketch the nets shown in Q2 of the 'Let's learn' section on p.184 on blank paper and to then cut them out to test whether they could be folded to make a shape. These nets can then be used to create a 2D and 3D display. Ask pupils to create labels to practice their vocabulary.

14 Angles, symmetry and transformation

Angles, symmetry and transformation 2A

Section overview

Textbook 2A units	Review units	Prior learning	Benchmarks/ Es&Os
Unit 14.1 Identifying angles	**1B** Unit 13.2 **1C** Unit 13.1	• I know that a right angle is 90°. • I know that 90° is a quarter turn and 180° is a half turn. • I can use my knowledge of right angles to estimate if an angle is bigger or smaller than 90°. • I can use a right-angle tester to check if my estimate is correct. • I can identify angles that are less than and more than a right angle (90°).	**MTH 2-17a** **MTH 2-17b** **MTH 2-17c** **MTH 2-17d** **MTH 2-18a** **MTH 2-19a** • Uses mathematical language including acute, obtuse, straight and reflex to describe and classify a range of angles identified within shapes in the environment. • Measures and draws a range of angles to within ± 2 degrees • Knows that complementary angles add up to 90° and supplementary angles add up to 180° and uses this knowledge to calculate missing angles. • Uses knowledge of the link between the eight compass points and angles to describe, follow and record directions. • Interprets maps, models or plans with simple scales; for example, 1 cm : 2 km.
Unit 14.2 Using an 8-point compass	**1C** Unit 13.3	• I know and can use north, south, east and west. • I know that there is a quarter turn (90°) between north/south and east/west. • I know there is a half turn (180°) between north and south, and between east and west. • I can turn clockwise.	
Unit 14.3 Plotting points using coordinates	**1C** Unit 13.4	• I can use grid references to locate objects.	
Unit 14.4 Lines of symmetry	**1A** Unit 13.3 **1B** Unit 13.4	• I know what horizontal means and can identify a horizontal line. • I know what vertical means and can identify a vertical line.	

Unit 14.5 Creating designs with lines of symmetry	**1A** Unit 13.3 **1B** Unit 13.4	• I can spot symmetry in patterns, pictures, nature and shapes. • Given a set of objects, I can create a symmetrical pattern with more than one line of symmetry. • I know the terms horizontal and vertical, and can draw horizontal and vertical lines. • I can spot symmetry in patterns, pictures, nature and shapes. • Given a set of objects, I can create a symmetrical pattern with more than one line of symmetry. • I can complete a symmetrical shape.	• Describes, plots and records the location of a point, in the first quadrant, using coordinate notation. • Identifies and illustrates line symmetry on a wide range of 2D shapes and applies this understanding to complete a range of symmetrical patterns, with and without the use of digital technologies.
Unit 14.6 Measuring angles up to 180°	**2A** Units 13.1 and 14.1	• I can identify angles on a protractor. • I can use a protractor to draw a square.	
Unit 14.7 Understanding scale	**1C** Units 8.2, 8.3 and 8.6	• I can measure a straight line accurately with a ruler. • I know there are 100 centimetres in a metre and 1000 metres in a kilometre.	

Conceptual context

Angles

Angle is a measurement, it is a measure of rotation. A key conceptual understanding of angles is that they can be both dynamic and static. For example, the movement involved in stretching the arms of a compass is the dynamic aspect of angle, which demonstrates the size of rotation, with the final position of the arms being a static representation of this turn. Thinking of angle as the difference in direction between two lines helps to link the dynamic view of angle with the static view.

When pupils are asked to define an angle, they may suggest it is a corner that is pointy or sharp, a place where two lines meet, the space between two lines, or the difference between the slope of the two lines. These are not correct. Angles can be defined as intersections of two or more lines, corners of shapes, bends, slopes, limited rotation (e.g. a door handle) and unlimited rotation (e.g. a wind turbine). To deepen conceptual understanding of angle, it is important that pupils have extensive experience of all five contexts.

Throughout the angles sections there should be a discussion and agreement with the pupils around the importance of accurate measurement and construction of angles. A tolerance of ±2 degrees is expected for the benchmark at second level.

Transformation

Pupils will learn about three inter-related transformations: line symmetry, rotational symmetry and translation. These can be considered as three types of congruence, meaning that although the shape has been transformed, the length of its sides remain the same, as do the size of the angles and the orientation of the sides and the angles.

Reflective, or line symmetry, is the first kind of symmetry that children will experience. In mathematics, it is a broader idea than just 'mirror images'. A shape that has reflective symmetry has two or more matching parts. It is important that pupils experience reflecting and folding shapes in many ways to understand the differences between horizontal, vertical and oblique (diagonal) lines of symmetry.

Rotational symmetry can occur around the centre of a shape, around a corner point and rotating around a point outside the shape. The number of times a shape will fit into its own outline during a full 360-degree rotation denotes the order of rotational symmetry. A shape that only fits into its own outline once, i.e. after a full rotation, does not have rotational symmetry.

Translation is the process of sliding a shape along a straight-line path to a new position. This aspect of transformation is closely linked to coordinate geometry in that the straight-line path can be described by a move horizontally and vertically.

A further form of transformation is to scale an object up or down to produce an enlargement or reduction. The angle sizes will remain the same, but the length of the sides will change according to the scale. This is not covered in second level; however, pupils will encounter scale in the topic of similarity in third and fourth levels.

Learning about and using scale as a context is a rich, real-life mathematics topic. In these textbooks, scale is used for learning about maps, plans and bearings. For this aspect of scale, as opposed to scaling an object, it is important to set out the purpose in learning about scale and its uses and make connections to multiplying, dividing, ratio and proportion.

Resources

- Protractors
- Interactive whiteboard protractor or large board protractor
- Strips of paper and Blu Tack or Geostrips
- Labels for the classroom walls: initially north, east, south, west then adding north-east, north-west, south-east and south-west
- Blank grid reference grids and blank coordinate grids
- Resource sheet 14a: shapes
- Rulers
- Centimetre squared paper
- Resource sheet 14b: Venn diagram and 26 capital letters of the alphabet
- 2D shapes, pegboards and beads/strings
- Mirrors
- Squared paper and colouring pencils
- Large board protractor or protractor for interactive whiteboard
- Selection of right-angled triangles on paper
- Maps of the area local to the school

Key vocabulary

- **8-point compass**
- **Acute**
- **Axes (x-axis, y-axis)**
- **Clockwise and anticlockwise**
- **Compass directions**
- **Coordinates**
- **Diagonally**
- **First quadrant**
- **Horizontal**
- **Line of symmetry**
- **Obtuse**
- **Origin**
- **Protractor**
- **Reflection**
- **Scale**
- **Straight angle**
- **Vertical**

Misconceptions

Potential misconception	Addressing the misconception
Units 14.1, 14.2 and 14.6 Pupils may have a misconception about what an angle is.	Children must experience both the static and dynamic notions of an angle as explained in the 'Conceptual context' section. Spotting dynamic and static angles in the environment and recognising these can help with this.
Units 14.1 and 14.6 Pupils may only recognise angles such as right angles when both arms of the angle are the same length and/or pointing in a standard direction.	Ensure that, when measuring and discussing angles, pupils experience them with unequal arm lengths and in various positions, e.g. the angle pointing to the right, left, up and down.
Unit 14.2 Pupils may confuse compass directions.	Have the pupils individually following directions before asking them to direct others. Reinforce going clockwise and remembering N, E, S, W in that order. You may want to connect to the position of the minute hand on a clock. Rotating clockwise will be relied upon for bearings later.
Unit 14.3 Pupils may have difficulty moving from locating an object using a letter and number to using coordinates. They may also have difficulty moving from labelling spaces to labelling lines.	As the mathematics becomes more sophisticated, pupils need to understand that changing from locating an object in a space to locating an object at a point is more accurate. To ensure that pupils do not struggle with this transition, extend the lines past the axes, before labelling the mark with a number.
Units 14.4 and 14.5 Pupils may struggle to use symmetry when the lines of symmetry are horizontal or diagonal.	Pupils are more able to identify and work with vertical lines of symmetry than horizontal and diagonal. Pupils should be encouraged to rotate their jotters, textbooks, etc. so that the line of symmetry is made vertical.
Unit 14.6 Pupils may use the bottom of the protractor as the zero rather than the internal zero line.	Introduce the different parts of a protractor to the pupils before you expect them to use them. Make sure that the scales and zero lines are clearly visible. Only use clear plastic protractors. Encouraging pupils to estimate an angle before they measure it will ensure they use the correct scale.
Unit 14.7 Pupils may have difficulty when converting between units of measure.	Encourage the pupils to notice the units of measurements in the question and talk about the units of measurement required in the answer. Constantly revisit converting between centimetres, metres and kilometres.

Teaching ideas

Unit 14.1 Identifying angles

- Ask pupils to create their own angles (using strips of paper and something to fix them in place or Geostrips). Ask pupils to sort these into 'less than a right angle – acute' and 'greater than a right angle – obtuse'. Can they then order these angles from smallest to largest? Ask: **How did you decide which angle was the biggest?** Encourage the pupils to focus on *the amount of turn about a point from one line to the other* and not the length of the arms.
- Discuss the different angle measurements that would be acute and obtuse. **What is the only angle measurement for a right angle?**

Unit 14.2 Using an 8-point compass

- Have labels on the walls of the classroom/open space: north, east, south, west.

  To review previous learning, give pupils compass directions to follow to go from a starting point to an end point. Start pairs of pupils off at different points within the space.

  Use the 'Before we start' instructions as a practical activity with the pupils.

  Ask: **What instruction would I have to give if I wanted you to face the corner rather than the wall? How could we make our instructions more precise?**

  Introduce north-east, north-west, south-east, and south-west and ask: **What do you notice about these directions and what do you think they mean?**
- To reinforce the pupils' understanding of the 8-point compass directions, give them instructions to turn in different directions – this can be done in the hall, playground or classroom. Emphasise rotating in a clockwise direction.

Unit 14.3 Plotting points using coordinates

- Show a blank grid reference grid and a blank coordinate grid with the axes labelled. Ask the pupils to describe the differences between the two grids. See if they notice that a coordinate grid has the numbers on the line and the origin marked and the grid reference grid uses numbers and letters along the axes between the lines.

  Put the shapes from Resource sheet 14a, on a projected grid reference grid – make the shapes so that they will comfortably fit inside the squares and could be moved within the square too.

Explain that the grid references describe the space where the object can be found. Put the objects in different positions within the grid squares. **What has stayed the same and what has changed?** The objective is for the pupils to understand that the grid reference is not an accurate description of location because the object can remain within the same grid even if it is moved.

Using the coordinate grid, put objects on points on the grid. **How can we describe where the objects are on this coordinate grid?** Emphasise that the horizontal number comes first, followed by the vertical.

What happens if the objects move to a different position on the coordinate grid? Can the coordinates stay the same?

Emphasise the importance of using a cross rather than a dot to plot a point. It is difficult to accurately position and see a dot at the intersection of the grid lines.

Unit 14.4 Lines of symmetry

- Using a Venn diagram with the labels 'Has horizontal symmetry' and 'Has vertical symmetry', sort the 26 capital letters of the alphabet.
- Ask pupils to match the full shapes from the halves on Resource Sheet 14b. Check their results using a mirror. Ask: **What do you see in the mirror that tells you there is a line of symmetry?**

Unit 14.5 Creating designs with lines of symmetry

- Allow pupils to explore reflective symmetry using 2D shapes, pegboards or by making necklaces using beads on a string. Discuss how they know their creations show symmetry. Ask: **How does this differ from the activities we have completed previously?** Draw pupils' attention to the colours they have used, the shapes they have chosen and the distance between the shapes.
- With the full shapes that were used in a Unit 14.4 activity, allow pupils to explore placing their mirrors across the shapes in different ways. How does the shape you see in the mirror differ from the paper shape? Ask: **Do lines of symmetry always have to be horizontal or vertical?**
- Mark a line of symmetry on a sheet of squared paper (this can also be done on an interactive whiteboard). In pairs or small groups, each pupil takes turns to colour one square and its reflection on the squared paper/interactive whiteboard grid. Pupils must ensure that when a square is coloured, the corresponding square that makes the shape symmetrical is also coloured. The pupils should agree that the reflection is correct. They should also be describing their strategies for placing the squares correctly.

Unit 14.6 Measuring angles up to 180°

It is suggested that, as this is the pupils' first encounter with measuring angles, these pages from the textbook are photocopied on to A3 paper.

- Pupils will be using protractors for the first time in this unit. Check protractors to ensure that the zero line and the centre point of the protractor can be clearly seen and have not been damaged by pencil or compass marks.

  A large protractor, or one for an interactive whiteboard, may be useful to demonstrate how to measure angles. Demonstration angles should be drawn so that pupils can see how to measure an angle from the left of the zero line *and* the right. Pupils should also be encouraged to use their angles vocabulary and knowledge of angles to check their measurements – **Is this an obtuse angle? Does it measure more than 90° and less than 180°?** This gives pupils the opportunity to consolidate their understanding of key vocabulary.
- After pupils complete the 'Challenge' section, ask them to share their thoughts and rules in a larger group. Having right-angled triangles available on paper may be helpful in demonstrating that the angles in a triangle equal 180°; therefore, if one is 90°, the other two must total 90°.

Unit 14.7 Understanding scale

In this Unit, scale is set in the context of maps and plans.

- To be able to work with scale, pupils must have a good grasp of converting between measures of length and hence the 'Before we start' section could be extended to further check and consolidate related prior knowledge.
- Before introducing 'Let's learn', discuss maps, plans and bird's eye views using a mapping tool, such as Google Maps (in satellite view and street view), OS Maps, etc. Perhaps print off maps of the area local to the school – if your school is in a rural setting, choose the map of a local town.

  Using the example on page 181 (or one devised from a real map), explain how 1 cm represents 1 km in real life. Devise other examples to reinforce this understanding and extend to using scales of 1cm for every 2 km, 1 cm for every 10 km, etc., making sure to make the connection to multiplying.
- For the 'Challenge', be aware that the pupils must measure to the nearest cm.

Angles, symmetry and transformation 2B

Section overview

Textbook 2B units	Review units	Prior learning	Benchmarks/Es&Os
Unit 14.1 Identifying and sorting angles	**1B** Unit 13.2 **1C** Unit 13.1 **2A** Units 14.1 and 14.6	• I know that a right angle is 90°. • I know that an acute angle is smaller than 90° and an obtuse angle is bigger than 90°. • I can use my knowledge of right angles to estimate if an angle is bigger or smaller than 90°. • I know that a straight angle is 180°.	**MTH 2-17a** **MTH 2-17b** **MTH 2-17c** **MTH 2-17d** **MTH 2-18a** **MTH 2-19a** • Uses mathematical language including acute, obtuse, straight and reflex to describe and classify a range of angles identified within shapes in the environment. • Measures and draws a range of angles to within ± 2 degrees. • Knows that complementary angles add up to 90° and supplementary angles add up to 180° and uses this knowledge to calculate missing angles. • Uses knowledge of the link between the eight compass points and angles to describe, follow and record directions. • Interprets maps, models or plans with simple scales; for example, 1 cm : 2 km.
Unit 14.2 Measuring and drawing angles	**2A** Unit 14.1 **2A** Unit 14.6	• I know that 360° is a full turn and 180° is a half turn. • I can identify acute, right, obtuse, reflex and straight angles. • I know that an acute angle is less than 90° and an obtuse angle is greater than 90°. • I can measure angles using a protractor.	
Unit 14.3 Finding missing angles	**2A** Units 14.1 and 14.6	• I know a right angle is 90 degrees and a straight angle is 180 degrees.	
Unit 14.4 Locating objects using bearings	**2A** Units 14.1, 14.2 and 14.6	• I can locate objects using an 8-point compass rose. • I can describe simple journeys using an 8-point compass rose.	
Unit 14.5 Reading coordinates	**2A** Unit 14.3	• I can plot points on a coordinate grid given the coordinates. • I know that the first number in a pair of coordinates is the horizontal distance and the second number is the vertical distance. • I know and can use the terms vertical and horizontal.	

Unit 14.6 Line symmetry	**2A** Units 14.4 and 14.5	• I can identify shapes with one line of symmetry. • I can draw the line of symmetry in shapes with one line of symmetry.	• Describes, plots and records the location of a point, in the first quadrant, using coordinate notation. • Identifies and illustrates line symmetry on a wide range of 2D shapes and applies this understanding to complete a range of symmetrical patterns, with and without the use of digital technologies.
Unit 14.7 Symmetrical pictures and diagrams	**2A** Units 14.4 and 14.5	• I can identify shapes with up to two lines of symmetry. • I can complete a symmetrical shape by reflecting it in one line of symmetry.	
Unit 14.8 Reading scale maps	**2A** Unit 14.7	• I can calculate a real-life distance using a simple scale. • I know that there are 100 cm in a metre and 1000 m in a kilometre. • I can measure a straight line accurately with a ruler.	

Conceptual context

Angles

Angle is a measurement, it is a measure of rotation. A key conceptual understanding of angles is that they can be both dynamic and static. For example, the movement involved in stretching the arms of a compass is the dynamic aspect of angle, which demonstrates the size of rotation, with the final position of the arms being a static representation of this turn. Thinking of angle as the difference in direction between two lines helps to link the dynamic view of angle with the static view.

When pupils are asked to define an angle, they may suggest it is a corner that is pointy or sharp, a place where two lines meet, the space between two lines, or the difference between the slope of the two lines. These are not correct. Angles can be defined as intersections of two or more lines, corners of shapes, bends, slopes, limited rotation (e.g. a door handle) and unlimited rotation (e.g. a wind turbine). To deepen conceptual understanding of angle, it is important that pupils have extensive experience of all five contexts.

Throughout the angles sections there should be a discussion and agreement with the pupils around the importance of accurate measurement and construction of angles. A tolerance of ±2 degrees is expected for the benchmark at second level.

Transformation

Pupils will learn about three inter-related transformations: line symmetry, rotational symmetry and translation. These can be considered as three types of congruence, meaning that although the shape has been transformed, the length of its sides remain the same, as do the size of the angles and the orientation of the sides and the angles.

Reflective or line symmetry is the first kind of symmetry that children will experience. In mathematics, it is a broader idea than just mirror images. A shape that has reflective symmetry has two or more matching parts. It is important that pupils experience reflecting and folding shapes in many ways to understand the differences between horizontal, vertical and oblique lines of symmetry.

Rotational symmetry can be around the centre of a shape, around a corner point and rotating around a point outside a shape. The number of times a shape will fit into its own outline during a full 360-degree rotation denotes the order of rotational symmetry. A shape that only fits into its own outline once does not have rotational symmetry.

Translation is the process of sliding a shape along a straight-line path to a new position. This aspect of transformation is closely linked to coordinate geometry in that the straight-line path can be described by a move horizontally and vertically.

A further form of transformation is to scale an object up or down to produce an enlargement or reduction. The angle sizes will remain the same, but the length of the sides will change according to the scale. This is not covered in second level; however, pupils will encounter scale in the topic of similarity in third and fourth levels.

Learning about and using scale as a context is a rich, real-life mathematics topic. In these textbooks, scale is used for learning about maps, plans and bearings. For this aspect of scale, as opposed to scaling an object, it is important to set out the purpose in learning about scale and its uses and make connections to multiplying, dividing, ratio and proportion.

Resources

- Selection of acute, right, obtuse and reflex angles
- A selection of 2D shapes to use to measure acute, obtuse and reflex angles
- Protractors (and one large or one on the interactive whiteboard)
- N, NE, E, SE, S, SW, W, NW labels on the walls and corners of the classroom
- Rulers
- Centimetre squared paper
- Resource sheet 14a
- Mirrors
- Access to a selection of logos and/or national flags
- Squared paper and coloured pens/pencils
- Paper and scissors or art materials
- Maps of the local area
- Centimetre squared paper
- Scissors
- For the re-designing their bedroom idea – Argos catalogues

Key vocabulary

- **Acute, obtuse, right angle, straight angle, reflex, full turn**
- **Bearing, three-figure bearing**
- **Clockwise**
- **Coordinates: x coordinate and y coordinate**
- **Coordinate grid**
- **Dimensions (length and breadth)**
- **Horizontal, vertical, diagonal**
- **Lines of symmetry, symmetrical**
- **Protractor**
- **Real-life distance**
- **Vertices**
- **Scale**

Misconceptions

Potential misconception	Addressing the misconception
Units 14.1 and 14.2 Pupils may only recognise angles such as right angles when both arms of the angle are the same length and/or pointing in a standard direction.	Ensure that, when drawing and measuring angles, pupils experience them with unequal arm lengths and in various positions, e.g. the angle pointing to the right, left, up and down.
Unit 14.4 Pupils may incorrectly name NE, NW, SE and SW and mix up where they are located on the 8-point compass.	Practise the order of the 8-point compass directions and link them to the associated bearings.
Unit 14.5 Pupils may have difficulty moving from labelling spaces to labelling lines / labelling zero on both axes / confusing x and y axis / remembering which way round the numbers are recorded.	When labelling lines, extend the lines past the axes, before labelling the mark with a number. It is important to label the origin (0,0) with either one 0 or a 0 on each axis. To help pupils remember the order of x and y coordinates, use 'along the corridor and up the stairs' or 'along the runway, up in the air'. This way of thinking about coordinates will not work with coordinates in four quadrants but by the time pupils encounter negative coordinates, they should have a conceptual understanding of x before y.
Units 14.6 and 14.7 The pupil struggles to use symmetry when the lines of symmetry are horizontal or diagonal.	Pupils are more able to identify and work with vertical lines of symmetry than horizontal and diagonal. Pupils should be encouraged to rotate their jotters, textbooks, etc. so that the line of symmetry is made vertical. Some may also think that a line dividing a shape into two parts or two equal parts must be line of symmetry. Experiences with mirrors being placed along lines of symmetry and exploring this is key to move pupils beyond this reasoning.
Unit 14.8 Pupils may use incorrect units of measure.	Highlight the units of measure of the scale, reinforcing the need to check every time. Encourage the pupils to check which units of measure the question is asking them to answer in.

Teaching ideas

Unit 14.1 Identifying and sorting angles

- Using a selection of acute, right, obtuse and reflex angles, ask pupils to sort these into 'What is *not* an acute angle', etc. Ensure that pupils are justifying why they think particular angles are not obtuse, etc.
- Use the 'Let's learn' section of the textbook on page 186 as a matching activity, photocopy the page and cut the sections into cards for the pupils to match the angle to the description.
- Issue the pupils with shapes or angles to sort into a table with four sections with the labels: acute, right, obtuse and reflex. Ensure the pupils agree on the position of the shapes. Some shapes may have two different angles and this can promote discussion.

Unit 14.2 Measuring and drawing angles

You may wish to enlarge Questions 1 and 2 on page 190 to help pupils measure angles more accurately.

- Ask: **Is it easier to measure an acute angle or a reflex angle? How could we measure a reflex angle more easily? How could we use our knowledge of a full turn to help us?** If needed, drawing an acute angle and the corresponding reflex angle may help support them in coming to their conclusion.
- Recap on the important features of a protractor and how they are used to measure and draw angles. Focus on how we decide which side of the zero line we measure from, how we line up the vertex of the angle with the centre-point of the protractor and how we measure/draw angles as accurately as possible.
- Practise estimating and drawing angles such as 100 degrees and 170 degrees whilst justifying/ verbally reasoning why they have been drawn like that. Pupils will hopefully have a mental image of angles such as 90 degrees, 180 degrees, 45 degrees, etc. and be able to use this to draw angles slightly more or slightly less.

Unit 14.3 Finding missing angles

- Use 'Before we start' to review learning from Units 14.1 and 14.2, concentrating on identifying whether the angle is acute or obtuse *before* using a protractor to measure and discussing the common errors when using a protractor.
- Draw a 90° angle on the board, with the right-angle symbol shown and a 180° angle on the board, with the vertex and the angle marked. **What can you tell me about the size of these angles? How do you know without measuring them?**

  On both the 90° angle and 180°angle introduce a third line, splitting the angles in two, labelling the known angle in degrees and the missing angle A, B, etc.

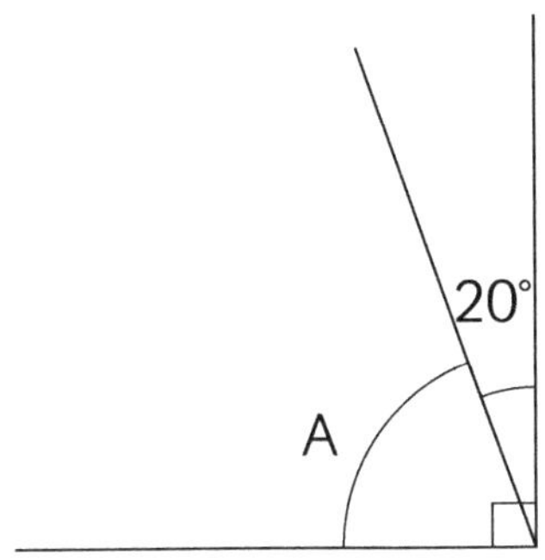

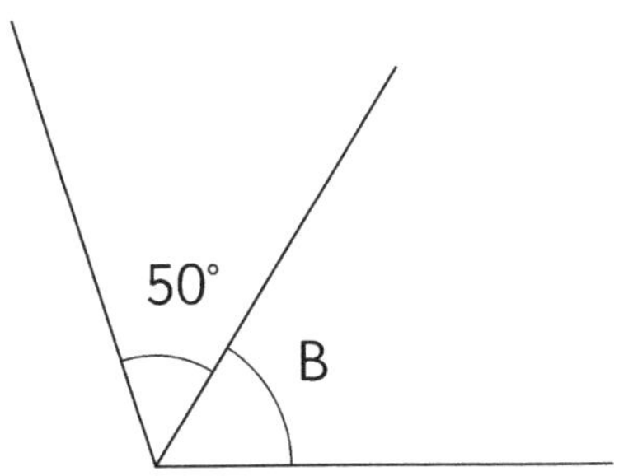

How do we calculate the missing angle A? How do we calculate the missing angle B? Etc. Invite the pupils to measure the missing angles to check that the calculation is correct. Introduce the language of *complementary* and *supplementary* angles and explore other examples of calculating complementary and supplementary angles, asking the pupils to check by measuring – this is important as a further review of their ability to use a protractor accurately.

You may wish to enlarge question 3 on page 193 to help pupils measure the angles more accurately.

Unit 14.4 Locating objects using bearings

Although bearings are not in CfE Second Level, their introduction at this point provides a useful link between, consolidation of learning of, and context for 8-point compass directions and angles.

- Pin up N, NE, E, SE, S, SW, W, NW labels on the walls and corners of the classroom, reinforce the learning from Unit 14.2 in Textbook 2A by asking the pupils to turn to these directions. Insist on the pupils turning clockwise to get to the correct compass direction.

  After each new instruction, reset the pupils to facing north and once they have turned, ask: **How many degrees have you turned?** Begin with north to east, north to south, etc. then introduce north to north east, etc. Record their answers on a diagram like this:

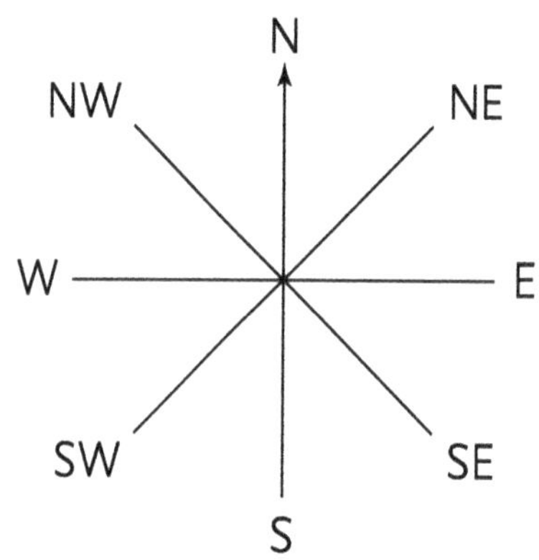

- Use the diagram above to discuss the position of objects on bearings of 100°, 190°, 275°, etc.
- When the pupils have finished the 'Challenge', ask: **What is the connection between the bearing the ship must take to investigate the flare and the bearing it needs to take to go back to the start?**

Key things to reinforce about bearings: always start from north, always rotate clockwise and always write bearings using three figures.

Unit 14.5 Reading coordinates

- To combine the skills of plotting and reading coordinates, and as a whole-class consolidation activity, play coordinate bingo. Ask the pupils to draw a four-by-four coordinate grid. Pupils then plot four points. The teacher calls out coordinates and, if the pupil has a point at that coordinate, they put a circle around the cross. When a pupil has circled all four points, they call 'bingo'. The pupil must then read back the coordinates of the points to win.
- To be able to do 'Let's practise' question 3b, the pupils must have encountered reflection in a horizontal line. See Textbook 2A, Units 14.4 and 14.5.

Emphasise the importance of using a cross rather than a dot to plot a point. It is difficult to accurately position and see a dot at the intersection of the grid lines.

Unit 14.6 Line symmetry

- Use a selection of paper shapes (see Resource sheet 14a) to allow pupils to check which shapes have two lines of symmetry by folding. Use mirrors to 'see' the whole shape when the mirror is placed along the fold showing the line of symmetry. Ask: **Do the shapes have horizontal or vertical symmetry? What happens when we place the mirror across the shape but not on a line of symmetry?**
- State: **A diagonal is a line of symmetry on a square and on a rectangle. Am I correct?** Support the pupils to use folding and lines of symmetry to prove whether this statement is true or false.
- Set the pupils an inquiry task: can the pupils find famous logos that have exactly two lines of symmetry? What other logos use symmetry?

Unit 14.7 Symmetrical pictures and diagrams

- Set the pupils an inquiry task: can the pupils find logos or national flags that have exactly two lines of symmetry? Are there flags with different symmetry?
- Using squared paper, create designs with two lines of symmetry by using horizontal and vertical folds in the paper and colouring pens/pencils.
- Horizontal and vertical symmetry can also be explored through art. Paint or folding and cutting paper are especially good mediums for this, e.g. by creating paper snowflakes.

Unit 14.8 Reading scale maps

- Use the 'Before we start' section to review work from Textbook 2A. Ask: **Why is the answer NOT a, c or d?** If the pupils are unsure, provide a further question using the same scale.
- If you have printed off maps of the local area, in an appropriate scale, use these to discuss scaled and real-life distances.
- Using 'Let's learn', ask the pupils to check that the distance from A to B is 5 cm and discuss how to write the scale and how to interpret the scale key. If you have started work on ratio, the pupils should recognise the colon ':'. If you have not, then introduce this as a shorthand for 'to' and remember to make this connection again when discussing ratio. Ask: **How would you write the scale using 1 cm : _____ metres?** Identify the steps for answering scale questions.
- For question 3 in 'Let's practise', the pupils will need to remember what a diameter is.
- You could use the 'Challenge' as a whole-class activity and get the children to re-design the classroom. Alternatively, you could ask the children to measure their bedroom and main pieces of bedroom furniture, draw a scale drawing of their rooms and re-arrange the furniture. You could combine this activity with a budgeting activity to buy new furniture for their rooms at home given a specified budget.

Angles, symmetry and transformation 2C

Section overview

<table>
<tr><th>Textbook 2C units</th><th>Review units</th><th>Prior learning</th><th>Benchmarks/Es&Os</th></tr>
<tr><td>Unit 14.1
Drawing triangles</td><td>2A
Unit 14.1
2B
Unit 14.2</td><td>• I know the properties of triangles.
• I can identify a scalene, an isosceles and an equilateral triangle.</td><td rowspan="5">MTH 2-17a
MTH 2-17b
MTH 2-17c
MTH 2-17d
MTH 2-18a
MTH 2-19a
• Uses mathematical language including acute, obtuse, straight and reflex to describe and classify a range of angles identified within shapes in the environment.
• Measures and draws a range of angles to within ± 2 degrees.
• Knows that complementary angles add up to 90° and supplementary angles add up to 180° and uses this knowledge to calculate missing angles.
• Uses knowledge of the link between the eight compass points and angles to describe, follow and record directions.
• Interprets maps, models or plans with simple scales; for example, 1 cm : 2 km.</td></tr>
<tr><td>Unit 14.2
Reflex angles</td><td>2B
Units 14.1 and 14.2</td><td>• I can identify a reflex angle.
• I know that there are 360° in a full turn.</td></tr>
<tr><td>Unit 14.3
Finding missing angles</td><td>2B
Unit 14.3</td><td>• I can find complementary angles.
• I can find supplementary angles.
• I know that there are 360° in a full turn, 180° in a half turn and 90° in a quarter turn.</td></tr>
<tr><td>Unit 14.4
Using bearings 1</td><td>2B
Unit 14.4</td><td>• I know and can use the 8-point compass rose.
• I can use three-figure bearings to locate objects on a map or plan.</td></tr>
<tr><td>Unit 14.5
Using bearings 2</td><td>2B
Units 14.4 and 14.8
2C
Unit 14.4</td><td>• I know and can use the 8-point compass rose.
• I can use three-figure bearings to locate objects on a map or plan.
• I can use three-figure bearings to describe journeys.</td></tr>
</table>

Unit 14.6 Using coordinates	**2B** Unit 14.5 **2C** Unit 13.2	• I can read and plot points in the first quadrant. • I can calculate the area of a rectangle, square. • I can name a square, rectangle, rhombus and kite. • I know and can use the term *congruence*.	• Describes, plots and records the location of a point, in the first quadrant, using coordinate notation. • Identifies and illustrates line symmetry on a wide range of 2D shapes and applies this understanding to complete a range of symmetrical patterns, with and without the use of digital technologies.
Unit 14.7 Symmetry 1	**2A** Units 14.4 and 14.5 **2B** Units 14.6 and 14.7	• I can identify two lines of symmetry in a shape. • I can create designs with two lines of symmetry. • I can complete shapes that have two lines of symmetry.	
Unit 14.8 Symmetry 2	**2A** Units 14.4 and 14.5 **2B** Units 14.6 and 14.7	• I can identify lines of symmetry in shapes. • I can reflect shapes and designs in up to two lines of symmetry.	
Unit 14.9 Using scale	**2B** Unit 14.8	• I can calculate a real-life distance using a scale given as 1 cm :. • Given a real-life distance and a scale, I can calculate how far it should be on a map or plan. • I know that there are 100 cm in a metre and 1000 m in a kilometre. • I can measure a straight line accurately with a ruler.	

Conceptual context

Angles

Angle is a measurement, it is a measure of rotation. A key conceptual understanding of angles is that they can be both dynamic and static. For example, the movement involved in stretching the arms of a compass is the dynamic aspect of angle, which demonstrates the size of rotation, with the final position of the arms being a static representation of this turn. Thinking of angle as the difference in direction between two lines helps to link the dynamic view of angle with the static view.

When pupils are asked to define an angle, they may suggest it is a corner that is pointy or sharp, a place where two lines meet, the space between two lines, or the difference between the slope of the two lines. These are not correct. Angles can be defined as intersections of two or more lines, corners of shapes, bends, slopes, limited rotation (e.g. a door handle) and unlimited rotation (e.g. a wind turbine). To deepen conceptual understanding of angle, it is important that pupils have extensive experience of all five contexts.

Throughout the angles sections, there should be a discussion and agreement with the pupils around the importance of accurate measurement and construction of angles. A tolerance of ±2 degrees is expected for the benchmark at second level.

Transformation

Pupils learn about three inter-related transformations: line symmetry, rotational symmetry and translation. These can be considered as three types of congruence, meaning that although the shape has been transformed, the length of its sides remain the same, as do the size of the angles and the orientation of the sides and the angles.

Reflective or line symmetry is the first kind of symmetry that children will experience. In mathematics, it is a broader idea than just mirror images. A shape that has reflective symmetry has two or more matching parts. It is important that pupils experience reflecting and folding shapes in many ways to understand the differences between horizontal, vertical and oblique lines of symmetry.

Rotational symmetry can be around the centre of a shape, around a corner point and rotating around a point outside a shape. The number of times a shape will fit into its own outline during a full 360-degree rotation denotes the order of rotational symmetry. A shape that only fits into its own outline once does not have rotational symmetry.

Translation is the process of sliding a shape along a straight-line path to a new position. This aspect of transformation is closely linked to coordinate geometry in that the straight-line path can be described by a move horizontally and vertically.

A further form of transformation is to scale an object up or down to produce an enlargement or reduction. The angle sizes will remain the same, but the length of the sides will change according to the scale. This is not covered in second level; however, pupils will encounter scale in the topic of similarity in third and fourth levels.

Learning about and using scale as a context is a rich, real-life mathematics topic. In these textbooks, scale is used for learning about maps, plans and bearings. For this aspect of scale, as opposed to scaling an object, it is important to set out the purpose in learning about scale and its uses and make connections to multiplying, dividing, ratio and proportion.

Resources

- Protractors
- Rulers
- Paper for drawing
- Mini whiteboards and pens
- 8-point compass labels for the classroom wall
- Street map of the local area around the school
- Centimetre squared paper
- Squared paper and scissors
- Pre-cut squares or squared paper to allow pupils to cut their own squares
- Access to logos and/or national flags
- Various 2D shapes and 2 mirrors per pupil
- Calculators
- Pairs of compasses

Key vocabulary

- **3-figure bearing**
- **Acute, obtuse, reflex angles**
- **Coordinates: x coordinate and y coordinate**
- **Coordinate grid**
- **Congruent**
- **Diameter**
- **Horizontal, vertical and diagonal lines of symmetry**
- **Isosceles**
- **Interior and exterior angles**
- **Rectangle, square, rhombus, kite**
- **Scale**
- **Vertices**
- **x-axis and y-axis**

Misconceptions

Potential misconception	Addressing the misconception
Unit 14.2 Pupils may only recognise angles such as right angles when both arms of the angle are the same length and/or pointing in a standard direction.	Ensure that, when measuring and discussing angles, pupils experience them with unequal arm lengths and in various positions, e.g. the angle pointing to the right, left, up and down.
Unit 14.3 Pupils may confuse the different types of angle calculations.	Consolidate complementary, supplementary angles and angles around a point to ensure the pupils remember and can use the appropriate totals for each.
Unit 14.4 Pupils may not take bearings from a new north line when plotting a new leg of a journey.	Encourage pupils to draw a new north line every time they take a new bearing.
Units 14.7 and 14.8 Pupils may struggle to use symmetry when the lines of symmetry are horizontal or diagonal.	Pupils are more able to identify and work with vertical lines of symmetry than horizontal and diagonal. Pupils should be encouraged to rotate their jotters, textbooks, etc. so that the line of symmetry is made vertical. Some pupils will give horizontal or vertical images as answers for any objects placed in a horizontal or vertical orientation despite the orientation of the mirror line. This is particularly important when there is an oblique mirror line. Pupils should be given experiences exploring what happens when the mirror line is placed horizontally, vertically and on the oblique to aid this understanding.

Teaching ideas

Unit 14.1 Drawing triangles

- This is the first time pupils will have encountered naming angles and triangles using mathematical naming conventions. This will need to be explained to pupils. The 'Let's learn' section gives an example of using the letter at the vertex to name the angle.
- Pupils can work in pairs or individually with a mini whiteboard to test out Isla's theory in 'Before we start' of an isosceles triangle before sharing their work and theories with the rest of the group.
- Ask pupils to use their previous knowledge of drawing shapes, such as a square, to describe to the teacher/another pupil how to draw a triangle when the length of the sides and size of the angles are known. How will they use the protractor and ruler?
- It may need explained to pupils that, in order to show the sides of a shape are the same length, a small line can be drawn perpendicular to the side. The first two shapes in the 'Let's practise' section on page 187 have this feature.
- During the 'Challenge' section on page 188, pupils may decide to work in groups and draw one triangle each, then collaborate to work out the pattern, or choose to do this individually.

Unit 14.2 Reflex angles

- Discuss what the smallest and largest acute, obtuse and reflex angles would be in order to recap on the properties of these angles. Challenge the pupils to sketch these angles and share their results with the group.
- Recap on the link between acute and reflex angles. Can the pupils make the link that angles in a full turn add up to 360° and angles on a straight line (on a protractor) add up to 180°? A secure understanding in measuring the acute/obtuse angle to calculate the reflex angle is important to allow the pupils success in the calculations in this section.
- After completing the 'Challenge', the pupils can create their own pathways or maze and challenge a partner to correctly measure the angles.

Unit 14.3 Finding missing angles

- Review complementary and supplementary angles, the learning from Unit 14.3 in Textbook 2B, before pupils attempt the 'Before we start' section. Ask the pupils to work in pairs to discuss this section, leading to a whole-class discussion. In this section the pupils should be identifying the right-angle square symbol denoting a 90-degree angle.
- Extend their ability to calculate missing angles by introducing missing angles around a point with some worked examples. Making sure that the angles are accurately drawn means the pupils can measure them to check the angles are correct. How to measure using a protractor will be required for question 2 in the 'Let's practise' section. Remember that the smallest inaccuracy when measuring several angles can lead to an expected right angle measuring more or less than 90° or a full turn being greater than 360° and so on. This is an opportunity to remind pupils about accuracy, but also that measurement can have small errors. The pupils should use estimation when measuring to predict or check their solutions are correct.
- Use question 1 of the 'Let's practise' section to assess their understanding, collecting whole-class feedback using mini whiteboards. Discuss the pupils' answers to D and E. **Is this always the same? How could we check?**
- Before or after the pupils attempt the 'Challenge' section, ask: **What assumptions will you need to/did you make to calculate the missing angles?**

You may wish to enlarge question 2 on page 192 to help pupils measure the angles more accurately.

Unit 14.4 Using bearings 1

- This unit introduces advanced skills in using bearings to describe a journey and requires a secure understanding of reflex angles, bearings and how to measure bearings using a protractor.
- Recap the 8-point compass directions and associated bearings. You can put up signs around the room or in a large space for the 8-point compass. Then challenge the pupils to rotate to the correct direction. Remind the pupils to always start at north and rotate clockwise. This activity can be further enhanced by changing to bearings as well as compass directions.
- Work together through an example, similar to question 1 on page 194, to identify the key steps involved in describing a journey using measured bearings. Use the maps of the local streets around the school to make describing a journey more relevant and personal to the pupils. Avoid complicating the situation by measuring the distances at this stage.
- For question 1 in 'Let's practise', tell the pupils to round to the nearest 5°; unless you decide to provide an enlarged version of the 'Let's practise' section.

Unit 14.5 Using bearings 2

- This unit, like Unit 14.4, relies upon advanced understanding of reflex angles, bearings, how to use a protractor to measure bearings and brings in the pupils' understanding of scale to plot journeys.
- The instructions on page 195 in the 'Let's learn' section clearly explain the steps required to plot a journey given bearings and distances and using an appropriate scale. Encourage the pupils to work in pairs and use the instructions to 'coach' each other through questions 2 and 3 in 'Let's practise'.
- Consider using a map of the local area around the school to contextualise the 'Challenge', this time including measuring the distances, knowing the map's scale.

Unit 14.6 Using coordinates

- This unit combines the pupils' knowledge and understanding of how to plot points, properties of 2D shapes and congruency.
- Using the same idea as the 'Before we start' section as a whole-class activity, read out coordinates for the pupils to plot and ask the pupils to predict what shape the coordinates will make. Make up examples giving three vertices of a quadrilateral and ask: **To complete the shape to make a quadrilateral, what are the coordinates of the missing vertex? Are there other possibilities?** Throughout your questioning and explanations you can remind the pupils of the shape and angles vocabulary.

- Although the 'Let's learn' section explains how to *calculate* the coordinates of missing vertices of 2D shapes, some pupils may find this difficult. It is acceptable for pupils to plot the given points, complete the shapes and read the coordinates of the missing vertex. However, once they have completed 'Let's practise', ask the pupils to tabulate the coordinates of the points and look for things that have stayed the same and things that have changed about the x and y coordinates. **Why has this happened? If points have the same x coordinate, what does that tell you about the points? If points have the same y coordinate, what does that tell you about the points?**
- To combine reading, plotting and predicting coordinates to complete 2D shapes, play shape coordinate battleships. Pupils work in pairs. Each pupil plots four 2D shapes on a coordinate grid. They take it in turns to call out coordinates. If the coordinate falls inside a shape, this is a 'hit'. If a shape is hit three times, it is destroyed.
- Before the pupils tackle the 'Challenge' section, ask: **Without drawing the 4th rectangle, how will you predict what the x and y coordinates of its vertices will be? How do the x and y coordinates relate to whether they are in the 1st, 2nd or 3rd rectangle? What strategies will help you to look for patterns and predict the x and y coordinates of the vertices of the 10th rectangle? How can you check that you are correct?**

Unit 14.7 Symmetry 1

- In preparation for the 'Before we start' and 'Let's learn' section, encourage pupils to create shapes on squared paper that can be folded in half, twice. By creating two of these, the pupils can match the folded shape to the original. Encourage discussion of how others managed to match the folded shape to the original shape and use of vocabulary such as: quarters, half, length of sides, angles.
- During the 'Let's learn' section, it would be beneficial for pupils to have paper squares that they can fold to show lines of symmetry in different ways. It is worth noting that a *diagonal line of symmetry* is relative to the orientation of the shape.
- Ask pupils to find logos and/or national flags that have various lines of symmetry. These can then be placed in the correct section of a Venn diagram – *One or more diagonal line of symmetry* and *vertical and horizontal lines of symmetry.*
- It would be worthwhile asking pupils to share their results of their tables in the 'Challenge' section before predicting how many lines of symmetry shapes (f) and (g) have. Features such as regular shapes having sides of equal length and equal angles linking to the number of lines of symmetry should be explored. Again, explore links between the etymology, e.g. the 'hex' in 'hexagon' and the number of sides.

Unit 14.8 Symmetry 2

- A fun way to introduce this topic is to show the pupils a kaleidoscope. There are short videos online that show how to construct a simple kaleidoscope using a toilet roll and shiny card. There are also free online websites that can create kaleidoscope pictures.
- Allow pupils time to explore working with one 2D shape and two mirrors and share what happens when there are two horizontal mirror lines below the shape. This will help visualise the problem in the 'Before we start' section.
- Having two mirrors available for each pupil will help support them during the 'Let's learn' and 'Let's practise' sections. As the pupils tackle symmetry problems, remind them that colour is important. They should notice that the mirror image reflects the line, shape and colour.

Unit 14.9 Using scale

- Understanding and using scale is an advanced skill. Here, the pupil is using the scale to find the distance they will have to draw on a map given the scale and the actual distance and this requires them to have a secure knowledge and understanding of multiplication, division, ratio and proportion.
- 'Before we start' reviews the work encountered previously and checks the pupils' understanding of how to use a scale, although as it is 1·5 cm : 30 km, discussion about what 1 cm would measure in real life should be explored.
- Before moving on to 'Let's learn', use a diagram drawn to scale, like the one on page 182, question 1 of Textbook 2A, annotate the diagram and put the real-life distances on it and discuss how, **if you know the real-life distance and the scale, how can you calculate the length you would have to draw on the map or plan?**
- Do several examples of working from real-life distances to map/plan lengths using a variety of scales, including 1 cm : 2 km; 1 cm : 5 km; 1 cm : 20 000 m; 1 cm : 350 km. Include examples where the measurement will have to be converted from km to m or vice versa before the scale can be applied. Ask: **What do you notice about the scale and the units of measurement? What will we need to remember when we are calculating the map/plan length when the measurements are not in the same units?**
- For question 1 in the 'Let's practise' section, explain to the pupils that they are drawing lines to represent the distance on a map and not the direction.
- For question 3, pupils will require a pair of compasses.
- The 'Challenge' section can be used as a whole-class activity and used to produce a map to use for orienteering.

15 Data handling and analysis

Data handling and analysis 2A

Section overview

Textbook 2A units	Review units	Prior learning	Benchmarks/Es&Os
Unit 15.1 Reading and interpreting information	**1C** Unit 4.7 Unit 10.1 Unit 14.1 Unit 14.4	• I can read and interpret information from block graphs, bar graphs, tables, Carroll diagrams and Venn diagrams to display information. • I can include a suitable title, labelling on both axes, and an appropriate scale. • I can calculate halves and quarters of amounts.	**MNU 2-20a** **MNU 2-20b** **MTH 2-21a** • Devises ways of collecting data in the most suitable way for the given task. • Collects, organises and displays data accurately in a variety of ways including through the use of digital technologies; for example, creating surveys, tables, bar graphs, line graphs, frequency tables, simple pie charts and spreadsheets. • Analyses, interprets and draws conclusions from a variety of data. • Draws conclusions about the reliability of data taking into account, for example, the author, the audience, the scale and sample size used. • Displays data appropriately making effective use of technology and chooses a suitable scale when creating graphs.
Unit 15.2 Organising and displaying data	**1C** Unit 13.2 Unit 14.4	• I can identify and measure a right angle and a straight angle. • I can find half and quarter of an amount.	
Unit 15.3 Reading and interpreting pie charts	**1C** Unit 5.2 Unit 5.8	• I can make a tally. • I can make a block graph. • I can include a suitable title and axis labels. • I can use an appropriate scale.	
Unit 15.4 Collecting data	**1C** Unit 14.1 Unit 14.3 Unit 14.4	• I can read and interpret simple graphs, charts and tables.	

Conceptual context

The ability to read, interpret, collect and analyse data is an intrinsic skill for all learners. Pupils must be equipped with the necessary skills to question the world around them, to deepen their understanding, and apply clear knowledge from many organisers across numeracy and mathematics.

It is vitally important that pupils are supported to read a variety of scales with a range of intervals. This can be done initially using a counting stick vertically and horizontally. Pupils must have the opportunity to draw conclusions from their data through valid and supported questioning. For example, pupils may rely on addition to calculate how many brought fruit or vegetables to school. However, they must then be able to compare how many more pupils brought sweets or biscuits than fruit. Pupils rely on a concrete understanding of addition and subtraction strategies to calculate and justify their interpretations.

Pupils must be given the opportunity to *collect data* of their own choice, relevant to their lives or within the context of the school. This inquiry usually starts with a 'big question', and with the phrase 'I wonder…'. Pupils must choose a recording table or graph that suits the data that they wish to collect, and this should centre around the life and ethos of the school. For example, collecting information on favourite sports would be more suitable to a frequency table or bar graph, whereas a table or line graph is more suitable when collecting time series data. Pupils should be encouraged to have a recording system when collecting data and organising data prior to collecting, analysing and displaying findings.

It is important that pupils discuss the data collected and have opportunities to interpret what it means. Does their data answer the 'big question'. Pupils must progress through:

1. Ask the question (what are we finding out?)
2. Make a plan (what do we need to know?)
3. Collect the data (how do we do this?)
4. Present the data
5. Interpret the data

Bar graphs can be used to display data that have discrete values and are not continuous. The length of the bars is proportional to the number of items in each category. For one-to-one correspondence, the length would equal the number of items. Bar graphs are displayed with clear gaps between the bars. Pupils begin to use their knowledge of multiplication and division to create bar graphs with intervals of more than one, and to include an effective scale to present their data.

Pictographs use icons or pictures to represent data. With one-to-one correspondence, each picture represents one item of data, making it very similar to a block graph. Each picture or icon can represent more than one item. It is important that pupils understand the key chosen and how this changes the value of each individual picture or icon.

Line graphs illustrate the frequency by drawing a line instead of a bar. This type of graph is very good for showing trends over time, but the x-axis must be drawn carefully to avoid creating a misleading impression. Points on line graphs are usually joined with straight lines.

Pie charts are used to represent discrete data within a circle and are used to display and compare data. Pie charts are split into sectors to represent the categories of data. Each sector must be labelled, or a key provided to aid interpretation. Pie charts can be used to display data clearly, and must not have too many sectors, to ensure clarity.

Resources

- squared paper
- rulers
- protractors
- coloured pencils
- pencils
- clipboards
- 100 square
- Resource sheets 15a, 15b and 15c: examples of bar graph, line graph, pictograph
- pair of compasses
- laptops/computers
- access to spreadsheets through digital technology

Key vocabulary

- **analysis**
- **axes**
- **bar graph**
- **bias**
- **block graph**
- **category data**
- **continuous data**
- **discrete data**
- **frequency table**
- **histogram**
- **interpreting**
- **key**
- **line graph**
- **pie chart**
- **spreadsheet**
- **survey**
- **tables**
- **tally**
- **time series data**

Misconceptions

Potential misconception	Addressing the misconception
Unit 15.1 Pupils may read information from the wrong axis.	Provide opportunities for pupils to explore bar graphs, posing questions to initially encourage the identification of data. Highlight the importance of reading the x-axis prior to the y-axis before developing this concept by encouraging the pupils to compare data.
Unit 15.1 Pupils may not be able to count in 2, 5, 10 or 100 to calculate the axis or from pictographs.	It is important that pupils understand the scale of the axis they are working within. Provide opportunities for pupils to work on scales, asking them to identify the intervals. Promote skip counting in 2, 5, 10 and model by colour coding on a 100 chart.
Unit 15.1 Pupils may not know that they could use subtraction as a strategy to calculate the difference between two amounts.	Pupils must be supported to understand the vocabulary necessary to draw conclusions and analyse data. They need lots of opportunities to explore word problems and statements and discuss which operation they would use and what their calculation would look like.

Potential misconception	Addressing the misconception
Units 15.2 and 15.4 Pupils may not be able to count in 5s when using tally marks. For example, displaying the number 18 relies on three equal groups and 3 individual lines.	Pupils must be provided with the opportunity to use their multiplicative knowledge practically. Provide practical resources to allow pupils to use repeated addition before promoting the use of an individual 100 square.
Units 15.2 and 15.4 Pupils may struggle to identify the place of numbers that are not within intervals on the axis. For example, finding the placement of an odd number on an axis with intervals of 2.	Provide a table sized empty bar graph with a y-axis in intervals of 2, and a tray of cubes or counters in 5 colours. Ask learners to group the counters within colours, asking how we could show an odd number.
Units 15.2 and 15.4 Pupils may not know how to read information from a line graph. For example, knowing to read along the x-axis first and then read up the y-axis to identify the point the lines meet. Pupils may find it difficult to present data accurately. For example, display bars unequally or place bars together (histogram).	Provide opportunities for pupils to explore a range of completed bar graphs. It is important that they understand the key features before applying this to their own learning.
Unit 15.3 Pupils may not relate their knowledge of fractions to the sectors of a pie chart.	Provide pupils with a circle, cut from white paper. Ask pupils to fold their circle in half once and then open it up. Discuss what fraction is shown $\left(\frac{1}{2}\right)$. Repeat this process to demonstrate quarters and eighths. Allow pupils to then colour in segments of their circle and show their data as fractions.
Unit 15.3 Pupils may not relate knowledge of fractions to angles. For example, knowing that one quarter of the pie chart is equal to 90°.	Provide practical opportunities for pupils to explore rotating in 45°, 90° and 180° to relate this knowledge to fractions. (See Angles, symmetry and transformation Units 14.1 and 14.2)
Units 15.1, 15.2, 15.3 and 15.4 Pupils may not include correct labelling, including a title or key.	Provide pupils with clear examples of effective bar graphs, line graphs, pictographs and pie charts.

Teaching ideas

Unit 15.1 Reading and interpreting information - reading bar charts

- Provide pupils with an example of a bar graph (Resource sheet 15a: Favourite type of pet).
- Explain that bar graphs are used to display information to compare different groups or to track changes over time.
- Explain that one axis displays a choice of category and the other displays a discrete value.
- Ask pupils to discuss what information the bar graph displays and what they can tell from it.
- Ask questions to encourage pupils to interpret the bar graph.
  - o What is the topic of the bar graph? How do you know?
  - o Which pet is the most popular?
  - o Which pet is the least popular?
  - o How many more people like dogs than rabbits?
  - o How many people took part in the survey altogether?

  Repeat with further examples of bar graphs before applying these skills when creating their own.
- Provide pupils with opportunities to skip count in 2, 5 and 10 through circle counting games.

  Provide pupils with an example of a pictograph (Resource sheet 15b: Number of goals scored).

  Explain that pictographs use images to represent amount and that this is shown by a key.

  Ask pupils to discuss what information the pictograph displays and what they can tell from it.
  - o What is the topic of the pictograph? How do you know?
  - o On which day were the most goals scored? How do you know? How did you calculate it?
  - o On which day were the least goals scored? How do you know? How did you calculate it?
  - o How many more goals were scored on day 2, than day 3?
  - o How many people took part in the survey altogether?
- Repeat with further examples of pictographs before pupils apply these skills when creating their own.
- Provide pupils with an example of a line graph (Resource sheet 15c: Visitors to Edinburgh Castle).

  Explain that line graphs are used to display continuous data (data gathered over time).

  Ask pupils to discuss what information the line graph displays and what they can tell from it.

  Ask questions to encourage pupils to interpret the line graph.
  - o What is the topic of the line graph? How do you know?
  - o What is the scale of the line graph?
  - o Which month had the greatest number of visitors to Edinburgh Castle? How do you know?
  - o Which month had the least visitors to Edinburgh Castle? How do you know?

- o How many people visited Edinburgh castle in September?
- o What is the difference between the highest and lowest values? How do you know?
- o What other information gathered in a real-life context could be displayed on a line graph?
- o Repeat with further examples of line graphs before applying these skills when creating their own.
- Provide opportunities for pupils to create and display their data using digital technology.

Unit 15.2 Organising and displaying data

- Display the word 'frequency' on the board for pupils – what does the word 'frequency' mean?
- Agree that frequency means the rate that something occurs over a period or within a sample.
- When in school might we wish to know how frequent something was?
- Choose a focus from the pupils' answers: for example, method of travel to school.
- Display table on the board and ask pupils to respond with how they travelled to school that day.

Method of travel	Tally	Frequency
Walk		
Bus		
Cycle		
Scooter		
Park and Stride		
Other		

- Record pupils' responses in gated tally marks (blocks of 5) to ease totalling, before asking pupils to use repeated addition or multiplication to calculate the frequency.
- Which method of travel was the most frequent?
- Which method of travel was the least frequent?
- How many pupils cycled to school?
- How many more pupils walked to school than travelled by bus?
- Ask pupils how this data could be displayed and support pupils as they display the data gathered.
- Provide opportunities for pupils to create and display their data using digital technology.

Unit 15.3 Reading and interpreting pie charts

- Ask pupils to recap the ways that they have read and gathered data so far.
- Ask pupils what their favourite fruit is from: banana; strawberry; apples; oranges; or other.
- Record their responses using tally marks on the board within a table:

Type of fruit	Favourite
Banana	
Strawberry	
Apple	
Orange	
Other	

- What would be the most effective way to display this data?
- Is there another way that this data could be displayed?
- Introduce a pie chart as another way of displaying the gathered information.
- Explain that pie charts represent the whole amount and are then separated into different sectors to represent categories or groups to compare data visually.
- Draw a circle on the board and ask pupils to consider how the pie chart could be used to represent the data gathered.
- What would we need to do to the circle to display our information?
- How can we make sure that everyone is represented on the pie chart?
- Agree that the pie chart must be split into the exact amount of responses for pupils who have contributed to the fruit question.
- Split the pie chart into equal sectors representing each individual pupil.
- How many liked bananas? Code the correct number of sectors for bananas using yellow. If six pupils responded that they liked bananas the most, colour in six sectors.
- Repeat the question for the other remaining fruit responses.
- What could I title my pie chart?
- What labels do I need to add to make the data in my pie chart clear?
- Provide pupils with a frequency table and encourage them to go through the above steps to create their own pie chart.
- Draw a frequency table on the board showing:

Type of vegetable	Number planted (frequency)
Potato	32
Carrot	16
Onion	16

- Ask pupils to consider how they could display the numbers 32, 16 and 16 without having to draw out 64 sectors.
- What number can we divide 32 and 16 equally by?
- Encourage pupils to create a pie chart to scale by finding the common factor as 4 so that the pie chart could be drawn as 8 sectors for potatoes, 4 for carrots and 4 for onions.
- Provide opportunities for pupils to create and display their data using digital technology.

Unit 15.4 Collecting data

- Ask pupils to consider a big question to gather data on for their own inquiry.
- Ask pupils to consider how they would like to display their data: recording table or type of graph.
- Depending on the type of question, pupils must decide if they are gathering category data (answers fitting into pre-agreed categories) or time-series data (measurements collected regularly over time).
- Pupils can then create their question and a table to collect their data, considering headings and units of measurement.
- Pupils carry out their survey or investigation to complete their table.
- Make a bar graph or line graph to display data, considering the title and labelling the axes.
- Pupils should write up what they have found out from carrying out their investigation, interpreting the data to analyse if it has answered their big question.
- Provide opportunities for pupils to create and display their data using digital technology.

Data handling and analysis 2B

Section overview

Textbook 2B units	Review units	Prior learning	Benchmarks/Es&Os
Unit 15.1 Working with a range of graphs	**1C** Unit 14.1 Unit 14.2 Unit 14.4 **2A** Unit 15.1 Unit 15.2 Unit 15.3	• I can read and interpret information from block graphs, bar graphs, tables, Carroll diagrams and Venn diagrams to display information. • I can choose an appropriate graph to display data. • I can talk about the difference between category data and time series data. • I can include a suitable title, labelling on both axes, and an appropriate scale.	**MNU 2-20a** **MNU 2-20b** **MTH 2-21a** • Devises ways of collecting data in the most suitable way for the given task. • Collects, organises and displays data accurately in a variety of ways including through the use of digital technologies; for example, creating surveys, tables, bar graphs, line graphs, frequency tables, simple pie charts and spreadsheets. • Analyses, interprets and draws conclusions from a variety of data. • Draws conclusions about the reliability of data taking into account, for example, the author, the audience, the scale and sample size used. • Displays data appropriately making effective use of technology and chooses a suitable scale when creating graphs.
Unit 15.2 Using pie charts	**2A** Unit 14.1 Unit 14.6 Unit 15.3	• I can measure angles accurately. • I know that there are 360° in a full turn. • I can read and interpret simple pie charts (straight, 90° and 45° angles only).	
Unit 15.3 Creating and interpreting graphs	**2A** Unit 15.1 Unit 15.4	• I can draw a bar graph or line graph including a suitable title and scale, and label my axes. • I can think up a big question for an inquiry, starting with 'I wonder…' • I can create a recording template for my inquiry.	
Unit 15.4 Drawing conclusions from graphs	**2A** Unit 15.4	• I can interpret a range of simple graphs. • I can make statements about data displayed in graphs.	

Conceptual context

The ability to read, interpret, collect and analyse data is an intrinsic skill for all learners. Pupils must be equipped with the necessary skills to question the world around them, to deepen their understanding, and apply clear knowledge from many organisers across numeracy and mathematics.

It is vitally important that pupils are supported to read a variety of scales with a range of intervals and this can be done initially using a counting stick both vertically and horizontally. Pupils must have the opportunity to draw conclusions from their data through valid and supported questioning. For example, pupils may rely on addition to calculate how many brought fruit or vegetables to school. However, they must then be able to compare how many more pupils brought sweets or biscuits than brought fruit. Pupils rely on a concrete understanding of addition and subtraction strategies to calculate and justify their interpretations.

Pupils must be given the opportunity to *collect data* of their own choice, relevant to their lives or within the context of the school. This inquiry usually starts with a 'big question', and with the phrase 'I wonder…'. Pupils must choose a recording table or graph that suits the data that they wish to collect, and this should centre around the life and ethos of the school. For example, collecting information on favourite sports would be more suitable to a frequency table or bar graph, whereas a table or line graph is suited when collecting time series data. Pupils should be encouraged to have a recording system when collecting data and organising data prior to collecting, analysing and displaying findings. In this guide we have given suggested data, but please do change the context if it fits better with the experience of the pupils.

It is important that pupils discuss the data collected and have opportunities to interpret what it means. Does their data answer the 'big question'. Pupils must progress through:

1. Ask the question (what are we finding out?)
2. Make a plan (what do we need to know?)
3. Collect the data (how do we do this?)
4. Present the data
5. Interpret the data

Bar graphs can be used to display data that have discrete values and are not continuous. The length of the bars is proportional to the number of items in each category. For one-to-one correspondence, the length would equal the number of items. Bar graphs are displayed with clear gaps between the bars. Pupils begin to use their knowledge of multiplication and division to create bar graphs with intervals of more than one, and to an effective scale to present their data.

Pictographs use icons or pictures to represent data. With one-to-one correspondence, each picture represents one item of data, making it very similar to a block graph. Each picture or icon can represent more than one item. It is important that pupils understand the key chosen and how this changes the value of each individual picture or icon.

Line graphs illustrate the frequency by drawing a line instead of a bar. This type of graph is very good for showing trends over time, but the x-axis must be drawn carefully to avoid creating a misleading impression. Points on line graphs are usually joined with straight lines.

Pie charts are used to represent discrete data within a circle and are used to display and compare data. Pie charts are split into labelled sectors to represent categories of data. Each sector in the pie chart is a portion of 360°. The angle needed to represent each individual category is calculated by dividing 360° by the number of responses. For example: 360 ÷ 10, if there were 10 categories or responses to be shown.

Once pupils have learned how to gather and read data from a variety of graphs, and have created their own graphs, they must be encouraged to interpret data.

Pupils must now develop 'I notice' statements to *interpret* and draw conclusions data:

- 'I notice that there were fewer goals scored by Aberdeen than Edinburgh.'
- 'I notice that video games were three times more popular than painting.'
- 'I notice that the temperature never fell below 12°C in Spain'.

After writing their *I notice* statements they can summarise their conclusions about the data they have collected on the big question.

Pupils must then develop 'I wonder' statements from data gathered to allow them to create predictions based on data gathered and to suggest other data they would like to collect to help deepen their understanding of the big question.

This process will support pupils to demonstrate not only the technical skills of graph reading and drawing, but also the ability to *enquire* and use data to answer questions they have or are presented with.

Resources

- squared paper
- rulers
- protractors
- coloured pencils
- pencils
- clipboards
- 100 square
- Resource sheets 15a, 15b and 15c: examples of bar graph, line graph, pictograph
- pair of compasses
- laptops/computers
- access to spreadsheets through digital technology

Key vocabulary

- analysis
- axes
- bar graph
- bias
- block graph
- category data
- conclusion
- continuous data
- degrees
- discrete data
- double bar graph
- double line graph
- frequency table
- full turn
- histogram
- interpreting
- line graph
- pie chart
- spreadsheet
- survey
- tables
- tally
- time series data

Misconceptions

Potential misconception	Addressing the misconception
Units 15.1, 15.2, 15.3 and 15.4 Pupils may not include correct labelling, including a title or key.	Provide pupils with clear examples of effective bar graphs, line graphs, pictographs and pie charts.
Units 15.1, 15.2, 15.3 and 15.4 Pupil may not understand the most appropriate graph to display data on.	Revisit Textbook 2A, Unit 15.1 Provide pupils with a range of bar, line and pictographs to explore the most appropriate methods to display.
Unit 15.2 Pupils may not be able to calculate the correct size of angle to draw a sector of a pie chart.	Each sector in the pie chart is a portion of 360°. The degrees to represent for individual categories is calculated by dividing 360° by the number of responses. For example: 360 ÷ 10, if there were 10 categories or responses to be shown.
Unit 15.2 Accurate use of protractor.	Revisit Textbook 2B, Unit 14.2 Offer pupils different baselines from which to measure and draw angles so that it is not always horizontal to the page.

Teaching ideas

Unit 15.1 Working with a range of graphs

- Provide pupils with the following tables and data:

Height of a sunflower	
Months of the year	**Height in cm**
May	4 cm
June	40 cm
July	160 cm
August	120 cm

- Ask the question, 'Which type of graph would be most suitable to display this data?'
- Discuss ways that the data could be presented (bar graph or line graph, continuous data).
- Provide pupils with the next data set and ask how this could be presented.

	Score in test 1	Score in test 2
Anwar	14	16
Bob	18	18
Charlie	12	15
Duncan	15	20
Erica	20	20

- Discuss with pupils that there are two sets of data gathered for each pupil.
- Explain that the data could be displayed in a double bar chart with each set of data colour coded.
- If we only use column 1, how could we display the data? Give pupils the opportunity to create a bar or line graph using the data from column 1.
- If we were to add in column 2, how could we add the data to your graph? Provide opportunities for pupils to create and display their data using digital technology.

Unit 15.2 Using pie charts

- Present pupils with pie chart image, for example:

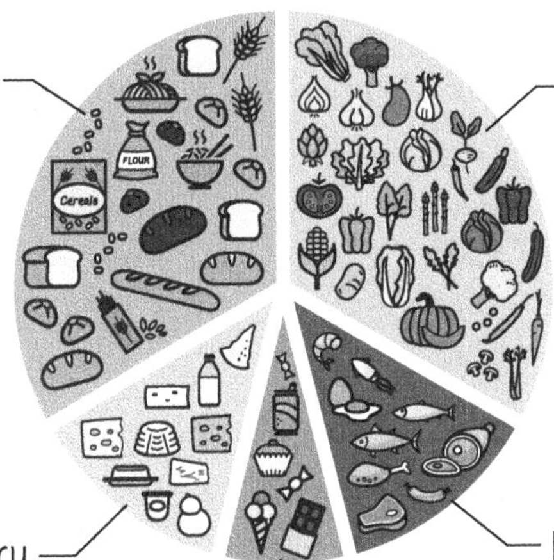

- What information can you tell from the data displayed?
- Ask pupils to stand up and turn 90°, 180°, 270° and 360° and establish that 360° is a full turn or rotation.
- Present a circle on the board.
- Explain to pupils that 36 pupils were asked their favourite flavour of ice cream and establish how each pupil's response would be shown within the pie chart (360° ÷ 36 = 10°).
- Display on the board that one response = 10° by measuring 10° as an angle within the circle drawn and make this into a segment to demonstrate the size of segment for one response.
- How could I show three responses?
- How would I show five responses?
- Work together to create a class pie chart for ice cream flavours, ensuring pupils can calculate the angle needed.
- Explain to pupils that they must use addition to ensure that they have added up all of the responses and that all angles must add to 360.

Ice cream flavour	Responses	Angle
Strawberry	10	10 × 10° = 100°
Vanilla	4	
Chocolate	14	
Raspberry	8	
Total	36	

- Provide opportunities for pupils to create and display their data using digital technology.

Unit 15.3 Creating and interpreting graphs

- Provide pupils with two tables and ask them to consider how they would display the data in the tables.
- Big question: 'I wonder what the most popular flower is.'

Type of flower	Number
Daffodil	6
Tulip	2
Pansy	1
Daisy	14
Lily	10

- Big question: 'I wonder how the height of a ramp affects the distance a toy car travels.'

Height of ramp (cm)	Distance travelled (cm)
10	25
20	32
30	38
40	43
50	47

- Pupils must then create a range of 'I notice' questions about the data in each table.
- Pupils must then create an 'I wonder…' statement to lead them on to their own inquiry question.
- Ask pupils to consider a big question to gather data on for their own inquiry.
- Ask pupils to consider how they would like to display their data: recording table or type of graph.
- Depending on the type of question, pupils must decide if they are gathering category data (answers fitting into pre-agreed categories) or time-series data (measurements collected regularly over time).
- Pupils can then create their question and a table to collect their data, considering headings and units of measurement.
- Pupils carry out their survey or investigation to complete their table.
- Make a bar graph or line graph to display data, considering the title and labelling the axes.
- Pupils should write up what they have found out from carrying out their investigation, interpreting the data to analyse if it has answered their big question.
- Provide opportunities for pupils to create and display their data using digital technology.

Unit 15.4 Drawing conclusions from graphs

- Share the following bar graph with with pupils.

  The bar graph shows us information from an inquiry into favourite activities.

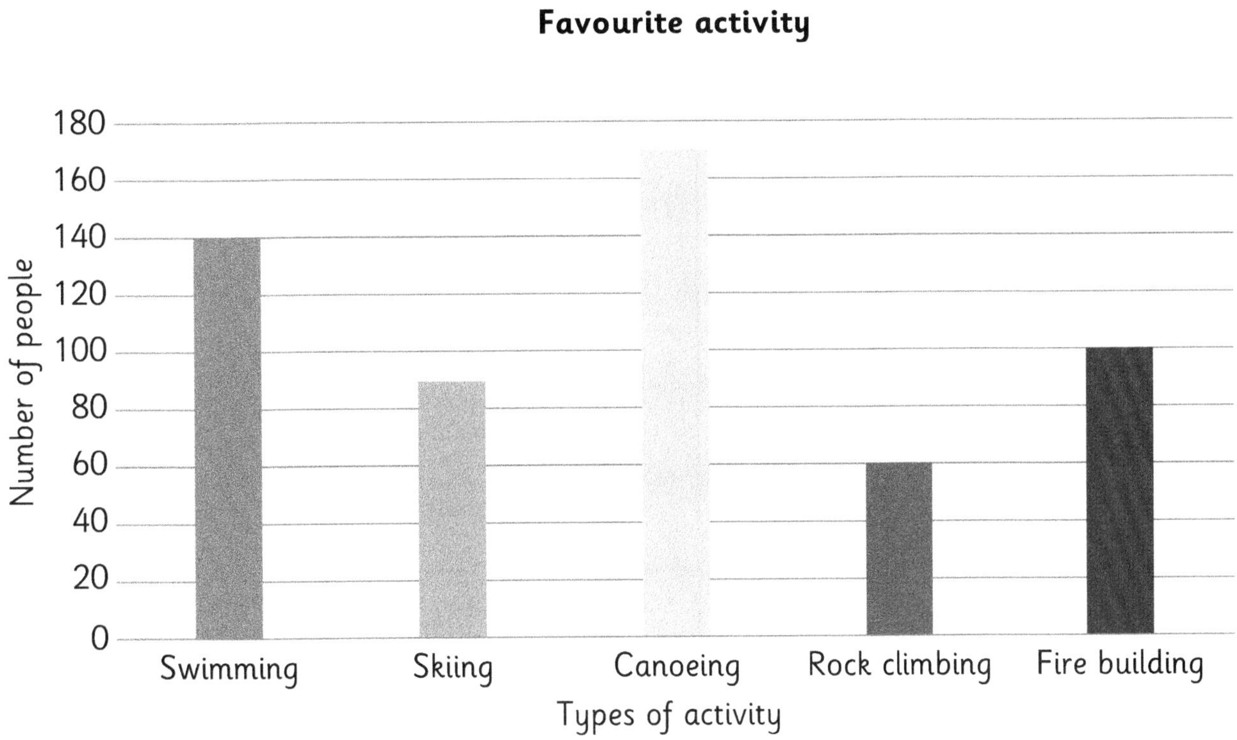

I found that people had lots of hobbies, but skiing was the least popular.
I wonder why so many people dislike rock climbing?
I wonder why canoeing was so popular?
I found out that swimming was the second most popular activity.

- Which statements are appropriate? Which statements are not appropriate?
- What other statements could you create based on the evidence that you have?
- Provide pupils with a range of bar, line and pictographs and encourage pupils to create their own conclusion statements based on the information that they have. What do they notice? What do they wonder?

Data handling and analysis 2C

Section overview

Textbook 2C units	Review units	Prior learning	Benchmarks/Es&Os
Unit 15.1 Sampling	**2A** Unit 15.1 **2B** Unit 15.1 Unit 15.2	• I can interpret, analyse and draw conclusions from a range of graphs. • I can represent data in a range of graphs and charts.	**MNU 2-20a** **MNU 2-20b** **MTH 2-21a** • Devises ways of collecting data in the most suitable way for the given task. • Collects, organises and displays data accurately in a variety of ways including through the use of digital technologies; for example, creating surveys, tables, bar graphs, line graphs, frequency tables, simple pie charts and spreadsheets. • Analyses, interprets and draws conclusions from a variety of data. • Draws conclusions about the reliability of data taking into account, for example, the author, the audience, the scale and sample size used. • Displays data appropriately, making effective use of technology and chooses a suitable scale when creating graphs.
Unit 15.2 Interpreting graphs	**2A** Unit 15.2 **2B** Units 15.1, 15.2 and 15.3	• I can interpret information from a range of graphs, charts, tables and diagrams. • I can display information in a range of graphs, selecting a suitable scale.	
Unit 15.3 Understanding sampling bias	**2B** Units 15.1, 15.2 and 15.3	• I can interpret, analyse and draw conclusions from a range of graphs. • I can say whether a survey has asked a whole population or a sample. • I can suggest a suitable sample.	
Unit 15.4 Pie charts	**2A** Unit 15.3 **2B** Unit 1.2 Units 4.3 – 4.8 Unit 4.10 Unit 4.11 Units 6.1 – 6.3 Unit 6.7 Units 6.9 – 6.11 Unit 13.2 Units 14.1 – 14.3 Unit 15.3	• I can multiply and divide. • I can interpret information from simple pie charts. • I can calculate percentages. • I can round to the nearest whole number.	

Conceptual context

The ability to read, interpret, collect, and analyse data is an intrinsic skill for all learners. Pupils must be equipped with the necessary skills to question the world around them, to deepen their understanding, and apply clear knowledge from many organisers across numeracy and mathematics.

It is vitally important that pupils are supported to read a variety of scales with a range of intervals and this can be done initially using a counting stick both vertically and horizontally. Pupils must have the opportunity to draw conclusions from their data through valid and supported questioning. For example, pupils may rely on addition to calculate how many brought fruit or vegetables to school. However, they must then be able to compare how many more brought sweets or biscuits than brought fruit. Pupils rely on a concrete understanding of addition and subtraction strategies to calculate and justify their interpretations.

Pupils must be given the opportunity to *collect data* of their own choice, relevant to their lives or within the context of the school. This inquiry usually starts with a 'big question', and with the phrase 'I wonder…'. Pupils must choose a recording table or graph that suits the data that they wish to collect, and this should centre around the life and ethos of the school. For example, collecting information on favourite sports would be more suitable to a frequency table or bar graph, whereas a table or line graph is suited when collecting time series data. Pupils should be encouraged to have a recording system when collecting data and organising data prior to collecting, analysing and displaying findings. In this guide we have given suggested data, but please do change the context if it fits better with the experience of the pupils.

It is important that pupils discuss data collected and have opportunities to interpret what it means. Does their data answer the 'big question'. Pupils must progress through:

1. Ask the question (what are we finding out?)
2. Make a plan (what do we need to know?)
3. Collect the data (how do we do this?)
4. Present the data
5. Interpret the data

Bar graphs can be used to display data that has discrete values and is not continuous. The length of the bars is proportional to the number of items in each category. For one-to-one correspondence, the length would equal the number of items. Bar graphs are displayed with clear gaps between the bars. Pupils begin to use their knowledge of multiplication and division to create bar graphs with intervals of more than one, and to an effective scale to present their data.

Pictographs use icons or pictures to represent data. With one-to-one correspondence, each picture represents one item of data, making it very similar to a block graph. Each picture or icon can represent more than one item. It is important that pupils understand the key chosen and how this changes the value of each individual picture or icon.

Line graphs illustrate the frequency by drawing a line instead of a bar. This type of graph is very good for showing trends over time, but the x-axis must be drawn carefully to avoid creating a misleading impression. Points on line graphs are usually joined with straight lines.

Pie charts are used to represent discrete data within a circle and are used to display and compare data. Pie charts are split into labelled sectors to represent categories of data. Each sector in the pie chart is a portion of 360°. The angle needed for each individual category is calculated by dividing 360° by the number of responses. For example: 360 ÷ 10, if there were 10 categories or responses to be shown.

Once pupils have learned how to gather and read data from a variety of graphs, and have created their own graphs, they must be encouraged to interpret data.

Pupils must now develop 'I notice' statements to *interpret* and draw conclusions data:

- 'I notice that there were fewer goals scored by Aberdeen than Edinburgh.'
- 'I notice that video games were three times more popular than painting.'
- 'I notice that the temperature never fell below 12°C in Spain'.

After writing their *I notice* statements, they can summarise their conclusions about the data they have collected on the big question

Pupils must then develop 'I wonder' statements from data gathered to allow them to create predictions and to suggest other data they would like to collect to help deepen their understanding of the big question.

This process will support pupils to demonstrate not only the technical skills of graph reading and drawing, but also the ability to *enquire* and use data to answer questions they have or are presented with.

Resources

- squared paper
- rulers
- protractors
- coloured pencils
- pencils
- clipboards
- 100 square
- examples of bar graph, line graph, pictograph
- pair of compasses
- laptops/computers
- access to spreadsheets through digital technology
- examples of lists of data (such as class registers) for pupils to sample from
- calculators
- protractors or angle measurers

Key vocabulary

- analysis
- axes
- bar graph
- bias
- block graph
- category data
- conclusion
- continuous data
- degrees
- discrete data
- double bar graph
- double line graph
- frequency table
- full turn
- histogram
- inquiry
- interpreting
- line graph
- pie chart
- population
- sample
- scale
- spreadsheets
- survey
- tables
- tally
- time series data

Misconceptions

Potential misconception	Addressing the misconception
Units 15.1, 15.2 and 15.4 Using the wrong axes and scale – misleading graphs.	Revisit Textbook 2B, Units 15.3 and 15.4 (creating, interpreting and drawing conclusions from graphs). Show the graphs starting at zero on the vertical axis and then ask pupils to use the data to re-plot the graphs starting at different points. Explore and compare how the graphs appear.
Unit 15.3 Sampling – appropriate size of sample.	Use a big question such as 'What is your favourite animal?' Create a simple data collection sheet for pupils to tally responses. In groups of three, ask one pupil to collect ten responses, another to collect ten from each class, and another to ask everyone in the school. Compare results to see that ten pupils are too few and the whole school is too many but ten from each class is a fair representation of the results from the whole school.
Unit 15.3 Bias when sampling.	Provide pupils with some samples and discuss which are a fair representation of the whole population. This could include asking only the football team when the inquiry question is 'Which sports do pupils in this school prefer?' and asking ten pupils from each class when the inquiry question is 'How many hours TV do you watch on a school night?' Encourage discussion and ask pupils to create inquiry questions, choosing a biased and an unbiased sample.

Potential misconception	Addressing the misconception
Unit 15.4 Calculation of percentage.	Revisit Textbook 2B, Units 6.9, 6.10 and 6.11 on early percentage work and Textbook 2C, Units 6.9 and 6.10 for further practice.
Unit 15.4 Calculation of angle on pie chart.	Revisit Textbook 2C, Unit 14.3 for angles around a point and missing angle calculations. Ensure pupils can understand equivalent fractions are related to the proportion of 360^0 needed in the pie chart (see Textbook 2C, Unit 15.4 for example)
Unit 15.4 Linking 100% to 360^0 in a pie chart.	Revisit Textbook 2C, Unit 6.9 for further practice on equivalence.
Unit 15.4 Accurate use of protractor.	Revisit Textbook 2C, Units 14.1, 14.2 and 14.3. Offer pupils different baselines from which to measure and draw angles so that it is not always horizontal to the page.

Teaching ideas

Unit 15.1 Sampling

- Big question '**I wonder which subject pupils like best in my school'?**
- Ask pupils: **What should we do to find the answer to this question?**
- Ask pupils to discuss in groups of three.
- Remind them about the inquiry process:
    - o Start with a big question. 'I wonder….'
    - o Design a data collection table with suitable headings (in this case, the subjects covered in the school day).
    - o Choose an appropriate graph to display the data (likely to be a bar graph for discrete data).
    - o Write as many 'I notice….' statements as you can about the data and the graph.
    - o Write a conclusion for your inquiry. 'I found that….'
- Pupils should then design their data collection table.
- A model can be made available to support pupils.

Subject	Maths	English	History	PE	Geography	Art	Science
Number of pupils (tally)							
Frequency							

- Show your teacher-version of the data collection table (note that the pupils may have come up with different versions, but your version is best to use in order to direct whole class discussion) and ask pupils: **How can we collect this data?**
- Ask: **Do we need to ask every pupil in this school, or could we just ask some of them to give ourselves an idea of the answer to our big question?**
- Discuss as a group or class.
- Pupils (in threes) can then carry out the survey in their own class and draw a bar graph to show the data. They should compare results as a class, perhaps presenting posters or creating a display using a software package.
- Now remind pupils that we need to think about the whole school.
- **'Does our class represent the whole school?'**, **'Have we asked enough people?'**
- Pupils may:
  - o conclude that their graphs would be too big if they had to plot data from the whole school;
  - o decide that it would take a long time to collect all the data; or
  - o think it would be too difficult to be accurate and they might forget who they have asked.
- Explain how a sample of pupils would be enough and we can then compare our samples to see if we get similar results.
- Think about:
  - o The larger the sample, the closer the result will be to the whole population.
  - o A very small sample will not give us enough information to see the bigger picture.
  - o Decide on a reasonable sample size that is easy to collect, does not introduce too many points to plot on the graph, and includes a way of checking that we haven't asked the same person twice.
- Pupils, with teacher modelling, may decide on a sample size of 10 pupils from each year group. This means that:
  - o The data represents all ages in the school.
  - o The data represents all classes across the school.
  - o Ten pupils are easy to keep track of when collecting the data.
  - o The final result is a manageable number for graphing.
- Pupils should then agree as a class how they will conduct their survey and then carry it out.
- They should present data and compare their graphs and conclusions.
- Pupils discuss how accurate the sample was, compared to when they asked the whole class.

- Offer pupils some matching cards with 'Big questions', 'Whole population' and 'Sample' on them; ask them to match these (the first four examples are taken from 15.1, you can create others that are suited to your class). Pupils can also make up some of their own, once the concept is understood.

Big question	**Whole population**	**Sample**
Where do the pupils in our school like to go on holiday?	All the pupils in the school	10 pupils from every class
Which newspapers do the parents of pupils in our school read?	All the parents of the pupils in the school	6 parents from every class
How do the people who work in the local shops travel to work?	All the people in the local shops	3 people from each shop
What do librarians in Scotland read in their spare time?	All the librarians in Scotland	1 librarian from a sample of 50 libraries spread across Scotland
What is the favourite film of the teachers in the schools in our town?	All the teachers in our town	3 teachers from each school in our town

Unit 15.2 Interpreting graphs

- Show the pupils a range of number lines that are all the same length, but have different scales, for example: 0–1, 0–10, 0–100, 500–1000, etc.
- Ask pupils to discuss which numbers are represented.
- Ask pupils: **What is the same and what is different about these number lines?** to encourage them to look closely at the different scales.
- Give the pupils a 0–10 and a 0–100 number line of the same length.
    - o Ask them to show where zero is. Ask them to show where 10 is. Ask them to show where 5, 1 and 8 are.
- Discuss how much easier it is to show these numbers on the 0–10 scale because they look very small on the 0–100 number line.
- Give the pupils a data set of scores in a maths test, for example:

Pupil	A	B	C	D	E	F	G	H	I
Score	43	47	41	50	52	47	42	45	40

- The pupils will create a bar graph for this data but will be given different axes to use so they can see how the scale can change how the data looks. Give the pupils one of three different sets of axes for them to plot the scores onto. The three different sets of axes could be:
    - o the y-axis (vertical) 0–100, going up in 10s; between 40 and 60, going up in 2s; between 20 and 80, going up in 20s. In each case, it should be labelled 'Score'.

- The x-axis (horizontal) should be 'A' to 'I' for each. It should be labelled 'Pupils' on all three axes.
- Ask pupils: **What is the same and what is different about the graphs you have plotted?**
- Ask pupils: **Which graph gives us information about the test scores, and which are misleading or difficult to interpret?**
- Repeat the task with continuous data. For example, use the following data about the number of apples sold in a small grocery shop over 6 months:

Month	March	April	May	June	July	Aug
Apples sold	150	180	220	230	210	250

- Give pupils one of three different sets of axes for them to plot the number of apples sold. This will create a line graph.
- The y-axis (vertical) could be 0–1000, going up in 100s; 100–300, going up in 50s; 150–250, going up in 10s. It should be labelled 'Apples sold' in each case.
- The x-axis (horizontal) should be 'March' to 'August' for each. It should be labelled 'Months' on all three versions of the axes.
- Ask pupils: **What is the same and what is different about the graphs you have plotted?**
- Ask pupils: **Which graph presents the information about the number of apples sold most clearly? Which graphs are misleading or difficult to interpret?**
- Ask pupils to use data collected in the sampling section (15.1) to create some graphs using different scales. The pupils can present their graphs as a poster to show how the different scale can make the graph helpful or misleading. They can draw by hand or use graphing software to create them.

Unit 15.3 Understanding sampling bias

- Remind pupils about the work done on sampling in section 15.1.
- Big question: **'What category of TV programmes do people in my school like best?'**
- Ask pupils this question and then show the data collection table.

TV show	Sports	Drama	Comedy	Music	Quiz	Nature	Other
Tally							

- Tell the pupils that you will ask a sample from the class.
- Now ask for responses from only the boys in the class.
- Collect results and discuss whether this represents the whole school or not. Remind pupils that this is called the 'population' in a data inquiry.

- Explain the meaning of the word 'bias' (a preference or prejudice for, or against, one person or group or category).
- Ask pupils to think about the population matching cards they used/created in 15.1.
- Organise pupils into small groups. Ask them to create a fourth column for each 'Big question' that describes a biased sample. For example:

Big question	Whole population	Sample	Biased sample
Where do the pupils in our school like to go on holiday?	All the pupils in the school	10 pupils from every class	All the girls in one class
Which newspapers do the parents of pupils in our school read?	All the parents of the pupils in the school	6 parents from every class	10 mums from the youngest pupils' class
How do the people who work in the local shops travel to work?	All the people in the local shops	3 people from each shop	All the people in one shop only
What do librarians in Scotland read in their spare time?	All the librarians in Scotland	1 librarian from a sample of 50 libraries spread across Scotland	10 male librarians from Glasgow
What is the favourite film of the teachers in the schools in our town?	All the teachers in our town	3 teachers from each school in our town	Headteachers only

Unit 15.4 Pie charts

- Revisit calculating angles around a point and make sure that all pupils know that they sum to 360°.
- Give each pupil a circle of paper. Ask them to fold it in half and then fold one of the halves into quarters. They should label one section 'half' and two sections 'quarter'.
- Ask pupils to identify the centre of the circle and the number of degrees they have folded to make half a circle and a quarter of a circle (180° and 90°).
- Ask pupils to add the labels '50%' in the half and '25%' in each of the quarters.
- Agree that this sums to 100% and also to 360°.
- Agree that for a circle 100% = 360°; 50% = 180° = $\frac{1}{2}$; 25% = 90° = $\frac{1}{4}$.
- Ask pupils to think about how they could work out the percentage and the circle degrees for different simple fractions $\left(\frac{3}{4}, \frac{1}{8}, \frac{5}{8}, \frac{1}{10}, \frac{1}{3}\right)$.
- Model using a radius to construct different angles and label these on a new paper circle (pupils can fold into eighths and check with a protractor to start with).
- On a new circle, ask pupils to show 90°, 75°, 15° and 180° in the same circle.
- You may need to show how each new angle is measured from the last radius drawn.
- Check that all pupils agree that they have all (100%) of the circle's degrees because the angles sum to 360°.

- Remind pupils about pie charts (Unit 15.2 in 2B).
- Give pupils some data about the colour of the socks 90 people are wearing in an office:

Sock colour	Black	Red	Green	Other
Frequency	45	15	15	15

  - o Ask pupils to work out how they could show this in a pie chart.
  - o Agree that we need to work out how many degrees each colour should get in the pie chart.
  - o We can reason like this: 45 is half of 90, so this is worth 180° since this is half of 360°.
  - o Pupils can then reason about equally sharing the other 180° between three to get 60° each.
  - o Ask: **What fraction of the pie chart is each colour sock?**
  - o Ask: **How many degrees is each sock worth?**
  - o Model $60 \div 15 = 4°$ each.
  - o Note that $4 \times 90 = 360$ and that $\frac{45}{90} \times 360° = 180°$ and that $\frac{45}{90} \times 100 = 50\%$.
- Give pupils data about the sales of different types of vegetables in a small grocery shop in one week:

Vegetable	**Number sold**	**Number of degrees**	**Percentage of the total**
Potato	105		
Carrot	158		
Runner beans	82		
Turnip	15		
Total	360		

- Ask them to show this data in a pie chart.

Vegetable	**Number sold**	**Number of degrees**	**Percentage of the total**
Potato	105	105°	29%
Carrot	158	158°	44%
Runner beans	82	82°	23%
Turnip	15	15°	4%
Total	360	360°	100%

- Discuss the need to round to the nearest whole number.

- Now give the pupils a second data set for a different week's sales:

Vegetable	Number sold	Number of degrees	Percentage of the total
Potato	99		
Carrot	68		
Runner beans	235		
Turnip	18		
Total	420		

- Ask them **to show this data in a pie chart**.
- Remind pupils that 420 is 100% of the amount sold.
- Remind the pupils how to find a fraction of a whole amount by writing, e.g. $\frac{99}{420}$.
- To find the degrees, they need this fraction of 360°, so will need to calculate $\frac{99}{420} \times 360$.
- To find the percentage, they need this fraction of 100%, so will need to calculate $\frac{99}{420} \times 100$.

Vegetable	Number sold	Number of degrees	Percentage of the total
Potato	99	85°	24%
Carrot	68	58°	16%
Runner beans	235	201°	56%
Turnip	18	15°	4%
Total	420	359°	100%

- Discuss that the need for rounding to whole numbers has created a slight rounding error for the number of degrees that is not significant. Remind pupils that this is modelling real-life and so the answer is sometimes not exact!
- Ask pupils to compare the two pie charts and write some 'I notice…' statements about them.

16 Ideas of chance and uncertainty

Ideas of chance and uncertainty 2A

Section overview

Textbook 2A units	Review units	Prior learning	Benchmarks/Es&Os
Unit 16.1 Predicting and explaining simple chance situations	**1C** Unit 16.1	• I can use the terms: probable, likely, unlikely, certain, uncertain, never, possible, impossible, fair and unfair to describe the likelihood of events occurring. • I can represent the chance of an event occurring on a number line from 0 to 1 including $\frac{1}{2}$. • I can interpret data gathered through everyday experiences to make reasonable predictions.	**MNU 2-22a** • Uses the language of probability accurately to describe the likelihood of simple events occurring; for example: equal chance; fifty-fifty; one in two; two in three; percentage chance; and $\frac{1}{6}$. • Plans and carries out simple experiments involving chance with repeated trials; for example: What is the probability of throwing a six if you throw a die fifty times? • Uses data to predict the outcome of a simple experiment.

Conceptual context

In mathematics, probability is a measurement just like any other measurement; the difference is that it is applied to the likelihood of whether or not an event will occur. We are teaching pupils to explore doubt and uncertainty by understanding the vocabulary that is used and the different situations where they might use this as a measurement.

In probability, we discuss 'events' and 'outcomes', with an event being anything that happens and an outcome being the result of what happens. For example, when you toss a coin it can have two possible results: heads or tails. Therefore, tossing the coin is the event and the outcome is either heads or tails.

Using probability, you can measure whether an outcome is more or less likely to occur as the result of any given event. By dividing the number of times that the outcome has occurred by the total number of events you can start to predict the probability of the outcome. To work out the probability that a coin when tossed will land one way or the other, you might toss the coin 20 times. If the coin turns up heads 10 times, then the probability that the coin will land heads up the next time is $\frac{10}{20}$, or $\frac{1}{2}$ or 50%.

Resources

- Dice
- Coins
- Counters
- Playing cards
- Probability scale (Resource sheet 16a)
- Probability cards (Resource sheet 16b)

Key vocabulary

- **Certainty**
- **Chance**
- **Doubt**
- **Likelihood**
- **Likely**
- **Predicting**
- **Probability**
- **Uncertainty**
- **Unlikely**

Misconceptions

Potential misconception	Addressing the misconception
Unit 16.1 Events are thought of together and not individually.	If you ask pupils how likely it is to roll a 6 on a dice, they may make their mind up based on their most recent experience. So, if they have just rolled the dice 4 times and each time it was a 6, they will say very likely. However, if they have rolled the dice 4 times and got a different number each time, they will have a different answer despite the probability remaining the same in both cases. It needs to be clearly explained to pupils that the probability is $\frac{1}{6}$ regardless.
Personal experience and belief can influence an individual's ability to think about probability and chance.	If a pupil's frame of reference is that you win every time you play a game, their perception of the likelihood that an event will occur will be very different from a pupil who has experienced losing every game they have played. This needs to be explored to explain that the chances are the same for each game: this might be a difficult concept for some pupils to take on board.

Teaching ideas

Unit 16.1 Predicting and explaining simple chance situations

- Start by introducing the key vocabulary to pupils so they understand what each of the terms means to ensure that, when they use them, they are clear why they have selected a specific word to describe the event.
- Put the following words up on the board and ask the pupils to think individually, then discuss with a partner what each of them means:
    - Certain
    - Likely
    - Even
    - Unlikely
    - Chance
    - Possible
    - Impossible
- Open the discussion up to the whole class and allow pupils to share their understanding with each other. Next, ask pupils (either individually or in pairs) to write down an example of an event for each one, telling them that they will need to be able to explain their answer. This will provide an opportunity to clarify understanding and any misconceptions that pupils may have.
- Give pupils a copy of the probability scale (see Resource Sheet 16a) and a set of the probability cards (see Resource Sheet 16b) and ask them to work with a partner to place each card along the scale. Tell them that they will have to have a reason for placing each card. When asking pupils to explain their answers, ensure that they are using the correct vocabulary to reinforce their understanding. Also use the opportunity to question their reasons – 'It will rain tomorrow' – this may be placed on a number of sections of the scale from unlikely to certain depending on what the pupils have recently experienced. Possible questions to ask are:

  What makes you say that?

  What evidence do you have to support this?

  What weather would you expect at this time of the year?

  Is there any circumstance in which you would give me a different answer?

Ideas of chance and uncertainty 2B

Section overview

Textbook 2B units	Review units	Prior learning	Benchmarks/Es&Os
Unit 16.1 Investigating the possible outcomes of random events	**1C** Unit 16.1	• I can use the terms: probable, likely, unlikely, certain, uncertain, never, possible, impossible, fair and unfair to describe the likelihood of events occurring. • I can represent the chance of an event occurring on a number line from 0 to 1 including 0, 0·25, 0·5, 0·75 and 1. • I can represent the chance of an event occurring on a number line from 0 to 1 including 0, 25%, 50%, 75% and 100%. • I can interpret data gathered through everyday experiences to make reasonable predictions.	**MNU 2-22a** • Uses the language of probability accurately to describe the likelihood of simple events occurring; for example: equal chance; fifty-fifty; one in two; two in three; percentage chance; and $\frac{1}{6}$. • Plans and carries out simple experiments involving chance with repeated trials; for example: What is the probability of throwing a six if you throw a die fifty times? • Uses data to predict the outcome of a simple experiment.

Conceptual context

On a daily basis, we may be interested in finding the chance of occurrence of many different events. This forms the basis of the concept of probability. Prior to understanding what probability is, pupils need to understand what events are and the different kinds of events.

A random event is one where the outcome is unknown; for example, when we toss a coin, we do not know if it will land heads-up or tails-up, so tossing a coin is considered a random experiment.

Events can be independent, with each event being unaffected by other events; dependent or conditional, where an event is affected by other events; and mutually exclusive, where events can't happen at the same time.

When an event is independent, it is not affected by any other events; for example, a coin does not remember that it came up tails previously, therefore each toss of a coin is an isolated event and the past outcomes will not affect the next one.

Some events are dependent, which means that they are affected by previous events; for example, when you select a card from a deck. Once you have taken one card from the deck there are fewer cards available and so consequently, the probabilities change. However, if we put each card back after drawing it, or replace it, then the chances don't change, as the events are independent. Therefore, without replacement the chances will change, and the events will be dependent.

Resources

- Dice
- Coins
- Counters
- Playing cards
- Numbered cards (Resource sheet 16c)

Key vocabulary

- **Certainty**
- **Chance**
- **Dependent**
- **Doubt**
- **Independent**
- **Likelihood**
- **Likely**
- **Predicting**
- **Probability**
- **Random**
- **Uncertainty**
- **Unlikely**

Misconceptions

Potential misconception	Addressing the misconception
Unit 16.1 Events are thought of together and not individually.	If you ask pupils how likely it is to roll a 6 on a dice, they may make their mind up based on their most recent experience. So, if they have just rolled the dice 4 times and each time it was a 6, they will say very likely. However, if they have rolled the dice 4 times and got a different number each time, they may have a different answer despite the probability remaining the same in both cases. It needs to be clearly explained to pupils that the probability is $\frac{1}{6}$ regardless.

Potential misconception	Addressing the misconception
Personal experience and belief can influence an individual's ability to think about probability and chance.	If a pupils' frame of reference is that you win every time you play a game, their perception of the likelihood that an event will occur will be very different from a pupil who has experienced losing every game they have played. This needs to be explored to explain that the chances are the same for each game: this might be a difficult concept for some pupils to take on board.

Teaching ideas

Unit 16.1 Predicting and explaining the outcomes of simple chance situations and experiments

- Introduce this concept by explaining to pupils that there are some things that we know are certain and do not rely on the help, influence or impact of anything else to happen. Things that are certain can include: there are 7 days in a week; there are 31 days in December; and it will get dark. Things that are uncertain can include: it will rain today; there will be a fire drill; or I will laugh today.
- Ask pupils if they know what a random event might be. Extend the conversation by asking them to explain why they think that.
- Ask pupils: **What is the probability of someone in their class being the same age as them?** Ask them why they have given that answer. Extend this by asking them what the probability of someone in their school being the same age as them is. Have the pupils given a different answer? Look for pupils to recognise that, within their own class, the probability is higher as they are all around the same age but that, across the whole school, there is a wider range of ages and so the probability is lower. Tell pupils that this is an example of the outcome being dependent on the information being considered.
- Give pupils a set of cards numbered 1 to 6 (see Resource Sheet 16c) and ask them to shuffle the cards and place them face down. Write the following questions on the board and ask pupils to discuss their answers with a partner:

  What is the possibility that they pick a card with an even number?

  What is the chance that they will pick a card with an odd number?

  What is the likelihood that they pick a card with 7 on it?

  How likely is it that they select a card with 4 on it?
- For each of these questions, although the selection is random, there is a specific set of parameters within which pupils have to work as there are only 6 cards, there is only one of each number, there are only numbers 1 to 6 and this will not change. Therefore, they have a 50% chance of selecting either an odd or an even number, a 0% chance of selecting a 7, and a 1 in 6 chance of selecting a 4. Explore the vocabulary of chance and the different numerical ways that chance can be expressed. Use the probability scale to highlight the use of decimal notation, fractions and percentages to express this.

Ideas of chance and uncertainty 2C

Section overview

Textbook 2C units	Review units	Prior learning	Benchmarks/Es&Os
Unit 16.1 Predicting and explaining the outcomes of simple chance situations and experiments	**2B** Unit 16.1	• I can understand and can use the terms: probable, likely, unlikely, certain, uncertain, never, possible, impossible, fair and unfair to describe the likelihood of events occurring. • I can confidently represent the chance of an event occurring using appropriate vocabulary and numerical notation. • I can interpret data gathered through everyday experiences to make reasonable predictions. • I can explain data gathered through everyday experiences to make reasonable predictions.	**MNU 2-22a** • Uses the language of probability accurately to describe the likelihood of simple events occurring; for example, equal chance; fifty-fifty; one in two; two in three; percentage chance; and $\frac{1}{6}$. • Plans and carries out simple experiments involving chance with repeated trials; for example: What is the probability of throwing a six if you throw a die fifty times? • Uses data to predict the outcome of a simple experiment.

Conceptual context

Predicted probability has to do with how the probability is calculated and what the outcomes mean. A predicted probability is, essentially, the probability of an event that is calculated from available data.

There is a wealth of information available (e.g. about natural occurrences, such as the weather; or statistics, such as football results) that can be used to develop this concept. But there are also many opportunities to use practical experiments to deepen pupils' understanding. For example, posing the question: If I drop a drawing pin, will it land point up or not? is difficult to reason, but lends itself to an experiment. Collecting local weather information can lead to questions such as: How likely is it to rain in June? However, as children develop this concept, they should also appreciate its limitations. We can say if it is likely to rain more in June or January and we can predict how many days rain we might get in June, but we cannot say if it will rain on sports day!

When we want to make predictions, it is a good strategy to make an organised list. If we can identify what could happen, we can begin to work out how likely it is to happen. If children are able to make a list of all the possible outcomes, they can then make predictions and give numerical values to the probability of an event occurring.

Children can find making an organised list difficult, particularly in ensuring that they have all the possibilities listed. The initial focus should be on ensuring that all the possible outcomes have been found, rather than any numerical analysis. Take the example: **What are the possible outcomes when we toss two coins at the same time?**

Children will often say there are three possible outcomes, HEAD HEAD, TAIL TAIL and a HEAD and a TAIL. Therefore, all the outcomes are equally likely. However, there are four possibilities, HH, TT, HT, TH so you are, in fact, less likely to get HH than a head and a tail.

Resources

- Dice
- Coins
- Counters
- Playing cards

Key vocabulary

- **Certainty**
- **Chance**
- **Doubt**
- **Likelihood**
- **Likely**
- **Predicting**
- **Probability**
- **Uncertainty**
- **Unlikely**

Misconceptions

Potential misconception	Addressing the misconception
Unit 16.1 Events are thought of together and not individually.	If you ask pupils how likely it is to roll a 6 on a dice, they may make their mind up based on their most recent experience. So, if they have just rolled the dice 4 times and each time it was a 6, they will say very likely. However, if they have rolled the dice 4 times and got a different number each time, they may have a different answer despite the probability remaining the same in both cases. It needs to be clearly explained to pupils that the probability is $\frac{1}{6}$ regardless.
Personal experience and belief can influence an individual's ability to think about probability and chance.	If a pupils' frame of reference is that you win every time you play a game, their perception of the likelihood that an event will occur will be very different from a pupil who has experienced losing every game they have played. This needs to be explored to explain that the chances are the same for each game: this might be a difficult concept for some pupils to take on board.

Teaching ideas

Unit 16.1 Predicting and explaining the outcomes of simple chance situations and experiments

- Tell pupils that they are predicting what is going to happen every day: they choose their clothes based on what they think the weather will be, they choose what table to sit at in the lunch hall based on where they think their friends will sit. Every choice is a prediction of how likely they think an event is to happen.
- Recap that probability is the ratio of the times an event is likely to occur divided by the total possible events. Using a dice, explain that there are six possible events, and there is one likely event for each number with each roll, or a $\frac{1}{6}$ chance. Discuss the scenario if there was nothing/were no dots on any of the sides: the probability of rolling a 3 would be zero because there would be no 3 and no other dots either, giving us this ratio $\frac{0}{0}$. If every side had one dot, then the probability of rolling a 1 would be 1 because it would be $\frac{6}{6}$, or 1.
- Ask pupils to think about the lunch menu. If they come to school Monday to Friday and they know that pizza is on the menu two days that week: **What is the probability that pizza will be served on any given day?** (you can adapt this for your own school menu).

- Write on the board:

Monday	Tuesday	Wednesday	Thursday	Friday
?	?	?	?	?

Discuss with pupils how they could work this out…

It would be $\frac{2}{5}$, because there are two desired outcomes (pizza) and five possible outcomes (days of the school week).

As the week continues however, do the chances change? If pizza is not served on Monday, what is the probability of it being served on Tuesday? Is it still $\frac{2}{5}$? Ask pupils to discuss this with a partner.

Monday	Tuesday	Wednesday	Thursday	Friday
X	?	?	?	?

If there is no pizza served on Tuesday, how will this change their prediction?

Monday	Tuesday	Wednesday	Thursday	Friday
X	X	?	?	?

If pizza is then served on Wednesday, what can they predict for the rest of the week?

Monday	Tuesday	Wednesday	Thursday	Friday
X	X		?	?